BIOENGINEERED FRUIT AND VEGETABLES

Trends in Postharvest Quality Improvement

Postharvest Biology and Technology Series

BIOENGINEERED FRUIT AND VEGETABLES

Trends in Postharvest Quality Improvement

Edited by

Mohammad Irfan, PhD
Pankaj Kumar, PhD

AAP | APPLE ACADEMIC PRESS

First edition published 2025

Apple Academic Press Inc.
1265 Goldenrod Circle, NE,
Palm Bay, FL 32905 USA

760 Laurentian Drive, Unit 19,
Burlington, ON L7N 0A4, CANADA

CRC Press
2385 NW Executive Center Drive,
Suite 320, Boca Raton FL 33431

4 Park Square, Milton Park,
Abingdon, Oxon, OX14 4RN UK

Library and Archives Canada Cataloguing in Publication

CIP data on file with Canada Library and Archives

Library of Congress Cataloging-in-Publication Data

CIP data on file with US Library of Congress

ISBN: 978-1-77491-748-0 (hbk)
ISBN: 978-1-77491-749-7 (pbk)
ISBN: 978-1-00350-195-4 (ebk)

DEDICATION

To our most respected parents and
PhD and PDF mentors.

ABOUT THE BOOK SERIES: POSTHARVEST BIOLOGY AND TECHNOLOGY

Editor-in-Chief

Mohammed Wasim Siddiqui, PhD
Scientist-cum-Assistant Professor, Department of Food Science and Technology, Bihar Agricultural University, Sabour, Bhagalpur, Bihar, India
AAP Acquisitions Editor | Horticultural Science
Founding/Managing Editor | *Journal of Postharvest Technology*
Email: wasim_serene@yahoo.com

As we know, preserving the quality of fresh produce has long been a challenging task. In the past, several approaches were in use for the postharvest management of fresh produce, but due to continuous advancement in technology, the increased health consciousness of consumers, and environmental concerns, these approaches have been modified and enhanced to address these issues and concerns.

The Postharvest Biology and Technology series features edited books that address several aspects related to postharvest technology of fresh produce. The series presents existing and novel management systems that are in use today or that have great potential to maintain the postharvest quality of fresh produce in terms of microbiological safety, nutrition, and sensory quality.

The books are aimed at professionals, postharvest scientists, academicians researching postharvest problems, and graduate-level students. This series is intended to be a comprehensive venture that provides up-to-date scientific and technical information focusing on postharvest management for fresh produce.

Books in the series address the following themes

- Nutritional composition and antioxidant properties of fresh produce
- Postharvest physiology and biochemistry
- Biotic and abiotic factors affecting maturity and quality

- Preharvest treatments affecting postharvest quality
- Maturity and harvesting issues
- Nondestructive quality assessment
- Physiological and biochemical changes during ripening
- Postharvest treatments and their effects on shelf life and quality
- Postharvest operations such as sorting, grading, ripening, de-greening, curing etc.
- Storage and shelf-life studies
- Packaging, transportation, and marketingVase life improvement of flowers and foliage
- Postharvest management of spice, medicinal, and plantation crops
- Fruit and vegetable processing waste/byproducts: management and utilization
- Postharvest diseases and physiological disorders
- Minimal processing of fruits and vegetables
- Quarantine and phytosanitory treatments for fresh produce
- Conventional and modern breeding approaches to improve the postharvest quality
- Biotechnological approaches to improve postharvest quality of horticultural crops

BOOKS IN THE SERIES

Postharvest Biology and Technology of Horticultural Crops: Principles and Practices for Quality Maintenance
Editor: Mohammed Wasim Siddiqui, PhD

Insect Pests of Stored Grain: Biology, Behavior, and Management Strategies
Ranjeet Kumar, PhD

Postharvest Management of Horticultural Crops: Practices for Quality Preservation
Editors: Mohammed Wasim Siddiqui, PhD, and Asgar Ali, PhD

Innovative Packaging of Fruits and Vegetables: Strategies for Safety and Quality Maintenance
Editors: Mohammed Wasim Siddiqui, PhD,
Mohammad Shafiur Rahman, PhD, and Ali Abas Wani, PhD

Advances in Postharvest Technologies of Vegetable Crops
Editors: Bijendra Singh, PhD, Sudhir Singh, PhD, and
Tanmay K. Koley, PhD

Plant Food By-Products: Industrial Relevance for Food Additives and Nutraceuticals
Editors: J. Fernando Ayala-Zavala, PhD, Gustavo González-Aguilar, PhD, and Mohammed Wasim Siddiqui, PhD

Emerging Technologies for Shelf-Life Enhancement of Fruits
Editors: Basharat Nabi Dar, PhD, and Shabir Ahmad Mir, PhD

Sensor-Based Quality Assessment Systems for Fruits and Vegetables
Editors: Bambang Kuswandi, PhD, and Mohammed Wasim Siddiqui, PhD

Emerging Postharvest Treatment of Fruits and Vegetables
Editors: Kalyan Barman, PhD, Swati Sharma, PhD, and
Mohammed Wasim Siddiqui, PhD

Packaging and Storage of Fruits and Vegetables: Emerging Trends
Editor: Tanweer Alam, PhD

Postharvest Physiological Disorders of Fruits
Editors: Ghulam Khaliq, PhD, and Sajid Ali, PhD

Bioengineered Fruit and Vegetables: Trends in Postharvest Quality Improvement
Editors: Mohammad Irfan, PhD, and Pankaj Kumar, PhD

Novel Packaging Systems for Fruits and Vegetables
Editors: Kirtiraj K. Gaikwad, PhD, and Suman Singh, PhD

Ozone Applications in the Food Industry: Trends and Future Prospects
Editors: Ravi Pandiselvam, PhD, Anjineyulu Kothakota, PhD, and
Kaavya Rathnakumar, PhD

Nonthermal Food Processing Technologies: Impact on Color Profile
Editors: Kaavya Rathnakumar, PhD, and R. Pandiselvam, PhD

Ready-to-Eat Snacks: Emerging Technologies for Production and Safety
Editors: R. Pandiselvam, PhD and Gülşah Çalışkan Koç, PhD

ABOUT THE SERIES EDITOR

Mohammed Wasim Siddiqui, PhD
Scientist-cum-Assistant Professor, Department of Food Science and Technology, Bihar Agricultural University, Sabour, Bhagalpur, Bihar, India

Mohammed Wasim Siddiqui, PhD, is an Assistant Professor and Scientist in the Department of Food Science and Post-Harvest Technology, Bihar Agricultural University, Sabour, India, and author or co-author of peer-reviewed research articles, book chapters, two manuals, and conference papers. He has over 20 edited and an authored books to his credit, published by Elsevier, CRC Press, Springer, and Apple Academic Press. Dr. Siddiqui has established the international peer-reviewed *Journal of Postharvest Technology*. He is Editor-in-Chief of two book series (*Postharvest Biology and Technology and Innovations in Horticultural Science*), published by Apple Academic Press. Dr. Siddiqui is also a Senior Acquisitions Editor for Apple Academic Press, USA, for Horticultural Science. He has been serving as an editorial board member and active reviewer of several international journals.

Dr. Siddiqui has received several grants and respected awards for his research work by a number of organizations. He is an active member of the organizing committees of several national and international seminars, conferences, and summits. He is one of key members in establishing the World Food Preservation Center (WFPC), LLC, USA, and is currently an active associate and supporter.

Dr. Siddiqui acquired a BSc (Agriculture) degree from Jawaharlal Nehru Krishi Vishwa Vidyalaya, Jabalpur, India. He received his MSc (Horticulture) and PhD (Horticulture) degrees from Bidhan Chandra Krishi Viswavidyalaya, Mohanpur, Nadia, India, with specialization in Postharvest Technology. He is a member of Core Research Group at the Bihar Agricultural University (BAU), which provides appropriate direction and assistance in prioritizing research.

ABOUT THE EDITORS

Mohammad Irfan, PhD
Postdoctoral Research Associate,
Plant Biology Section, School of Integrative Plant Science, College of Agriculture and Life Science, Cornell University, New York, USA

Mohammad Irfan, PhD, is a Postdoctoral Research Associate at the Plant Biology Section, School of Integrative Plant Science, College of Agriculture and Life Science, Cornell University, New York, USA. He is a well-established researcher with over 14 years of experience in plant and agricultural biotechnology, specializing in fruit biotechnology. Prior to joining Cornell, he worked at the Department of Biochemical and Chemical Engineering, Rutgers, The State University of New Jersey, USA. He earned his PhD in Agricultural Biotechnology at the National Institute of Plant Genome Research, New Delhi, India. Dr. Irfan has received several awards in recent years, including the Cornell Atkinson Small Grant Award by Cornell Atkinson Center for Sustainability, New York, USA (2022); the American Society of Plant Biologists-Early Career Plant Scientists-PB2021 Attendance Award (2021); the CSIR-Senior Research Associateship Award, India (2016); as well as having received international travel grant awards given by DBT, Government of India (2015) and the Society of Experimental Biology, UK (2016). He has also been granted various research fellowship awards by the Government of India, such as CSIR-UGC-JRF (2007), ICMR-JRF (2007), and ICAR-ASRB-NET (2013).

Dr. Irfan is a member of several international professional societies, including the American Society of Plant Biologists, the American Society for Horticultural Science, the New York Academy of Sciences, Hong Kong Chemical, Biological, and Environmental Engineering Society, the Indian Science Congress Association, and The Society of Biological Chemists, India. He has published over 40 research and review articles in high-impact factor international peer-reviewed journals and has contributed numerous book chapters. In addition, Dr. Irfan serves on the editorial boards of prestigious international journals such as the *Journal of Plant*

Growth Regulation, Frontiers in Plant Science, PeerJ, PLoS ONE, and the *Journal of Postharvest Technology*. He is also a review panel member of the Cornell University Schmittau-Novak Integrative Plant Science Small Grant Program and Boyce Thompson Institute Undergraduate Research Internship Program, funded by NSF and USDA. Dr. Irfan has presented his work at various international conferences and organizations in France, Sweden, Japan, India, and the USA.

Pankaj Kumar, PhD
Assistant Professor (Biotechnology),
Department of Biotechnology, Dr. Yashwant Singh Parmar University of Horticulture and Forestry, Solan, Himachal Pradesh, India

Pankaj Kumar, PhD, currently serves as an Assistant Professor (Biotechnology) at the Department of Biotechnology, Dr. Yashwant Singh Parmar University of Horticulture and Forestry, Solan, Himachal Pradesh, India. He completed his postdoctoral studies at the National Institute of Plant Genome Research, New Delhi, and the Council of Scientific and Industrial Research, Institute of Himalayan Bioresource Technology, Palampur, Himachal Pradesh, India. Dr. Kumar was awarded the DST INSPIRE JRF/SRF Fellowship by the Department of Science and Technology, Ministry of Science and Technology, Government of India for his PhD program. With eight years of expertise in plant genetic engineering and molecular biology, Dr. Kumar has focused on improving horticultural crops, particularly fruits and vegetables, using biotechnological tools. He has qualified for the ICAR AICE-SRF (PGS) and ICAR ASRB National Eligibility Test (NET). Dr. Kumar has published 65 research/review papers in various internationally and nationally recognized journals as well as contributed to 21 book chapters, five popular articles, two edited books (*Crop Improvement: Biotechnological Advances* (Routledge, Taylor and Francis Group (CRC Press) and *Agricultural Biotechnology: Latest Research and Trends* (Springer Nature) and two textbooks. He has received numerous awards from various scientific societies and organizations for his contributions, including the Young Scientist Award in Biotechnology (2018) by the Society for Plant Research, Young Scientist Award (2019) by the Society of Tropical Agriculture, New Delhi, India, and Excellence in Research Award (2019) by the Agro-Environmental Development Society, India.

CONTENTS

CONTRIBUTORS

Diksha Abrol
Department of Entomology, Dr. Y. S. Parmar University of Horticulture and Forestry, Nauni, Solan, Himachal Pradesh, India

Sekhu Ansari
Division of Pathology, Cincinnati Children's Hospital Medical Center, Cincinnati, Ohio, USA

Pranita Bhatele
Assistant Professor, Mata Gujri Mahila Mahavidyalaya, Jabalpur, Madhya Pradesh, India

Dharmendra Nath Bhatt
Laboratory of Immunogenetics, National Institute of Allergy and Infectious Diseases, National Institutes of Health, Bethesda, Maryland, USA

Rajat Chandel
Department of Food Science and Technology, Punjab Agricultural University, Ludhiana, Punjab, India

Kajal Choudhary
Research Scholar, Department of Plant Science, Central University of Himachal Pradesh, India

Mani Kant Choudhary
Department of Biology, University of Arkansas, Little Rock, Arkansas, USA

Neha Gautam
Department of Basic Sciences, Dr. Y. S. Parmar University of Horticulture and Forestry, Nauni, Solan, Himachal Pradesh, India

Sunakshi Gautam
Department of Food Technology, Dr. Y. S. Parmar University of Horticulture and Forestry, Nauni, Solan, Himachal Pradesh, India; M.S. Swaminathan School of Agriculture, Shoolini University, Bhajol, Solan, Himachal Pradesh, India

Mohammad Irfan
Plant Biology Section, School of Integrative Plant Science, Cornell University, Ithaca, New York, USA

Bhuvnesh Kapoor
University Institute of Biotechnology (UIBT), Chandigarh University, Gharuan, Mohali, Punjab, India

Ramandeep Kaur
Department of Food Science and Technology, Punjab Agricultural University, Ludhiana, Punjab, India

Alexander M. Khan
Tropical Marine Fisheries Undergraduate Program for Pangandaran Campus, Universitas Padjadjaran, Pangandaran, Indonesia

Ehasanullah Khan
Department of Biotechnology, Jamia Millia Islamia, New Delhi, India

Ankit Kumar
Center of Food Science and Technology, CCS-HAU, Hisar, Haryana, India

Anshul Kumar
Department of Fruit Science, Dr. Y. S. Parmar University of Horticulture and Forestry, Nauni, Solan, Himachal Pradesh, India; M.S. Swaminathan School of Agriculture, Shoolini University, Bhajol, Solan, Himachal Pradesh, India

Jitender Kumar
Assistant Professor, Department of Plant Science, Central University of Himachal Pradesh, India

Pankaj Kumar
Department of Biotechnology, Dr. Yashwant Singh Parmar University of Horticulture and Forestry, Nauni, Solan, Himachal Pradesh, India

Ritesh Kumar
Department of Agronomy, Kansas State University, Manhattan, Kansas, USA

Satish Kumar
Department of Food Science and Technology, Dr. Y. S. Parmar University of Horticulture and Forestry, Nauni, Solan, Himachal Pradesh, India

Surender Kumar
Department of Biotechnology, Dr. Yashwant Singh Parmar University of Horticulture and Forestry, Nauni, Solan, Himachal Pradesh, India

Vikas Kumar
Department of Food Science and Technology, Punjab Agricultural University, Ludhiana, Punjab, India

Vinay Kumar
Department of Biotechnology, Dr. Y. S. Parmar University of Horticulture and Forestry, Solan, Himachal Pradesh, India

Vinay Kumar
Pennsylvania State University HMC, Hershey, Pennsylvania, USA

Mukesh Kumar Meena
National Institute of Plant Genome Research, New Delhi, India; Leibniz Institute of Plant Biochemistry, Halle (Saale), Germany

Megha
Amity University, Manesar, Gurugram, Haryana, India

Jueria Mustaqeem
Dr. B.R. Ambedkar Center for Biomedical Research, University of Delhi, New Delhi, India

Nitika Negi
Department of Entomology, College of Agriculture, CSK HPKV, Palampur, Himachal Pradesh, India

Buntora Pasaribu
Department of Marine Science, Faculty Fisheries and Marine Science, Universitas Padjadjaran, Bandung, Indonesia

Noir Primadona Purba
Department of Marine Science, Faculty Fisheries and Marine Science, Universitas Padjadjaran, Bandung, Indonesia

Riddhi Rathore
Department of Biotechnology, Dr. Yashwant Singh Parmar University of Horticulture and Forestry, Nauni, Solan, Himachal Pradesh, India

Shelja
Research Scholar, Department of Plant Science, Central University of Himachal Pradesh, India

Anupama Singh
Department of Biotechnology, Dr. Yashwant Singh Parmar University of Horticulture and Forestry, Nauni, Solan, Himachal Pradesh, India

Nisha Singh
Gujarat Biotechnology University, Gandhinagar, Gujarat, India

Nitesh Singh
Department of Biosciences, UIBT, Chandigarh University, India

Shubham Smakaria
Department of Food Technology, Dr. Y. S. Parmar University of Horticulture and Forestry, Nauni, Solan, Himachal Pradesh, India

Kamlesh Kumar Soni
Institute of Plant and Microbial Biology, Academia Sinica, Taipei, Taiwan

Anchal Tandon
Department of Biotechnology, Dr. Yashwant Singh Parmar University of Horticulture and Forestry, Nauni, Solan, Himachal Pradesh, India

Rajul Tayal
National Institute of Plant Genome Research, New Delhi, India

Aayushee Thakur
Department of Biotechnology, Dr. Yashwant Singh Parmar University of Horticulture and Forestry, Nauni, Solan, Himachal Pradesh, India

Shikha Tiwari
Department of Biochemistry, All India Institute of Medical Sciences, Bhopal, Madhya Pradesh, India

Parul Tomar
Department of Biotechnology, Dr. Yashwant Singh Parmar University of Horticulture and Forestry, Nauni, Solan, Himachal Pradesh, India

Anuradha Tyagi
Department of cBRN, Institute of Nuclear Medicine and Allied Sciences, Brig S.K. Mazumdar, Delhi, India

Vikrant
Department of Biotechnology, Dr. Yashwant Singh Parmar University of Horticulture and Forestry, Nauni, Solan, Himachal Pradesh, India

PREFACE

The book titled, *Bioengineered Fruit and Vegetables: Trends in Postharvest Quality Improvement,* aims to address issues related to post-harvest losses in fruit and vegetable crops using cutting-edge technologies such as bioengineering, gene silencing, genome editing with zinc-finger nucleases (ZFNs), transcription activator-like effector nucleases (TALENs), and clustered regularly interspaced short palindrome repeats (Cas9). We attempt to bring together the most recent developments in this field, including gene overexpression approaches (enzymes, transcriptional factors, or combinatorial engineering), promoter engineering, enhancing postharvest stability (shelf life), epigenetic regulation, antioxidant-regulated expression, metabolic engineering, and regulation of postharvest pathogen infection and decay using biotechnological tools, as well as the current status and future prospects.

This book provides much-needed and up-to-date information on the application of modern technology and research for post-harvest quality assurance and improvements. It offers the latest insights and understanding of modern biotechnological and genome engineering tools in improving postharvest quality attributes of fruits and vegetables. It serves as a reference book for readers, providing them with the latest references and research-based case studies in different fruit and vegetable crops.

—Editors

CHAPTER 1

BIOENGINEERING OF FRUITS AND VEGETABLES: CURRENT STATUS AND FUTURE PROSPECTS

VINAY KUMAR,[1] PANKAJ KUMAR,[1] and MOHAMMAD IRFAN[2]

[1]*Department of Biotechnology, Dr. Y. S. Parmar University of Horticulture and Forestry, Solan, Himachal Pradesh, India*

[2]*Plant Biology Section, School of Integrative Plant Science, Cornell University, Ithaca, New York, USA*

ABSTRACT

Fruits, vegetables, and ornamentals are crucial for human health and the global economy, but their susceptibility to postharvest losses significantly reduces their quality and quantity. Bioengineering technologies offer promising solutions to extend shelf life, improve quality, and enhance resistance to stress. This chapter explores the current status and future prospects of bioengineering in horticulture, focusing on genetic engineering, suppression technology, and genome editing. We discuss how these techniques target ripening processes, regulate ethylene production, and modify key genes to achieve desired traits. Recent advancements in these technologies have led to the development of commercially available GM crops with extended shelf life and improved quality. However, regulatory hurdles and public concerns remain challenges that need to be addressed. Future directions include the development of new regulatory frameworks, increased public acceptance, and continued exploration of omics technologies for a deeper understanding of postharvest responses in

Bioengineered Fruit and Vegetables: Trends in Postharvest Quality Improvement.
Mohammad Irfan and Pankaj Kumar (Eds.)

fruits and vegetables. By harnessing the power of bioengineering, we can significantly reduce postharvest losses and ensure a sustainable and secure food supply for the growing global population.

1.1 INTRODUCTION

All fruits, vegetables, and ornamentals are considered horticultural crops, and the majority of them have the largest economic value due to their significant contribution to global agricultural production. The essential components for human diets are minerals, vitamins, fibers, carbohydrates, proteins, and organic acids found in fruits and vegetables, whereas ornamentals are valued for their aesthetic appeal. The quick perishability of these crops, as well as their ability to respire and transpire even after harvest, causes excessive ripening-related softness during post-harvest storage, which is a serious problem (Shipman et al., 2021; Irfan et al., 2023). In addition to providing a variety of wholesome nutrients, fruits, and vegetables also offer us a number of enjoyable sensory experiences that enhance our quality of life. However, these products are highly perishable. New crop varieties with improved characteristics are expected to be easier to cultivate than ever before in a short amount of time and at a low cost due to modern technology. Post-harvest physiology is a field of study that examines the physiology of harvested agricultural products, particularly living plant tissues. The techniques applied to agricultural produce after harvesting for preservation, conservation, quality control/enhancement, processing, packaging, storage, and various other purposes are essentially post-harvest technologies (Sharma & Kumari, 2020). To meet consumer expectations, post-harvest technologies focus on enhancing the nutritional content of food items. Nearly one-third of agricultural products available for human consumption are wasted each year. Food loss occurs when food is available for human consumption but remains uneaten. Post-harvest losses (PHLs) are qualitative and quantitative food commodity losses that frequently occur during post-harvest processes.

Over the past few decades, as the world's population has increased, so has the need for food. As a result, in order to meaningfully alleviate the food crisis, reduce the demand for natural resources, end hunger, and raise farmer income, agricultural commodity losses due to post-harvest operations must be limited (Sarwicka, 2019). Weight loss, nutritional value and quality loss, viability loss, and cost loss can all come from PHLs. The quality and quantity of agricultural goods can frequently deteriorate as a result of PHLs. Weight loss, changes in color and apparent quality, changes in nutritional content, and modifications in flavor are examples of quality loss, whereas quantity loss refers to a drop in the volume of the product (Hegazy, 2013). Therefore, it can be said that postharvest loss is one of the most important elements influencing agricultural production, the global food supply chain, and economic growth. Numerous factors that have an impact on the quality of the output may contribute to PHLs of agricultural commodities like primary, biological, and environmental factors (Figure 1.1).

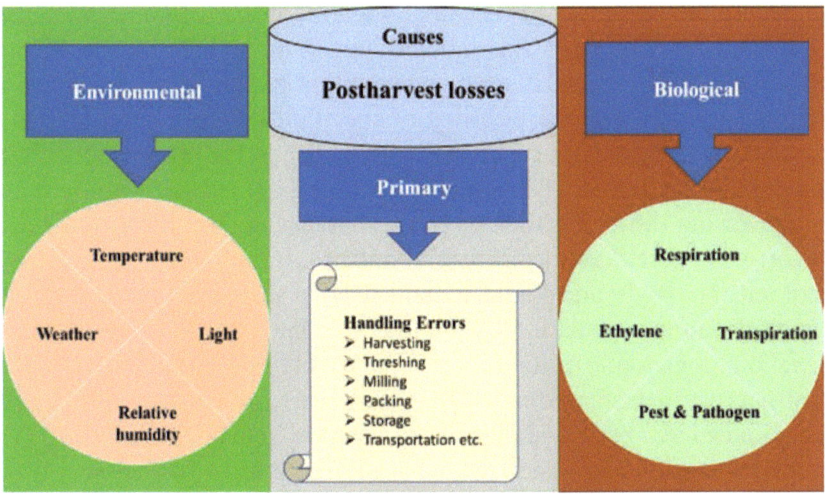

FIGURE 1.1 Various factors of post-harvest losses in horticultural crops.

Consuming fresh food is crucial for maintaining human health and has significant economic worth. However, physiological and pathological degradation causes 25–50% of fruits and vegetables to be lost or discarded after harvest. Production of crops places a lot of emphasis on improving postharvest storage. Increasing disease resistance, decreasing chilling susceptibility, and reducing water loss rate are the three key factors in fruits and vegetables (Kramer & Redenbaugh, 1994; Rommens et al., 2006). These are the losses of food in the food chain from harvest to consumption in terms of quality, nutrition, viability as seeds, and market value. For many years, extending crop shelf life using traditional methods such as low temperature, cold storage, and chemical treatment was the only approach to mitigate these losses. However, problems with low-temperature storage are brought on by insufficient storage capacity and organization, farmers' limited access to storage facilities, and inadequate transportation. Produce quality is also impacted by chemical processing, which shows health hazards due to its inappropriate use. So, for the enhancement of post-harvest features and their nutritional quality, breeders have made many attempts other than conventional breeding methods having low market value, labor-intensive, and time-consuming nature.

On the other hand, precision technologies like genetic and genomic editing have a huge potential to effectively address post-harvest losses and quality (Irfan & Datta, 2017). Since the emergence of civilization, plant biotechnology has improved plant production through crop domestication, breeding, and selection. Plant breeders adopt tactics such as the transformation of plants with the insertion of specific DNA sequences into plants broadening the gene pool. Food security is becoming more challenging as the population grows, yet many biotechnological techniques might assist in addressing difficult issues including hunger, malnutrition, disease, and poverty. Concerns have been raised about the virtues, societal repercussions, ecological hazards, and actual advantages of transgenic techniques in food plants (Rommens et al., 2006). On the other hand, novel technologies like genetic and genomic editing have a huge potential to effectively address post-harvest losses and quality. In recent years, genetic engineering (GE) techniques have been widely used to develop genetically modified (GM) crops by introducing genes for desirable features to reduce post-harvest losses and improve crop quality. The cause of post-harvest loss has been tackled in various crops using suppression technology such as asRNA (antisense RNA) and RNAi (RNA interference) (Savin et al., 1995). Utilizing complementary antisense sequences against the target sense RNA strand, asRNA technology uses antisense sequences to bind with the target sense strand and then inhibit gene expression. Understanding the biological mechanisms of fruit ripening and developing practical technologies, such as genome editing tools, is essential to enhance the quality, postharvest shelf life, marketability, processing attributes, and consumer appeal of these items. Ripening is a well-known irreversible process that gives the final commodity both good and bad qualities. Substantial, poor ripening features lead to significant expenditure and crop loss compared to decent ripening, which imparts good flavor, color, and texture (Owino & Ezura, 2011).

1.2 MOLECULAR BIOLOGY OF RIPENING

Research on the ethylene production route and the discovery and functional characterization of components of the ethylene signaling system in crop species has been motivated by the significance of ethylene in controlling features of agronomic interest (Kendrick & Chang, 2008). Due to its commercial importance, a rich source of biochemical and molecular data on the processes involved in ripening, relatively small genome, relative ease of genetic transformation, and availability of developmental mutants that are ripening impaired, the tomato has been the model system of choice for studying the changes during development and ripening of fleshy fruits and vegetables (Fei et al., 2006).

Extended shelf life is one of the most significant properties of tasty fruits, vegetables, and ornamentals, and it has a bigger influence on market potential. In breeding and genome engineering efforts, maintaining and extending shelf life is a significant difficulty. For many years, these crops have been maintained for a longer time in cold storage, but such a method is prohibitively expensive for smallholder farmers. As a result, it is critical to develop strategies for growing foods with a longer shelf life. To extend the shelf life of fruits and vegetables, the production of the plant hormone ethylene, commonly referred to as the ripening hormone, had to be managed (Jung et al., 2018; Li et al., 2018; Yu et al., 2017; Gao et al., 2019, 2020; Ezura & Owino, 2008; Lin et al., 2009). Senescence is the final stage of this ripening process under natural circumstances. Therefore, specific management techniques must be developed to postpone this ripening process during transit and storage in order to decrease the consequences of post-harvest losses. As a result of the horticultural produce's decline in quality from the ripening to the senescence stage, it is finally rejected by consumers, resulting in post-harvest losses. To retain the flavor, fragrance, and quality attributes, it is crucial to control the expression of genes associated with shelf life (Yu et al., 2017). Recently, sophisticated genome editing techniques have the potential to improve the quality and post-harvest features of horticultural crops. The tomato has a large platform of tools for functional genomics, including its genome sequence, which is continually being developed (Jung et al., 2018; Gao et al., 2020). However, other fruit species such as melon have been proposed as alternative models for studying fruit ripening and softening (Ezura & Owino, 2008).

The ethylene biosynthesis pathway, signal transduction genes, transcriptional control of fruit ripening, and genetic manipulation of ripening regulatory genes have been well characterized in tomatoes and other fruits (Lin et al., 2009; Irfan et al., 2022). The tomato's broad platform of functional genomics tools, which includes its genome sequence, is also constantly being improved. To examine how fruit ripens and softens, other fruit species, including the melon, have been suggested as alternatives (Ezura & Owino, 2008). In tomatoes and other fruits, the ethylene production process, signaling genes, transcriptional regulation of fruit ripening, and genetic modification of ripening regulatory genes have all been thoroughly characterized (Lin et al., 2009; Irfan et al., 2022). The model based on the formation of *MADS* protein complexes describes the *AGAMOUS C-type* lineage using the tomato *AGAMOUS (AG)* ortholog. It has been proposed that tomato MADS-box paralogs (*TAG1*, *TAGL2*, *TAGUI*, *TAGL12*, *TAGU*, *TDR6*, and *TDR4*) are involved in fruit development (Giovannoni, 2007). A putative *HD-Zip* homeobox protein that binds to the promoter of *ACC* (1-aminocyclopropane-1-carboxylic acid) oxidase (*LeACO1*), a crucial enzyme in the ethylene biosynthesis pathway, is encoded by the tomato *HB-1* gene (Lin et al., 2008). Ripening regulation requires the *MADS*-box and *SPBP* transcription factors (TFs) that are encoded by the rin and *CNR* (colorless non-ripening) genes, respectively (Manning et al., 2006). The fact that low ethylene, unripened fruit results from repression of *TAGU*, *MADS-RIN* (*MADS*-ripening inhibitor), *CNR-SBP*, or *HB-1* suggests that all four genes serve upstream of ethylene production and have some ethylene-independent roles. While *TAGU* regulates ethylene directly or indirectly) through regulation of *ACS2* (acetyl-coenzyme A synthetase 2) with no effect on *ACO1* (*ACC* oxidase), *HB1* regulates climacteric ethylene by direct regulation of *ACO1*. The rin mutant exhibits a decreased concentration of *CNR-SBP* mRNA, indicating that *MADS-RIN* positively regulates *CNR-SBP* expression (Vrebalov et al., 2009). It is also significant that (*rin/Rin*) hybrids serve as the foundation for the majority of the fresh market tomatoes produced today, proving the commercial significance of the *RIN* mutation. Arlequin (*Alq*) is a tomato mutant with a semi-dominant T-DNA that shows developmental alterations impacting flower and fruit ripening. The *Alq* T-DNA mutant's genetic, molecular, and functional investigations revealed that it corresponds to the *TAGL1* gene, a crucial *MADS*-box regulator of tomato reproductive development (Pineda et al., 2010; Gimenez et al., 2010). Since *ALQ/TAGL1* controls both flower and fruit development and functions upstream of ethylene-related genes, irrespective of the ripening pathway controlled by *RIN*, it cannot be regarded as a specialized fruit ripening gene. The *ALQ/TAGL1* could function as a connecting element between floral development and fruit ripening networks, enabling the effective completion of the reproductive process. However, the majority of these were hypothesized to be involved in early fruit development. Additional *MADS*-box genes have also been found for other fruits in recent years. Two *TAG* and *TAGL1*-like *MADS*-box genes from the peach have been cloned, and tomato expression of these genes suggests that they may be in charge of fruit development (Tadiello et al., 2009). Chinese pears *MAD*S-box genes have also been identified, and neither their expression in climacteric cultivars nor in cultivars that are not climacteric differed (Yamane et al., 2007). It has yet to be discovered if these genes are uniquely related to ripening in these fruits. Since *MaMADS2* is unaffected by ethylene, it is the most plausible candidate to act as an upstream regulator of ripening in Grand Nain bananas (Elitzur et al., 2010). It has been discovered that the strawberry RIN gene, which is associated with ripening, does not enhance respiration or ethylene output as the fruit ripens and is highly insensitive to ethylene action inhibitors such as 1-methylcyclopropene (*1-MCP*). This supports the idea that these regulators may be shared by both climacteric and non-climacteric fruits. Therefore, *NfN* and *CNR* may be effective alternatives to current ethylene control methods for reducing ripening and extending shelf life in the majority of produce. The research mentioned above shows that positive transcriptional regulators of ripening include *TAGL1, MADS-RIN, CNR-SBP,* or *HB-1*. The *APETALA 2 (AP2)* genes are a member of the *AP2/ERF* (Ethylene Response Element) family, primarily linked to the response to biotic and environmental stress, encoding putative TFs (Kim et al., 2006). The *APETALA2* gene *SlAP2a* from tomatoes was characterized, showing a detrimental effect on the ripening process by regulating the ethylene hormone and carotenoid pigmentation, crucial for tomato fruit color and nutritional quality (Chung et al., 2010).

The structural and chemical changes in the fruit cell wall during ripening are primarily responsible for the textural changes that take place in most fruits (Kumar et al., 2016). These modifications include adjustments to the constituent polysaccharides' structure and composition, which have been linked to the expression of a variety of hydrolases and transglycosylases, as well as the potential modification of covalent and non-covalent interactions between different polysaccharide classes (Vicente et al., 2007). One of the factors that may have impeded progress in identifying important individual drivers of textural alterations is the participation of a broader variety of enzymes. Other possible reasons include the fact that important enzyme activities are yet unknown, or that the textural changes are more complex than previously thought. However, it is evident that the control of texture and shelf life is much more intricate than previously believed, necessitating new strategies that involve more comprehensive models of the biochemical and physiological components that affect fruit firmness (Matas et al., 2009). The cuticle might be a biotechnological target to prolong fruit quality in addition to changing polysaccharide metabolism, for instance, as has been reported. Equivalent changes in cuticle architecture are also a crucial part of the ripening program. The capacity of the cuticle to maintain the integrity of the fruit's skin, to prevent cuticular transpiration, and to prevent microbial infection are only a few of their significant influences by other ripening-related structures, physiological processes, and biological processes that might have a significant influence on the quality and shelf-life of the fruit (Hovav et al., 2007). With the ultimate goal of enabling more thorough fruit ripening on the plant, promoting nutrient and flavor accumulation while maintaining fruit firmness and shelf-life, the integration of all these new approaches into a more comprehensive model of ripening-related textural changes should provide a host of new targets for gene manipulation.

1.3 DIFFERENT BIOENGINEERING TECHNOLOGY

1.3.1 GENETIC ENGINEERING (GE)

Crop yields are influenced more by adverse environmental circumstances than by pests and diseases, such as drought, floods, and severe heat. Plant scientists' principal goals are to find ways to sustain high productivity under stress and to develop crops with increased nutritional value. GM crops may prove to be successful complements to those cultivated using traditional methods in meeting the world's demand for high-quality meals.

GE, often known as genetic modification or genetic manipulation, is the direct alteration of an organism's DNA by biotechnology. It is a set of technologies used to change the genetic makeup of cells, including gene migration across and within species to generate better or altogether new animals. Artificial DNA synthesis is used to synthesize new DNA, whereas recombinant DNA procedures are used to isolate and copy the genetic material of interest. A vector construct is frequently used to insert this DNA into the host organism. Numerous steps that can be used for genetic manipulation in plants (Sharma & Kumari, 2020) (Figure 1.2).

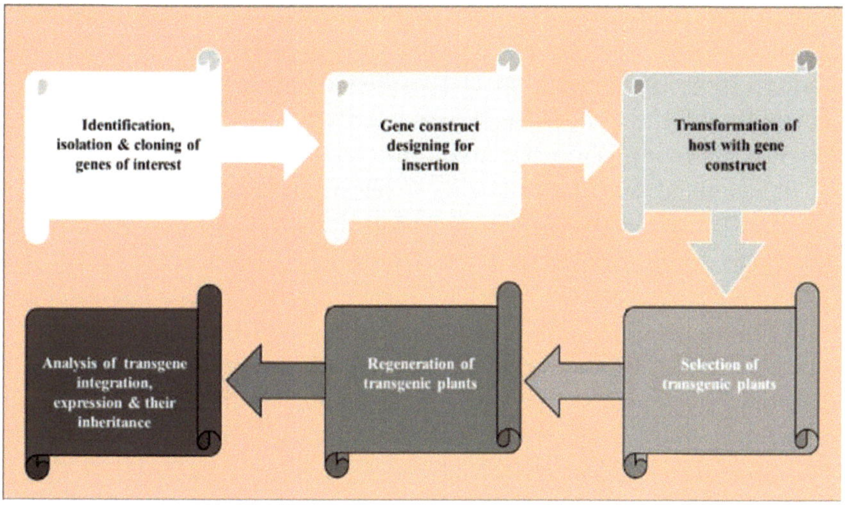

FIGURE 1.2 Steps used for genetic manipulation in plants.

GM crops can also be used to increase yields, nutritional quality, and resilience to a variety of biotic and abiotic stresses. Plants can also be genetically modified to produce recombinant proteins and specific chemicals, significantly enhancing their value. Despite some concerns about the biosafety and health risks associated with GM crops, the goal of achieving food security for both current and future generations can be attained through the sustainable integration of modern biotechnology with traditional agricultural practices. To meet the expanding worldwide need for food grains and agricultural commodities, biotechnology and genetic transformation technology will have a substantial impact on agricultural output and industry. Commonly used methods for reducing postharvest losses and improving shelf life, such as early harvesting, adjusting storage conditions, and selecting late ripening genotypes, may be inconsistent and insufficient for most commodities, leading to loss of taste, fragrance, and metabolites. Therefore, transgenic technologies are essential for developing improved crop varieties, enhancing plant resilience to various biotic and abiotic stresses, ensuring the nutritional quality of crop species, and various other applications. Agricultural postharvest degradation is a significant global issue. As previously mentioned, the key causes of shortened food shelf life include rapid rotting and softening of fruits and vegetables, low temperature, pathogen infection, and other factors. These factors have a negative impact on both product quality and availability. Biotechnological methods for enhancing postharvest factors in food products focus on various aspects: resilience to biotic and abiotic stresses, genetic transformation, secondary metabolites, simultaneous ripening for optimal harvesting, shelf life and improvement in organoleptic taste, and enhancement of nutritional quality (Sharma & Kumari, 2020; Owino & Ezura, 2011). Biotechnological techniques are also utilized to create new crop varieties that can adapt to current climate change, resist postharvest infections, and have an extended shelf life. The plant physiological, biochemical, and molecular mechanisms associated with postharvest attributes of fruits and vegetables, such as storage and shelf life, are well understood. Consequently, by employing molecular and transgenic approaches, these characteristics can be genetically analyzed and manipulated (Sharma & Kumari, 2020; Owino & Ezura, 2011).

1.3.2 SUPPRESSION TECHNOLOGY

Technology for suppression, such as RNA interference (RNAi) and antisense-mediated silencing, was more effective in modifying at the transcriptional or post-transcriptional level. These methods are used to reduce or stop a protein's production from its associated gene.

1.3.2.1 ANTISENSE RNA (ASRNA) TECHNOLOGY

A messenger RNA (mRNA) strand's complementary strand is a single-stranded RNA called an antisense RNA (asRNA). To create an antisense gene, a protein-coding region of a gene is inverted, or its orientation in relation to its promoter is reversed. Consequently, the gene's natural sense strand is oriented in the 3'-5' direction with respect to its promoter and initiates transcription. The gene's typical antisense strand, now oriented 5'-3', remains untranscribed. As tools for gene knockdown research, asRNAs can be synthetically produced and are widely utilized. By introducing single-stranded RNA into the cell that is complementary to an mRNA encoding proteins, gene expression is diminished. The asRNA binds to the mRNA strand, and the double-stranded RNA (dsRNA) molecules are targeted and degraded by the dsRNA-specific RNAse H, preventing gene translation into protein.

1.3.2.2 RNA INTERFERENCE (RNAI)

Other types of suppression technology include RNAi, which is a biological process in which RNA molecules such as siRNA and miRNA, collectively known as small RNAs, inhibit gene expression by neutralizing targeted mRNA molecules that share homology with small RNAs. Andrew Z. Fire and Craig C. Mello were awarded the Nobel Prize in Physiology or Medicine in 2006 for discovering the mechanism of gene silencing by dsRNA through research in the nematode worm *Caenorhabditis elegans*.

1. **Components of RNAi:** Dicer belongs to the RNase III family of endoribonucleases. It cleaves dsRNA into short dsRNA fragments known as small interfering RNA (siRNA), which are about 20–25 nucleotides long and have a two-base overhang on the 3′-end. MicroRNAs (miRNAs) are single-stranded RNA molecules with a length of 21–23 nucleotides that regulate gene expression. MiRNA molecules are partially complementary to one or more mRNA molecules and have the primary function of regulating gene expression.
2. **Small Interfering RNA (siRNA):** Also known as short interfering RNA or silencing RNA, siRNA is a type of dsRNA molecule with a length of 20–25 nucleotides that plays a variety of roles in biology. siRNA is an exogenous RNA that is directly introduced into the cytoplasm from the outside for gene suppression. The RNA-induced silencing complex (RISC) is a multiprotein complex that incorporates one strand of a siRNA or miRNA and uses it

as a template for recognizing complementary mRNA. The RISC complex eventually interrupts translation by cleaving a complementary piece of RNA with the argonaute (AGO) protein complex. Through suppression technology, a number of horticulture crops were modified, and their quality was related to PHLs (Figure 1.3).

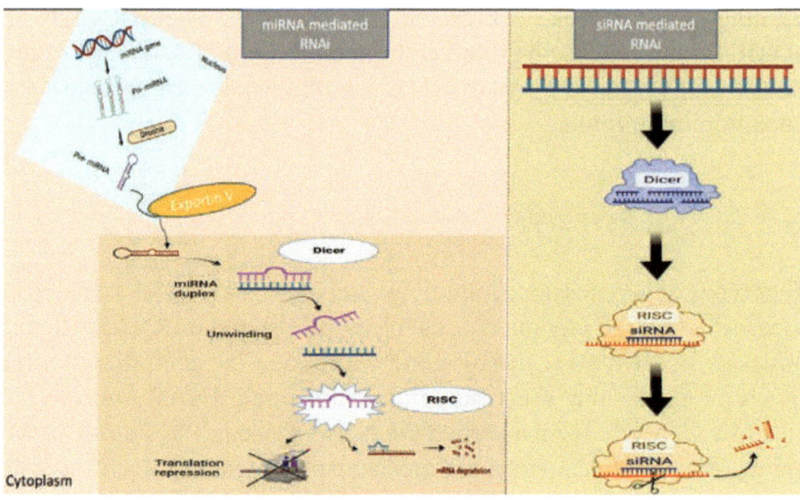

FIGURE 1.3 RNAi-mediated gene silencing in plants.

1.3.3 GENOME EDITING

The term "genome editing" refers to a class of advanced molecular biology techniques that allow for precise, effective, and targeted changes to genomic regions. Advanced biotechnological methods are made possible by genome-editing technologies, allowing for precise and effective targeted change of an organism's genome.

1.3.3.1 MAIN TOOLS/GENOME EDITORS

Various genome editing tools, such as CRISPR/Cas9 (clustered regularly interspaced short palindromic repeats (CRISPR)/CRISPR-associated protein 9 (Cas9) (Chen et al., 2019), TALENs (Christian et al., 2010), and ZFNs (Kim et al., 1996), are mostly used to improve the quality of various crops. Various genome editing tools like CRISPR/Cas9 (Chen et al., 2019), transcription activator-like effector nucleases (TALENs) (Christian et al., 2010), and zinc-finger nucleases (ZFNs) (Kim et al., 1996) are mostly used to enhance the quality of various crops. Site-specific double-stranded DNA breaks (DSBs) are created by these tools using endonucleases, and the internal DNA repair mechanisms of plants may then fix either through homology-directed/dependent recombination (HDR) or non-homologous end-joining (NHEJ). In the case of HDR, it results in the insertion of particular sequences specified by repair templates given in Trans, whereas repair by NHEJ results in allele knockout via random insertions or deletions (InDel) (Sharma & Sharma, 2021; Doudna & Charpentier, 2014).

1. **Zinc-Finger Nucleases (ZFNs):** These are synthetic restriction enzymes created by joining a DNA-cleaving domain to a zinc finger (ZF) DNA-binding domain (restriction endonuclease *FokI*). Zinc-finger nucleases may target distinct regions within complicated genomes because the zinc-finger domains are customized with target DNA sequences. Precisely, ZFNs are mostly used in the modification of higher species genomes by utilizing the native DNA repair mechanism (nonhomologous end joining and homologous recombination (HR)).
2. **Transcription Activator-Like Effector Nucleases (TALENs):** Restriction enzymes known as TALENs can be created to cut certain DNA sequences. They are created by joining a DNA cleavage domain to a TAL effector DNA-binding domain (restriction endonuclease *FokI*). When used in conjunction with a nuclease, transcription activator-like effectors (TALEs) develop to bind nearly to any desired DNA sequence. As a result, DNA can be cut at certain regions. These compounds can be used to accurately modify the genomes of higher species by utilizing the native DNA repair mechanism (nonhomologous end joining and HR).
3. **Clustered Regularly Interspaced Short Palindromic Repeats (CRISPR)/Cas:** A family of DNA sequences known as the clustered regularly interspaced short palindromic repeats (CRISPR)/Cas systems may be found in the genomes of prokaryotic species like bacteria and archaea. Cas9, also known as "CRISPR-associated protein 9," is an enzyme that recognizes and cleaves certain DNA strands that are complementary to CRISPR sequences by using such sequences as a guide strand. Providing a sort of acquired

immunity, the CRISPR-Cas system is a prokaryotic immune system that gives resistance to invading genetic material found in plasmids and phages. The spacer-containing RNA aids Cas (CRISPR-associated) proteins in identifying and slicing foreign pathogenic DNA. There are two domains, the DNA binding (DBD) domain of ZFNs or TALENs proteins and the nuclease domain based on type II restriction endonuclease FokI, which are united to form ZFNs and TALENs. DSBs (double strand breaks) are generated due to interactions between proteins and DNA. Conversely, the CRISPR/Cas9 system is founded on the defensive mechanism of bacteria (*Streptococcus pyogenes*). Bacteria utilize this system to combat invading bacteriophages and viruses by cleaving their DNA in a sequence-specific manner. The CRISPR/Cas9 system transitioned from a pivotal role in bacterial defense to a genome editing tool (sgRNA) by virtue of its capacity to be reprogrammed through modification of a few base pairs (20) in a single guide RNA (sgRNA). In this strategy, the nonspecific Cas9 endonuclease collaborates with the specific sgRNA/gRNA (single guide RNA). The protospacer adjacent motif (PAM) sequence, essential for Cas9's selective cleavage and induction of DSBs in the target sequence, is guided by the sgRNA/gRNA (Sharma & Sharma, 2021; Doudna & Charpentier, 2014; Jiang & Doudna, 2017). CRISPR/Cas9 surpasses ZFNs and TALENs as an editing technique due to its precision, cost-effectiveness, reprogramming capability, and versatility. Researchers employed a transformation process, akin to GE, to transfer CRISPR/Cas9 components into host plant cells. Commonly utilized biolistic weapons, protoplast fusion, or *Agrobacterium*-mediated techniques can be employed to accomplish this. While RDNA technology employs regulatory elements such as promoters, terminators, selectable marker genes, T-DNA, and so on, these approaches have limitations such as gene silencing and positional effects (Subburaj et al., 2016). Therefore, plants become eligible for GMO control when foreign recombinant DNA pieces are incorporated into their genomes. The inserted CRISPR/Cas9 complex has a short lifespan, which may be why the DNA-free preassembled Cas9/gRNA ribonucleoprotein (RNP) complex cleaves the target sites quickly after delivery and degrades swiftly. When compared to the aforementioned

DNA-based Cas9/gRNA complex delivery approaches, the frequent off-targeting may be attributed to the ongoing production of functional enzymes (Liu et al., 2019). Numerous RNP delivery techniques have been developed for the introduction of genome editing tools in target plants and animals. This encompasses a variety of methods spanning physical, biological, carrier-based, and chemical approaches, including microinjection, biolistic delivery, electroporation, microfluidics, filtroporation, nano-tubes, polyethylene glycol (PEG), protoplast transformation, lipofection, virus-like particles, lipid nanoparticles, lipopeptides, and various polymers. Among these techniques, biolistic delivery and PEG-mediated protoplast transfection have emerged as the primary methods for delivering ribonucleoproteins (RNPs) in plants (Zhang et al., 2021) (Figure 1.4).

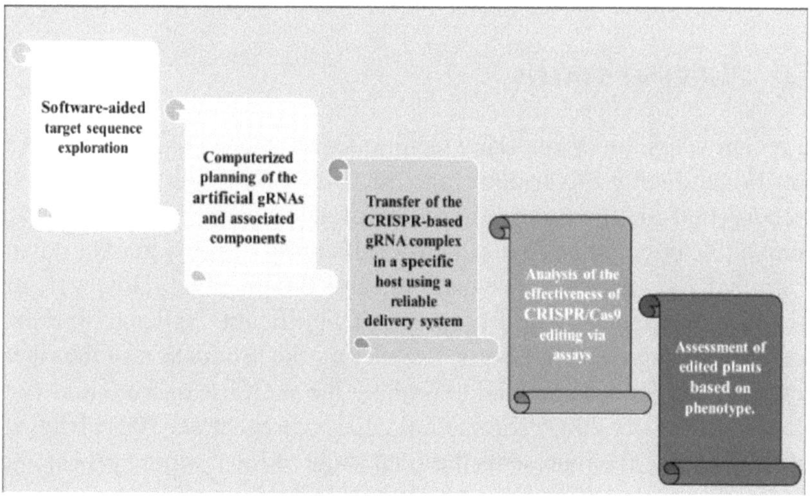

FIGURE 1.4 Steps involved in CRISPR/Cas9 genome editing system.

This technology has recently gained recognition on a worldwide scale and will receive the 2020 Nobel Prize. Numerous programs and websites, including Genome CRISPR, CRISPR-Plant, Cas-OFFinder, CRISPy-Web, CrisprGE, CRISPR-P, Prime Design, GT-Scan, pegFinder, CRISPRdirect, RGEN BE-designer, CRISPOR, SSFinder, RGEN Cas designer, CHOPCHOP, etc., have been developed to make this technology more effective and easy to use (Hassan et al., 2021; Wang et al., 2021). In addition to CRISPR/Cas9, other genome editing systems such as CRISPR/Cpf1 and Cas12a were developed from the bacterial species *Prevotella* and *Francisella* (Matje-Sprink et al., 2019). Crops with modified genomes are categorized into three groups based on mutation type/donor DNA editing behavior and repair mechanisms: SDN-1 (site-directed nucleases 1), SDN-2 (site-directed nucleases 2), and SDN-3 (site-directed nucleases 3). The use of SDN-1 results in gene knockout via random mutation in the form of insertions and deletions, which uses the endogenous NHEJ pathway to repair the DSBs (InDels). SDN-2 relies on HDR repair with targeted sequence change at the DSB target region, resulting in gene modification (gain of function) by using homologous template DNA (short). In contrast to SDN-1 and SDN-2, SDN-3 employs the HDR process by incorporating an additional gene or DNA sequence that results in gene insertion (Naegeli et al., 2020; Podevin et al., 2013; Entine, 2021).

1.4 CURRENT STATUS

In recent years, bioengineering techniques have been used to improve the nutritional quality of various crops, focusing on postharvest storage traits such as shelf life, texture, pathogen resistance, flavor, and nutrition development. A longer shelf life, which significantly impacts market potential, is crucial for soft fruits, vegetables, and ornamentals. However, maintaining and extending shelf life poses a significant challenge in breeding and genetic engineering efforts. Regulating the production of the ripening hormone ethylene is essential to prolong the shelf life of fruits and vegetables (Matas et al., 2009; Xiong et al., 2015; Tayal et al., 2021; Irfan et al., 2021). Senescence represents the final stage of the ripening process under natural conditions (Ghosh et al., 2013). In various climacteric and non-climacteric species, ripening inhibitor gene (*RIN*), Alcobaca gene (*ALC*), *ACC* (1-amino-cyclopropane-1-carboxylic acid) synthase gene, *PPO* (polyphenol oxidase), *Vinv* (Vacuolar invertase) gene, *Pg* gene (polygalacturonase) and *SAM* gene (S-adenosylmethionine hydrolase) responsible for the regulations of fruit ripening. By using these enzymes, bioengineering techniques were employed in various fruit crops to enhance their shelf life and quality (Albornoz et al., 2019). In recent years, developers such as Monsanto Company, Agritope Inc. (USA), Okanagan Specialty Fruits

Incorporated, J.R. Simplot Co., and Del Monte Fresh Produce Company have introduced a number of GM crops that were commercialized through GE and suppression technology in various crops, such as melon (Melon A and B), apple (Arctic™ "Golden Delicious" Apple, Arctic™, and Arctic™ Fuji Apple), potato (Innate® Acclimate, Simplot Innate, Innate® Hibernate, Innate® Cultivate, and Innate® Invigorate), and pineapple (Rosé), respectively (Table 1.1).

However, advanced genome editing methods may be used to improve the quality and post-harvest characteristics of horticultural crops. The tomato is a model climacteric crop that has been the subject of the most scientific research for genome editing studies because it was the first crop to be altered through GE and because molecular-level data on shelf-life processes are readily available. Editors used various genes and their regulated elements like *RIN, ALC, Vinv*, and *PPO* in various crops like mushrooms, tomatoes, and Potato (Lu et al., 2018; Jung et al., 2018; Yu et al., 2017; Waltz, 2016; Clasen et al., 2016) (Table 1.2).

In the case of horticultural crops, texture remained a crucial characteristic for post-harvest stability. These crops' hardness and softening processes are regulated by a variety of texture-related enzymes, including β-galactosidase (*β-gal*), N-glycoprotein-modifying enzymes, e.g., α-mannosidase (*α-Man*) and β-D-N-acetylhexosaminidase (*β-Hex*), polygalacturonase (*PG*), expansin (*EXPs*) endo-b-(1,4)-glucanase (*EGs*) and pectin methyl esterase (*PME*). The suppression of relevant gene expression in tomatoes and strawberries has been documented by several studies (Irfan et al., 2014, 2016). In tomato, the Monsanto company developed the first commercialized GM crop, FLAVR SAVR™, through antisense technology. However, this gene also somewhat affects the firmness of strawberries with greater Brix levels (Quesada et al., 2009). Furthermore, in tomatoes, another gene like the pectate lyase (*PL*) gene, which encodes a protein related to the cell wall, has been successfully silenced to improve fruit texture without affecting their other physical and biochemical parameters, ultimately affecting the sensory characteristics in strawberries (Jimenez-Bermudez et al., 2002) and tomatoes (Uluisik et al., 2016), respectively. CRISPR/Cas9 editing was used in the case of the *PL* gene, in which mutants had a favorable effect on fruit textures without altering their fruit's color, fragrance, and flavor parameters (Uluisik et al., 2016).

Fruits, vegetables, and ornamentals are most at risk from post-harvest infections at all stages of production, including ripening, harvesting, transit

TABLE 1.1 Applications of Genetic Engineering and Suppression Technology in Horticulture Crops to Improve Their Postharvest Quality and Shelf Life

Crop Species	Target Gene	Function	Event Name, Code, and Trade Name	Developer	Bioengineering Techniques	Transformation Method	Outcomes
Tomato	*PG* gene	Breakdown of pectin content in the cell wall.	B, SYN-ØØØØB-6 Da, SYN-ØØØDA-9 F (1401F, h38F, 11013F, 7913F), SYN-ØØØØF-1	Zeneca Plant Science and Petoseed Company	Antisense RNA technology.	*Agrobacterium tumefaciens* mediated transformation.	Delayed fruit softening.
			FLAVR SAVR™, CGN-89564-2 and *FLAVR SAVR*™	Monsanto Company			
	ACC (truncated)	A modified transcript of the 1-amino-cyclopropane-1-carboxylic acid synthase gene that reduced ethylene production and postponed fruit ripening.	1345-4	DNA Plant Technology Corporation (USA)	Genetic engineering.	*A. tumefaciens* mediated transformation.	Delayed ripening/ senescence.
	SAM-k	S-adenosylmethionine hydrolase enzyme which hydrolyzes SAM which used as a substrate for ethylene production.	35-1-N	Agritope Inc. (USA)			
	ACCd	1-amino-cyclopropane-1-carboxylic acid deaminase enzyme which metabolizes precursor of ethylene hormone that causes fruit to mature.	8338, CGN-89322-3	Monsanto Company			
	Anti-*efe*	Antisense RNA of the enzyme 1-amino-cyclopropane-1-carboxylate oxidase gene which delays ripening by reducing ethylene production by silencing the ACO gene.	Huafan No. 1	Huazhong Agricultural University (China)	Antisense RNA technology.	Microparticle bombardment.	Delayed ripening/ senescence.
Melon	*SAM-k*	S-adenosylmethionine hydrolase enzyme which hydrolyzes SAM which used as a substrate for ethylene production.	Melon A Melon B	Agritope Inc. (USA)	Genetic engineering	*A. tumefaciens*-mediated transformation	Delayed ripening/ senescence

TABLE 1.1 *(Continued)*

Crop Species	Target Gene	Function	Event Name, Code, and Trade Name	Developer	Bioengineering Techniques	Transformation Method	Outcomes
Pineapple	*Psy*	The phytoene synthase gene is responsible for the regulation of lycopene and beta-carotene levels.	EF2-114, FDP-ØØ114-5 and*Rosé*	Del Monte Fresh Produce Company	RNAi	*A. tumefaciens* mediated transformation.	Delayed ripening/ senescence.
	b-Lyc	Produce gamma-carotene involve in beta-carotene biosynthesis.					
	e-Lyc	Produce delta-carotene is involved in beta-carotene biosynthesis.					
	ACC	Responsible for the penultimate stage of the ethylene biosynthesis pathway.					
Potato	*asn1*	Asparagine synthetase gene that increases asparagine level.	Gen2-Z6, SPS-ØØØZ6-5, and *Simplot Innate*	J.R. Simplot Co.	RNAi	*A. tumefaciens* mediated transformation.	Decrease asparagine level, reduced black spot, decreased reducing sugars, foliar late blight resistance.
	ppo5	Polyphenol oxidase gene which oxidizes phenolic compounds.					
	PhL	Phosphorylase-L gene that increases the accumulation of reducing sugar levels.					
	VInv	Vacuolar invertase gene that can hydrolyze sucrose into glucose and fructose.					
	Rpi-vnt1	R gene confers resistance to potato late blight.					
Potato	*R1, Rpi-vnt1 asn1, ppo5, PhL,* and*VInv*	R1 gene regulates sugar levels while Rpi-vnt1 asn1, ppo5, Phl, andVInv showed resistance to potato late blight disease, asparagine synthesis, oxidized phenolic compounds, and hydrolyzed sugar level, respectively.	W8, SPS-ØØØW8-4 X17, SPS-ØØX17-5 and *Innate® Acclimate* Y9, SPS-ØØØY9-7 and*Innate® Hibernate*	J.R. Simplot Co.	RNAi	*A. tumefaciens*-mediated transformation	Decrease asparagine level, reduced black spot, decreased reducing sugars, foliar late blight resistance.

TABLE 1.1 *(Continued)*

Crop Species	Target Gene	Function	Event Name, Code, and Trade Name	Developer	Bioengineering Techniques	Transformation Method	Outcomes
Potato	*R1*, *asn1*, *ppo5* and*PhL*	R1 gene regulates sugar level while asn1, ppo5, andPhl genes are responsible for asparagine synthesis, oxidization of phenolic compounds, and hydrolyzed sugar level, respectively.	E12, SPS-ØØE12-8 and *Innate® Cultivate* E24, SPS-ØØE24-2 E56, SPS-ØØE56-7 F10, SPS-ØØF10-7 and*Innate® Generate* F37, SPS-ØØF37-7 G11, SPS-ØØG11-9 H37, SPS-ØØH37-9 H50, SPS-ØØH50-4 J3, SPS-ØØØJ3-4 and*Innate® Accelerate* J55, SPS-ØØJ55-2 J78, SPS-ØØJ78-7 V11, SPS-ØØV11–6 and *Innate® Invigorate*	J.R. Simplot Co.	RNAi	*A. tumefaciens* mediated transformation.	Decrease asparagine level, reduced black spot, decreased reducing sugars.
Potato	*GBSS* gene	Granule-bound starch synthase gene is responsible for amylase synthesis.	AM04-1020, BPS-A1Ø2Ø-5, andstarch potato EH92-527-1, BPS-25271-9 and *Amflora*™	BASF	Antisense RNA technology	*A. tumefaciens* mediated transformation.	Modified starch content
Apple	*PGAS PPO* suppression gene.	*PPO* suppression transgenes are responsible for the oxidization of phenolic compounds.	GD743, OKA-NBØØ1-8, and*Arctic*™ *"Golden Delicious" Apple* GS784, OKA-NBØØ2-9 and*Arctic*™ NF872, OKA-NBØØ3-1 and*Arctic*™ *Fuji Apple*	Okanagan Specialty Fruits Incorporated	RNAi	*A. tumefaciens* mediated transformation.	Reduction in the enzymatic browning.

TABLE 1.2 Applications of Genome Editing Tools in Horticulture Crops to Improve Their Postharvest Quality and Shelf Life

Crop Species	Target Gene	Function	Tool	Delivery Method	Repair Mechanisms	Outcome	References
Mushroom	*PPO*	Polyphenol oxidase catalyzed the oxidation of phenolic compounds into quinones.	CRISPR/Cas9	PEG-mediated transformation.	NHEJ	Reduced enzymatic browning.	Waltz (2016)
Tobacco	*BBL*	Berberine bridge enzyme used for biosynthesis of benzylisoquinoline alkaloids.	Mega nuclease	*A. tumefaciens* mediated transformation.	NHEJ	Reduced nicotine level.	Ku & Ha (2020)
Tomato	*Mutated ALC*	Alcobaca gene that inhibits ethylene biosynthesis.	CRISPR/Cas9	*A. tumefaciens* mediated transformation.	HDR	Longer shelf life	Lu et al. (2018); Yu et al. (2017)
	RIN	Ripening inhibitor gene used for inhibition of ethylene biosynthesis.	CRISPR/Cas9	*A. tumefaciens* mediated transformation.	NHEJ	Delayed ripening	Jung et al. (2018)
	CaMBD	Calmodulin-binding domain (CaMBD), by the removal of CaMBD, glutamic acid decarboxylase enzyme activity increases, which catalyzes the conversion of glutamate to GABA.	CRISPR/Cas9	–	NHEJ	Enhanced GABA content.	Waltz (2022)
	PL	The pectate lyase gene is responsible for plant cell wall degradation.	CRISPR/Cas9	–	NHEJ	Enhance fruit firmness.	Sonker et al. (2016)
	SlMAPK3	*Solanum lycopersicum* mitogen-activated protein kinases responsible for defense for biotic and abiotic stresses.	CRISPR/Cas9	–	NHEJ	Resistance to *Botrytis cinerea*	Zhang et al. (2018)
Apple	*IdnDH*	L-idonate dehydrogenase gene is responsible for tartaric acid biosynthesis.	CRISPR/Cas9	*A. tumefaciens* mediated transformation and PEG transformation system.	–	Enhance tartaric acid accumulation.	Osakabe et al. (2018)
Potato	*StGBSS*	Granule-bound starch synthase gene is responsible for amylase synthesis.	CRISPR/Cas9	*A. tumefaciens* mediated transformation and PEG transformation system.	NHEJ	Enhance amylopectin content.	Andersson et al. (2017); Kusano et al. (2018)

TABLE 1.2 *(Continued)*

Crop Species	Target Gene	Function	Tool	Delivery Method	Repair Mechanisms	Outcome	References
	SBE1 and *INV2*	Starch branching enzyme used for amylopectin formation by introducing α-1,6-linkages into starch, invertase gene responsible for irreversible hydrolysis of sucrose into glucose and fructose.	TALEN	*A. tumefaciens* mediated transformation.	NHEJ	Enhance amylopectin content and cold sweetening.	Ma et al. (2017)
	SBE1 and *SBE2*	Starch branching enzyme used for amylopectin formation by introducing α-1,6-linkages into starch.	CRISPR/Cas9	*A. tumefaciens* mediated transformation and PEG transformation system.	NHEJ	Improved amylopectin content.	Tuncel et al. (2019)
	Vinv	The vacuolar invertase gene is responsible for the irreversible hydrolysis of sucrose into glucose and fructose.	TALEN	*A. tumefaciens* mediated transformation and PEG transformation system.	NHEJ	Enhance cold storage and processing traits.	Clasen et al. (2016)
	PPO2	Polyphenol oxidase catalyzed the oxidation of phenolic compounds into quinones.	CRISPR/Cas9	PEG-mediated transformation.	NHEJ	Reduced enzymatic browning.	Gonzalez et al. (2020)
Citrus	*LOB1*	Lateral organ boundaries transcription factor, susceptibility gene of citrus canker.	CRISPR/Cas9	*A. tumefaciens*-mediated transformation	NHEJ	Enhance resistance to citrus bacterial canker disease.	Hu et al. (2014); Jia et al. (2017)
Chili pepper	*ERF28*	Ethylene-responsive factor gene, susceptibility gene of anthracnose disease.	CRISPR/Cas9	*A. tumefaciens* mediated transformation.	NHEJ	Enhance resistance to anthracnose disease.	Mishra et al. (2021)
Grape	*IdnDH*	L-idonate dehydrogenase gene is responsible for tartaric acid biosynthesis.	CRISPR/Cas9	*A. tumefaciens* mediated transformation.	NHEJ	Enhance tartaric acid accumulation.	Ren et al. (2016)

from field to farm, processing, and storage. Abiotic variables, such as temperature, relative humidity, and oxygen balance, have a significant impact on how susceptible horticultural crops are to pathological assaults during storage. Fungi, bacteria, yeast, and molds are the pathogens most frequently to blame for post-harvest losses (Sonker et al., 2016). The different infections, bacterial rots, and fungal rots (Ramirez-Gil et al., 2021) are the most destructive during post-harvest management of horticulture crops because they seriously impair perishables and canned goods. The majority of fruits and vegetables decay due to bacterial and fungal rot (Sonker et al., 2016). Various species of *Erwinia, Pseudomonas, Bacillus, Lactobacillus,* and*Xanthomonas* are the main causative bacterial soft rots agents. However, during different post-harvest procedures, fungal infections—which result in rot in fruits and vegetables—are more prevalent than bacterial ones. *Pythium, Sclerotium botrytis, Colletotrichum, Aspergillus, Dothiorella, Monilinia, Alternaria, Penicillium, Fusarium, Phomopsis, Rhizoctonia, Phytophthora, Diplodia, Rhizopus, Mucor*, etc., all contribute significantly to post-harvest losses. In recent years, bioengineering techniques like genetic transformation, suppression technology as well as genome editing tools have been widely used to minimize post-harvest loss in particular crops.

In potatoes, in addition to the shelf-life trait, J.R. Simplot Co. used the R gene through RNAi technology to develop a number of commercialized GM crops like Innate® Acclimate, Simplot Innate, and Innate® Hibernate to enhance resistance to Foliar late blight. Citrus canker, which manifests as severe symptoms in the stem, leaf, and fruit, is caused by the bacteria *Xanthomonas citri* subsp. Citri and is mostly responsible for economic losses in citrus. The occurrence of this illness both before and after harvest raises concerns. The CRISPR/Cas9 technique was used to overcome the losses caused by the pathogen by blocking LATERAL ORGAN BOUNDARIES 1 (*CsLOB1*), a key transcriptional factor responsible for disease manifestation (Hu et al., 2014; Jia et al., 2017). Similarly, in chili peppers, the CRISPR/Cas9 technique was used to knock out the expression of the ethylene-responsive factor gene (*ERF 28*), which is responsible for anthracnose disease manifestation (Mishra et al., 2021).

Numerous elements, including genetic makeup, field conditions, method of harvesting, and post-harvest treatment, have a significant impact on flavor and quality. Depending on the circumstances they are kept in for a lengthy period of time after harvesting until consumption, these perishable crops lose their distinctive edible characteristics (taste, fragrance, nutritional value, etc.), which results in an inedible state (off flavors). In order to retain the highest flavor and nutritional quality of horticulture products from harvest to consumption, it is crucial to cultivate the finest tasting genotypes utilizing better and updated techniques (Kumari et al., 2022). By altering *the GBSS* gene (granule-bound starch synthase gene) through antisense technology, BASF developed commercialized GM crops like Starch Potato and Amflora™ having modified starch content. On the other hand, nowadays genome editing tools in fruits and vegetable crops related to nutritional and flavor quality are an emerging field. GABA (gamma-aminobutyric acid) is a neurotransmitter with calming and blood pressure-lowering properties. By altering the GABA production route, the tomato variety "Sicilian Rouge" was created by Sanatech Seed with a high GABA content (4–5-fold) as the first commercialized genome-edited crops. To enhance the activity of the glutamic acid decarboxylase enzyme, which is involved in the decarboxylation reaction converting glutamate to GABA, the process includes the GABA shunt or the disruption of the calmodulin-binding domain (*CaMBD*) genes (Waltz, 2015). In potatoes, the CRISPR/Cas9 system was utilized to generate novel starch features by mutating the starch-branching enzymes (*SBE1 and SBE2*). Decreasing the expression of both *SBE1* and *SBE2* enzymes led to reduced amylopectin branching during granule growth, while *SBE2* influenced the initiation of starch granules (Tuncel et al., 2019). Additionally, TALENs-based targeting of the acid invertase (*INV2*) and the starch branching enzyme (*SBE1*) in the potato cultivars Russet Burbank and Shepody resulted in the corresponding mutations, which affected the level of cold-induced sweetening and starch branching (Ma et al., 2017). No off-target mutations were found when the CRISPR/Cas9 system altered the L-idonate dehydrogenase (*IdnDH*) gene, which is important for consistent tartaric acid (TA) buildup (Ren et al., 2016). This suggests that the grape is a good candidate for the use of this technology. This method subsequently altered the *IdnDH* gene in grapes and apples (Osakabe et al., 2018). Only four mutants exhibited knockout alleles of the potato CRISPR/Cas9-edited granule-bound starch synthase (*GBSS*) gene that is involved in the formation of amylopectin (Andersson et al., 2017). Similar to this, the same gene was altered using CRISPR/Cas9, and the mutants produced less amylose starch (Kusano et al., 2018). Here, we summarized some of the examples of bioengineered approaches to mitigate pre- and post-harvest losses in commercial tropical fruit and vegetable crops (Figure 1.5).

Crop(s)	Trait(s)	Biotechnological approaches
 Banana	Biotic Resistance (*Hrap, Pflp10, eBSV*)	CRISPR/Cas9 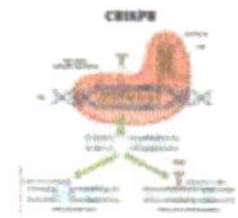
	Delayed Ripening (*MaMADS1* and *MaMADS2*)	Antisense RNA or RNAi 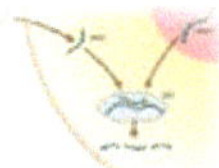
	Delayed Ripening (*MaACO1*)	CRISPR/Cas9 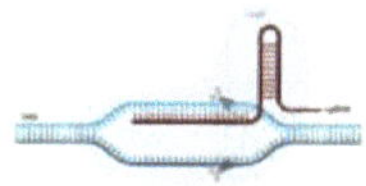
 Papaya	Delayed Ripening (*ACO*)	Biolistic gun method 
	Papaya Ringspot Virus (PRV) Resistance (*CP*)	Microprojectile-mediated transformation 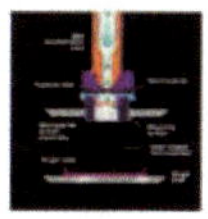
	Papaya Ringspot Virus (PRV) Resistance (*CP*)	*Agrobacterium*-mediated transformation 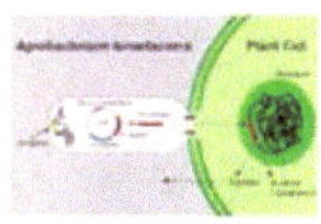
	Broader-spectrum Transgenic Resistance against PRSV (*CP*)	RNAi 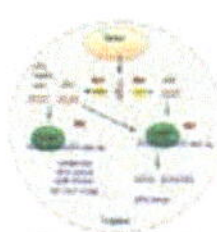
	Delayed Ripening (*ACO*)	RNAi 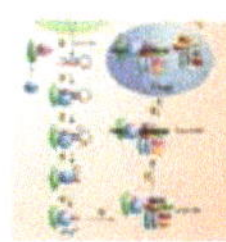

(Continued...)

Crop(s)	Trait(s)	Biotechnological approaches
 Citrus	Citrus Canker and Huanglongbing Resistance (Mthionin) Citrus Canker Resistance *(CsLOB1)*	Over-expression CRISPR/Cas9 
 Squash	Virus Resistance (CP)	*Agrobacterium*-mediated transformation 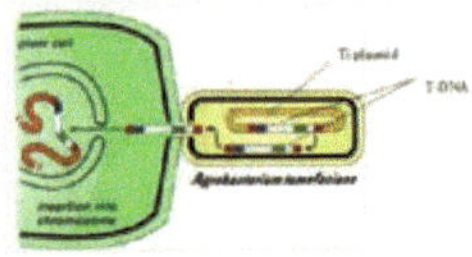
 Cucumber	Virus Resistance *(eIF4E)*	CRISPR/Cas9 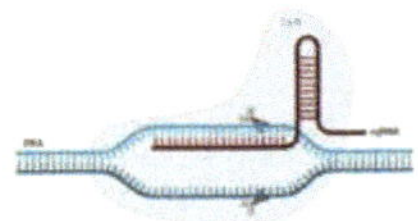
 Chili	Biotic Resistance (*CaERF28*)	CRISPR/Cas9 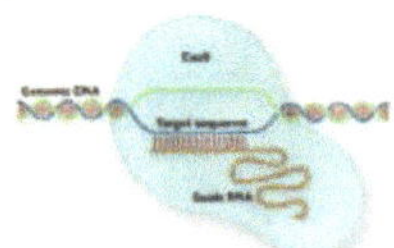
 Cassava	Cassava Brown Streak Disease (CBSD) (*IF4E, nCBP1, nCBP-2*) Gemini Virus Resistance (*AC2, AC3*)	CRISPR/Cas9 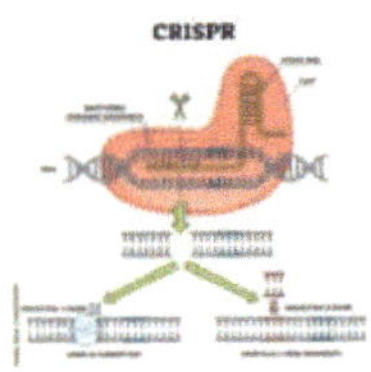

(Continued...)

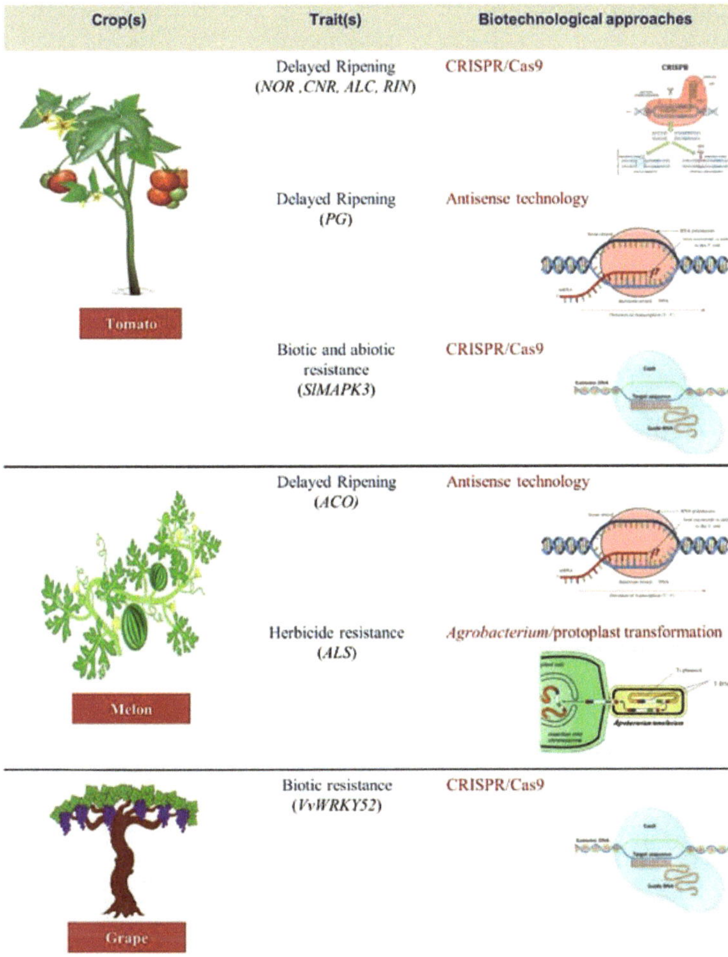

FIGURE 1.5 Bioengineered approaches to mitigate pre-and post-harvest losses in commercial tropical fruit and vegetable crops.

Abbreviations: Hrap: Hypersensitivity response-assisting protein (*Hrap*) gene, *Pflp10*: plant ferredoxin-like protein-10; *eBSV*: endogenous banana streak virus (eBSV); *MaMADS*: *Musa acuminata* MADS box gene; *MaACO1*: *Musa acuminata*1-aminocyclopropane-1-carboxylic acid oxidase-1; *ACO*: 1-aminocyclopropane-1-carboxylic acid oxidase; *CP*: coat protein; *VvWRKY52*: *Vitis vinifera* WRKY transcription factor 52; *ALS*: acetolactate synthase; *SlMAPK3*: *Solanum lycopersicum*-activated protein kinases-3; *PG*: polygalacturonase; *NOR*: non-ripening; *CNR:* colorless non-ripening; *ALC*: *Alcobaca*; *RIN*: ripening inhibitor; AC2: codes for transcription activation protein TrAP involved in gene activation, virus pathogenicity, and suppression of gene silencing; AC3: codes for the REn protein involved in replication enhancement; *IF4E*: translation initiation factor *4E*; *nCBP1*: novel cap-binding protein-1; *nCBP-2*:novel cap-binding protein 2; *CaERF28*: *Capsicum annuum* ethylene-responsive factor 28; *eIF4E*: eukaryotic translation initiation factor 4E; and *CsLOB1*: *Citrus sinensis* lateral organ boundaries 1.

1.5 FUTURE ASPECTS

Many nations have previously developed and implemented laws for GM crops. In each case, analysis of the regulatory status of genome-edited organisms shows that the majority of nations implement the GMO regulatory frameworks already in place (Eckerstorfer et al., 2019). Due to the regulatory triggers for the current biosafety laws differing between legislations and different countries pursuing various options to address/include genome-edited organisms, the global regulatory environment for genome-edited organisms is much more heterogeneous than for conventional GMOs. This reality was acknowledged by the OECD (Organization for Economic Co-operation and Development), an international organization that strives to harmonize the regulatory control of biotechnology. It is challenging to imagine how improved scientific collaboration and comprehension of these technologies may resolve this issue. In order to fully utilize this genome editing technology in the crop development program and meet the need for high-quality, nutrient-rich food that is accessible to the world's expanding population, the approval of relevant legislation must be expedited. To ensure that the current discussions and complexities surrounding the registration and commercialization of GM vegetables are adequately addressed, numerous industry stakeholders, private sectors, agriculturalists, policymakers, biotechnologists, scientists, extension agents, farmers, and the general public must be involved in the development of seed embodiments, products, and policies. The administration has given greater thought to the best approach to take to close the gap between supply and demand. The benefits of knowledge may be reaped if there is complete devolution of responsibility by all stakeholders regarding increased and maintained financing, greater agricultural R&D, and reduced costs and times for registration and commercialization of novel characteristics. Rapid and practical solutions are required in light of the changing global environment to supply food with enhanced nutritional content to the expanding population. A clear, consistent regulatory system and modern omics technology can be used to support its larger use with safety and public acceptance and address a number of related challenges.

Proteins are required for various plant processes such as development, growth, ripening, and senescence, as well as metabolic functions and resistance to abiotic and biotic stresses. Over the past few decades, research on proteomics and omics approaches has provided a new perspective on the ripening and senescence, development, and postharvest responses of perishable crops like apple, banana, citrus, grapes, strawberry, tomato, peach, papaya, mango, and others. They are crucial for comprehending how plants respond to pathogen infection. Proteins linked to defense systems, energy metabolism, and the antioxidant system are essential for food products in storage and trigger reactions, according to proteomics studies. Several postharvest investigations and experiments use gel-based approaches for protein measurement and identification. Furthermore, it is necessary to broaden the scope of proteome studies in fruits and vegetables to incorporate additional quantitative proteomic techniques such as iTRAQ, TMT, peptide demethylation labeling, ICAT, MRM, label-free methods, or the combination of the aforementioned techniques with LC-MS/MS. Postharvest proteomics is critical for reducing PHLs by selecting plants with desirable quality attributes. The selection of post-harvest food items with characteristics such as low sensitivity to bruising and chilling injury is crucial for changing the shelf life of processed commodities. Citrus storage quality has been studied using a mix of transcriptomics, metabolomics, and proteomics. Chilling injury is detrimental to production in storage because it affects not just protein and polysaccharide breakdown, but also organic acids and causes postharvest produce injury (Abdi et al., 2002).

1.6 CONCLUSION

The problem of horticultural loss and waste cannot be solved solely through genetic manipulation, as the issue is complex and multifaceted, requiring interdisciplinary breakthroughs that combine biological, engineering, and socio-cultural solutions. Implementing engineering solutions necessitates long-term investments in infrastructure, equipment, and energy. Agricultural produce is vulnerable to deterioration post-harvest, leading to losses in quality and quantity due to various factors. To mitigate post-harvest losses, research is ongoing into transgenic techniques and breeding methods. The advancement of genetically modified crops is closely linked to the significance of post-harvest physiology in agriculture.

KEYWORDS

- **antisense RNA technology**
- **bioengineering**
- **genetic engineering**
- **genome editing**
- **molecular biology**
- **ripening**
- **suppression technology**

REFERENCES

Abdi, N., Holford, P., & McGlasson, B., (2002). Application of two-dimensional gel electrophoresis to detect proteins associated with harvest maturity in stonefruit. *Postharvest Biology and Technology*, *26*(1), 1–13.

Albornoz, K., Cantwell, M. I., Zhang, L., & Beckles, D. M., (2019). Integrative analysis of postharvest chilling injury in cherry tomato fruit reveals contrapuntal spatio-temporal responses to ripening and cold stress. *Scientific Reports*, *9*(1), 1–14.

Andersson, M., Turesson, H., Nicolia, A., Fält, A. S., Samuelsson, M., & Hofvander, P., (2017). Efficient targeted multiallelic mutagenesis in tetraploid potato (*Solanum tuberosum*) by transient CRISPR-Cas9 expression in protoplasts. *Plant Cell Reports*, *36*(1), 117–128.

Chen, K., Wang, Y., Zhang, R., Zhang, H., & Gao, C., (2019). CRISPR/Cas genome editing and precision plant breeding in agriculture. *Annu. Rev. Plant Biol.*, *70*(1), 667–697.

Christian, M., Cermak, T., Doyle, E. L., Schmidt, C., Zhang, F., Hummel, A., Bogdanove, A. J., & Voytas, D. F., (2010). Targeting DNA double-strand breaks with TAL effector nucleases. *Genetics*, *186*(2), 757–761.

Chung, M. Y., Vrebalov, J., Alba, R., Lee, J., McQuinn, R., Chung, J. D., Klein, P., & Giovannoni, J., (2010). A tomato (*Solanum lycopersicum*) APETALA2/ERF gene, SlAP2a, is a negative regulator of fruit ripening. *The Plant Journal*, *64*(6), 936–947.

Clasen, B. M., Stoddard, T. J., Luo, S., Demorest, Z. L., Li, J., Cedrone, F., Tibebu, R., et al., (2016). Improving cold storage and processing traits in potato through targeted gene knockout. *Plant Biotechnology Journal*, *14*(1), 169–176.

Doudna, J. A., & Charpentier, E., (2014). The new frontier of genome engineering with CRISPR-Cas9. *Science*, *346*(6213). doi.org/10.1126/science.1258096.

Eckerstorfer, M. F., Engelhard, M., Heissenberger, A., Simon, S., & Teichmann, H., (2019). Plants developed by new genetic modification techniques—Comparison of existing regulatory frameworks in the EU and non-EU countries. *Frontiers in Bioengineering and Biotechnology*, *7*, 26.

Elitzur, T., Vrebalov, J., Giovannoni, J. J., Goldschmidt, E. E., & Friedman, H., (2010). The regulation of MADS-box gene expression during ripening of banana and their regulatory interaction with ethylene. *Journal of Experimental Botany*, *61*(5), 1523–1535.

Entine, J., Felipe, M. S. S., Groenewald, J. H., Kershen, D. L., Lema, M., McHughen, A., Nepomuceno, A. L., et al., (2021). Regulatory approaches for genome edited agricultural plants in select countries and jurisdictions around the world. *Transgenic Research*, *30*(4), 551–584.

Ezura, H., & Owino, W. O., (2008). Melon, an alternative model plant for elucidating fruit ripening. *Plant Science*, *175*(1, 2), 121–129.

Fei, Z., Tang, X., Alba, R., & Giovannoni, J., (2006). Tomato expression database (TED): A suite of data presentation and analysis tools. *Nucleic Acids Research*, *34*(suppl_1), D766–D770.

Gao, Y., Wei, W., Fan, Z., Zhao, X., Zhang, Y., Jing, Y., Zhu, B., et al., (2020). Re-evaluation of the nor mutation and the role of the NAC-NOR transcription factor in tomato fruit ripening. *Journal of Experimental Botany*, *71*(12), 3560–3574.

Gao, Y., Zhu, N., Zhu, X. F., Wu, M., Jiang, C. Z., Grierson, D., Luo, Y. B., et al., (2019). Diversity and redundancy of the ripening regulatory networks revealed by the fruit ENCODE and the new CRISPR/Cas9 CNR and NOR mutants. *Horticulture Research*, *6*, 39.

Ghosh, S., Singh, U. K., Meli, V. S., Kumar, V., Kumar, A., Irfan, M., Chakraborty, N., et al., (2013). Induction of senescence and identification of differentially expressed genes in tomato in response to monoterpene. *PLoS One*, *8*(9), e76029. https://doi.org/10.1371/journal.pone.0076029.

Giménez, E., Pineda, B., Capel, J., Antón, M. T., Atarés, A., Pérez-Martín, F., & Lozano, R., (2010). Functional analysis of the *Arlequin* mutant corroborates the essential role of the ARLEQUIN/TAGL1 gene during reproductive development of tomato. *PloS One*, *5*(12), e14427.

Giovannoni, J. J., (2007). Fruit ripening mutants yield insights into ripening control. *Current Opinion in Plant Biology*, *10*(3), 283–289.

González, M. N., Massa, G. A., Andersson, M., Turesson, H., Olsson, N., Fält, A. S., Storani, L., et al., (2020). Reduced enzymatic browning in potato tubers by specific editing of a polyphenol oxidase gene via ribonucleoprotein complexes delivery of the CRISPR/Cas9 system. *Frontiers in Plant Science*, *10*, 1649.

Hassan, M. M., Zhang, Y., Yuan, G., De, K., Chen, J. G., Muchero, W., Tuskan, G. A., et al., (2021). Construct design for CRISPR/Cas-based genome editing in plants. *Trends in Plant Science*, *26*(11), 1133–1152.

Hegazy, R., (2013). Post-harvest situation and losses in India. *Technical Report.* doi: 10.6084/m9.figshare. 3206851.v1.

Hovav, R., Chehanovsky, N., Moy, M., Jetter, R., & Schaffer, A. A., (2007). The identification of a gene (Cwp1), silenced during *Solanum* evolution, which causes cuticle microfissuring and dehydration when expressed in tomato fruit. *The Plant Journal*, *52*(4), 627–639.

Hu, Y., Zhang, J., Jia, H., Sosso, D., Li, T., Frommer, W. B., Yang, B., et al., (2014). Lateral organ boundaries 1 is a disease susceptibility gene for *Citrus* bacterial canker disease. *Proceedings of the National Academy of Sciences*, *111*(4), E521–E529.

Irfan, M., & Datta, A., (2017). Improving food nutritional quality and productivity through genetic engineering. *International Journal of Cell Science & Molecular Biology, 2*(1), 555576. https://doi.org/10.19080/IJCSMB.2017.02.555576.

Irfan, M., Ghosh, S., Kumar, V., Chakraborty, N., Chakraborty, S., & Datta, A., (2014). Insights into transcriptional regulation of β-DN-acetylhexosaminidase, an N-glycan-processing enzyme involved in ripening-associated fruit softening. *Journal of Experimental Botany, 65*(20), 5835–5848.

Irfan, M., Ghosh, S., Meli, V. S., Kumar, A., Kumar, V., Chakraborty, N., Chakraborty, S., & Datta, A., (2016). Fruit ripening regulation of α-mannosidase expression by the MADS box transcription factor ripening inhibitor and ethylene. *Frontier Plant Science, 7*, 10.

Irfan, M., Kumar, P., Ahmad, I., & Datta, A., (2021). Unraveling the role of tomato Bcl-2-associated athanogene (BAG) proteins during abiotic stress response and fruit ripening. *Scientific Reports, 11*(1), 1–15.

Irfan, M., Kumar, P., Ahmad, M. F., & Siddiqui, M. W., (2023). Biotechnological interventions in reducing losses of tropical fruits and vegetables. *Current Opinion in Biotechnology, 79*, 102850. doi.org/10.1016/j.copbio.2022.102850.

Irfan, M., Kumar, P., Kumar, V., & Datta, A., (2022). Fruit ripening specific expression of β-D-N-acetylhexosaminidase (β-hex) gene in tomato is transcriptionally regulated by ethylene response factor SlERF.E4. *Plant Science, 323*, 111380.

Jia, H., Zhang, Y., Orbović, V., Xu, J., White, F. F., Jones, J. B., & Wang, N., (2017). Genome editing of the disease susceptibility gene CsLOB1 in *Citrus* confers resistance to *Citrus* canker. *Plant Biotechnology Journal, 15*(7), 817–823.

Jiang, F., & Doudna, J. A., (2017). CRISPR-Cas9 structures and mechanisms. *Annu. Rev. Biophys., 46*(1), 505–529.

Jiménez-Bermudez, S., Redondo-Nevado, J., Munoz-Blanco, J., Caballero, J. L., López-Aranda, J. M., Valpuesta, V., & Mercado, J. A., (2002). Manipulation of strawberry fruit softening by antisense expression of a pectate lyase gene. *Plant Physiology, 128*(2), 751–759.

Jung, Y. J., Lee, G., Bae, S., & Kang, K. K., (2018). Reduced ethylene production in tomato fruits upon CRSPR/Cas9-mediated LeMADS-RIN mutagenesis. *Korean Journal of Horticulture Science and Technology, 36*, 396–405. doi: 10.12972/kjhst.20180039.

Kendrick, M. D., & Chang, C., (2008). Ethylene signaling: New levels of complexity and regulation. *Current Opinion in Plant Biology, 11*(5), 479–485.

Kim, S., Soltis, P. S., Wall, K., & Soltis, D. E., (2006). Phylogeny and domain evolution in the APETALA2-like gene family. *Molecular Biology and Evolution, 23*(1), 107–120.

Kim, Y. G., Cha, J., & Chandrasegaran, S., (1996). Hybrid restriction enzymes: Zinc finger fusions to *Fok I* cleavage domain. *Proceedings of the National Academy of Sciences, 93*(3), 1156–1160.

Kramer, M. G., & Redenbaugh, K., (1994). Commercialization of a tomato with an antisense polygalacturonase gene: The FLAVR SAVR™ tomato story. *Euphytica, 79*, 293–297.

Ku, H. K., & Ha, S. H., (2020). Improving nutritional and functional quality by genome editing of crops: Status and perspectives. *Frontiers in Plant Science, 11*, 577313.

Kumar, V., Irfan, M., Ghosh, S., Chakraborty, N., Chakraborty, S., & Datta, A., (2016). Fruit ripening mutants reveal cell metabolism and redox state during ripening. *Protoplasma, 253*, 581–594. https://doi.org/10.1007/s00709-015-0836-z.

Kumar, V., Neeraj, S. S., & Sagar, N. A., (2015). Post harvest management of fungal diseases in onion—A review. *International Journal of Current Microbiology and Applied Sciences*, *4*(6), 737–752.

Kumari, C., Sharma, M., Kumar, V., Sharma, R., Kumar, V., Sharma, P., Kumar, P., & Irfan, M., (2022). Genome editing technology for genetic amelioration of fruits and vegetables for alleviating post-harvest loss. *Bioengineering*, *9*(4), 176.

Kusano, H., Ohnuma, M., Mutsuro-Aoki, H., Asahi, T., Ichinosawa, D., Onodera, H., Asano, K., et al., (2018). Establishment of a modified CRISPR/Cas9 system with increased mutagenesis frequency using the translational enhancer dMac3 and multiple guide RNAs in potato. *Scientific Reports*, *8*(1), 1–9.

Li, T., Yang, X., Yu, Y., Si, X., Zhai, X., Zhang, H., Dong, W., et al., (2018). Domestication of wild tomato is accelerated by genome editing. *Nature Biotechnology*, *36*(12), 1160–1163.

Lin, Z., Hong, Y., Yin, M., Li, C., Zhang, K., & Grierson, D., (2008). A tomato HD-Zip homeobox protein, LeHB-1, plays an important role in floral organogenesis and ripening. *The Plant Journal*, *55*(2), 301–310.

Lin, Z., Zhong, S., & Grierson, D., (2009). Recent advances in ethylene research. *Journal of Experimental Botany*, *60*(12), 3311–3336.

Liu, W., Rudis, M. R., Cheplick, M. H., Millwood, R. J., Yang, J. P., Ondzighi-Assoume, C. A., Montgomery, G. A., et al., (2020). Lipofection-mediated genome editing using DNA-free delivery of the Cas9/gRNA ribonucleoprotein into plant cells. *Plant Cell Reports*, *39*(2), 245–257.

Lü, P., Yu, S., Zhu, N., Chen, Y. R., Zhou, B., Pan, Y., Tzeng, D., et al., (2018). Genome encode analyses reveal the basis of convergent evolution of fleshy fruit ripening. *Nature Plants*, *4*(10), 784–791.

Ma, J., Xiang, H., Donnelly, D. J., Meng, F. R., Xu, H., Durnford, D., & Li, X. Q., (2017). Genome editing in potato plants by agrobacterium-mediated transient expression of transcription activator-like effector nucleases. *Plant Biotechnology Reports*, *11*(5), 249–258.

Manning, K., Tör, M., Poole, M., Hong, Y., Thompson, A. J., King, G. J., Gio-Vannoni, J. J., & Seymour, G. B., (2006). A naturally occurring epigenetic mutation in a gene encoding an SBP-box transcription factor inhibits tomato fruit ripening. *Nature Genetics*, *38*(8), 948–952.

Matas, A. J., Gapper, N. E., Chung, M. Y., Giovannoni, J. J., & Rose, J. K., (2009). Biology and genetic engineering of fruit maturation for enhanced quality and shelf-life. *Current Opinion in Biotechnology*, *20*(2), 197–203.

Metje-Sprink, J., Menz, J., Modrzejewski, D., & Sprink, T., (2019). DNA-free genome editing: Past, present, and future. *Frontiers in Plant Science*, 1957. doi.org/10.3389/fpls.2018.01957.

Mishra, R., Mohanty, J. N., Mahanty, B., & Joshi, R. K., (2021). A single transcript CRISPR/Cas9 mediated mutagenesis of CaERF28 confers anthracnose resistance in chilli pepper (*Capsicum annuum* L.). *Planta*, *254*(1), 1–17.

Naegeli, H., Bresson, J. L., Dalmay, T., Dewhurst, I. C., & Epstein, M. M., (2020). Applicability of the EFSA opinion on site-directed nucleases type 3 for the safety assessment of plants developed using site-directed nucleases type 1 and 2 and oligonucleotide-directed mutagenesis. *EFSA Journal, 18*, e06299. e06299. EFSA Panel on Genetically Modified Organisms (EFSA GMO Panel). doi: 10.2903/j.efsa.2020.6299.

Osakabe, Y., Liang, Z., Ren, C., Nishitani, C., Osakabe, K., Wada, M., Komori, S., et al., (2018). CRISPR–Cas9-mediated genome editing in apple and grapevine. *Nature Protocols, 13*(12), 2844–2863.

Owino, W. O., & Ezura, H., (2011). Biotechnology in postharvest research of fruits: Prospects and achievements. *Stewart Postharvest Review*, *7*(1).

Pineda, B., Giménez-Caminero, E., García-Sogo, B., Antón, M. T., Atarés, A., Capel, J., Lozano, R., et al., (2010). Genetic and physiological characterization of the *Arlequin* insertional mutant reveals a key regulator of reproductive development in tomato. *Plant and Cell Physiology*, *51*(3), 435–447.

Podevin, N., Davies, H. V., Hartung, F., Nogué, F., & Casacuberta, J. M., (2013). Site-directed nucleases: A paradigm shift in predictable, knowledge-based plant breeding. *Trends in Biotechnology*, *31*(6), 375–383.

Que, Q., Chen, Z., Kelliher, T., Skibbe, D., Dong, S., & Chilton, M. D., (2019). Plant DNA repair pathways and their applications in genome engineering. *Plant Genome Editing with CRISPR Systems*, 3–24.

Quesada, M. A., Blanco-Portales, R., Posé, S., García-Gago, J. A., Jiménez-Bermúdez, S., Muñoz-Serrano, A., & Munoz-Blanco, J., (2009). Antisense down-regulation of the *FaPG1* gene reveals an unexpected central role for polygalacturonase in strawberry fruit softening. *Plant Physiology*, *150*(2), 1022–1032.

Ramírez-Gil, J. G., Henao-Rojas, J. C., & Morales-Osorio, J. G., (2021). Postharvest diseases and disorders in avocado cv. Hass and their relationship to preharvest management practices. *Heliyon*, *7*(1), e05905.

Ren, C., Liu, X., Zhang, Z., Wang, Y., Duan, W., Li, S., & Liang, Z., (2016). CRISPR/Cas9-mediated efficient targeted mutagenesis in chardonnay (*Vitis vinifera* L.). *Scientific Reports*, *6*(1), 1–9.

Rommens, C. M., Yan, H., Swords, K., Richael, C., & Ye, J., (2008). Low-acrylamide French fries and potato chips. *Plant Biotechnology Journal*, *6*(8), 843–853.

Rommens, C. M., Ye, J., Richael, C., & Swords, K., (2006). Improving potato storage and processing characteristics through all-native DNA transformation. *Journal of Agricultural and Food Chemistry*, *54*(26), 9882–9887.

Savin, K. W., Baudinette, S. C., Graham, M. W., Michael, M. Z., Nugent, G. D., Lu, C. Y., & Cornish, E. C., (1995). Antisense ACC oxidase RNA delays carnation petal senescence. *HortScience*, *30*(5), 970–972.

Sawicka, B., (2019). Post-harvest losses of agricultural produce. *Springer Nature*. doi: 10.1007/978-3-319-69626-3_40-1.

Sharma, B., & Kumari, A., (2020). *Role of Post Harvest Physiology in Evolution of Transgenics* Crops (pp. 1–21). IntechOpen.

Sharma, P., & Sharma, R., (2021). CRISPR/Cas9: An efficient tool for improving biotic stress in plants. In: *Plant-Microbial Interactions and Smart Agricultural Biotechnology* (pp. 185–198). CRC Press.

Shipman, E. N., Yu, J., Zhou, J., Albornoz, K., & Beckles, D. M., (2021). Can gene editing reduce postharvest waste and loss of fruit, vegetables, and ornamentals? *Horticulture Research, 8*, 121. doi: 10.1038/s41438-020-00428-4.

Sonker, N., Pandey, A. K., & Singh, P., (2016). Strategies to control post-harvest diseases of table grape: A review. *Journal of Wine Research, 27*(2), 105–122.

Subburaj, S., Chung, S. J., Lee, C., Ryu, S. M., Kim, D. H., Kim, J. S., Bae, S., & Lee, G. J., (2016). Site-directed mutagenesis in *Petunia* × *hybrida* protoplast system using direct delivery of purified recombinant Cas9 ribonucleoproteins. *Plant Cell Reports, 35*(7), 1535–1544.

Tadiello, A., Pavanello, A., Zanin, D., Caporali, E., Colombo, L., Rotino, G. L., Trainotti, L., & Casadoro, G., (2009). A PLENA-like gene of peach is involved in carpel formation and subsequent transformation into a fleshy fruit. *Journal of Experimental Botany, 60*(2), 651–661.

Tayal, R., Kumar, V., & Irfan, M., (2022). Harnessing the power of hydrogen sulphide (H_2S) for improving fruit quality traits. *Plant Biology, 24*(4), 594–601.

Tuncel, A., Corbin, K. R., Ahn-Jarvis, J., Harris, S., Hawkins, E., Smedley, M. A., Harwood, W., et al., (2019). Cas9-mediated mutagenesis of potato starch-branching enzymes generates a range of tuber starch phenotypes. *Plant Biotechnology Journal, 17*(12), 2259–2271.

Uluisik, S., Chapman, N. H., Smith, R., Poole, M., Adams, G., Gillis, R. B., Besong, T. M. D., et al., (2016). Genetic improvement of tomato by targeted control of fruit softening. *Nature Biotechnology, 34*(9), 950–952.

Vicente, A. R., Saladié, M., Rose, J. K., & Labavitch, J. M., (2007). The linkage between cell wall metabolism and fruit softening: Looking to the future. *Journal of the Science of Food and Agriculture, 87*(8), 1435–1448.

Vrebalov, J., Pan, I. L., Arroyo, A. J. M., McQuinn, R., Chung, M., Poole, M., Rose, J., et al., (2009). Fleshy fruit expansion and ripening are regulated by the tomato SHATTERPROOF gene TAGL1. *The Plant Cell, 21*(10), 3041–3062.

Waltz, E., (2016). Gene-edited CRISPR mushroom escapes US regulation. *Nature, 532*(7599), 293.

Wang, W., Liu, S., Deng, L., Ming, J., Yao, S., & Zeng, K., (2018). Control of *Citrus* post-harvest green molds, blue molds, and sour rot by the cecropin A-melittin hybrid peptide BP21. *Frontiers in Microbiology, 9*, 2455.

Wang, Y., Lecourieux, F., Zhang, R., Dai, Z., Lecourieux, D., Li, S., & Liang, Z., (2021). Data comparison and software design for easy selection and application of CRISPR-based genome editing systems in plants. *Genomics, Proteomics & Bioinformatics*.

Xiong, J. S., Ding, J., & Li, Y., (2015). Genome-editing technologies and their potential application in horticultural crop breeding. *Horticulture Research, 2*, 1–10.

Yamane, M., Abe, D., Yasui, S., Yokotani, N., Kimata, W., Ushijima, K., Nakano, R., et al., (2007). Differential expression of ethylene biosynthetic genes in climacteric and non-climacteric Chinese pear fruit. *Postharvest Biology and Technology, 44*(3), 220–227.

Yu, Q. H., Wang, B., Li, N., Tang, Y., Yang, S., Yang, T., Xu, J., et al., (2017). CRISPR/Cas9-induced targeted mutagenesis and gene replacement to generate long-shelf-life tomato lines. *Scientific Reports, 7*(1), 1–9.

Zhang, S., Shen, J., Li, D., & Cheng, Y., (2021). Strategies in the delivery of Cas9 ribonucleoprotein for CRISPR/Cas9 genome editing. *Theranostics, 11*(2), 614–648.

Zhang, S., Wang, L., Zhao, R., Yu, W., Li, R., Li, Y., Sheng, J., & Shen, L., (2018). Knockout of SlMAPK3 reduced disease resistance to *Botrytis cinerea* in tomato plants. *Journal of Agricultural and Food Chemistry*, *66*(34), 8949–8956.

CHAPTER 2

GENETIC TRANSFORMATION OF FRUIT AND VEGETABLE CROPS: CHALLENGES AND OPPORTUNITIES

BUNTORA PASARIBU,[1] NOIR PRIMADONA PURBA,[1]
ALEXANDER M. KHAN,[2] and RITESH KUMAR[3]

[1]*Department of Marine Science, Faculty Fisheries and Marine Science, Universitas Padjadjaran, Bandung, Indonesia*

[2]*Tropical Marine Fisheries Undergraduate Program for Pangandaran Campus, Universitas Padjadjaran, Pangandaran, Indonesia*

[3]*Department of Agronomy, Kansas State University, Manhattan, Kansas, USA*

ABSTRACT

With climate change exacerbating environmental stresses and increasing the incidence of pests and diseases, conventional breeding methods are insufficient to meet the demand for resilient crop varieties. Genetic transformation offers a viable alternative, enabling the introduction of foreign genetic material into plants to enhance their traits and resilience. This chapter provides a comprehensive overview of the various methods of genetic transformation, including *Agrobacterium tumefaciens*-mediated transformation, CRISPR/Cas9 genome editing, and nanoparticle-mediated transformation. While each method presents unique advantages, they also face challenges such as low transformation efficiency and off-target effects. Moreover, the chapter explores future prospects in genetic transformation,

Bioengineered Fruit and Vegetables: Trends in Postharvest Quality Improvement.
Mohammad Irfan and Pankaj Kumar (Eds.)

emphasizing the need for standardized protocols, improved regeneration techniques, and comprehensive genomic information for fruit and vegetable crops. Despite the existing challenges, genetic transformation holds immense potential for revolutionizing horticultural systems and ensuring the sustainability of fruit and vegetable crop production in the face of evolving environmental pressures.

2.1 INTRODUCTION

Global fruit production and supply will be significantly affected by climate change in the coming decades (Kumari et al., 2022). In a warming climate and erratic weather patterns, many fruit crops may struggle to sustain themselves and produce consistent crop yields (Sharma et al., 2022a). Extreme weather events such as heatwaves, droughts, floods, and storms are expected to increase as a result of climate change (Sharma et al., 2022a; Kapoor et al., 2023). Events like these can damage fruit crops, reduce yields, and disrupt supply chains (Kumari et al., 2022).

It was shown that increased atmospheric ozone levels resulted in a reduction in photosynthetic rate, growth, biomass accumulation, and a reduction in volatile ester levels in strawberries, and an increase in beta-carotene, lutein, and lycopene levels in tomatoes (Mattos et al., 2014). Moreover, pests and diseases are frequently major contributors to the decline of up to 20–30% in fruit and vegetable crops (Lei et al., 2016; Lobato-Gomez et al., 2021; Shivalingswami et al., 2002). One of the common infecting plant viruses, tomato yellow leaf curl virus (TYLCV), is transmitted by the whitefly *Bemisatabaci.* It causes stunting, leaf curling, and yellowing in tomato, pepper, bean, and cucurbit plants (Ghanim et al., 1998; Moriones & Navas-Castillo, 2000; Papayiannis et al., 2011; Anfoka et al., 2009). For over a decade, conventional breeding has contributed to developing an ideal variety of crops to meet current reduction factors. However, this method is very slow, difficult, and genotype-dependent (Wan et al., 2021). Hence, it is critical that horticultural systems of fruit and vegetable crops are continuously developed in order to maintain their productivity and sustainability.

Because conventional breeding is challenging in horticultural systems, genetic transformation in plants is an alternative method to increase crop production. Since its introduction at the Miami Winter Symposium in 1983 (O'Brien, 1983), genetic transformation has made significant progress. Plant genetic transformation experiments involve introducing foreign genetic material into a plant to modify its traits or characteristics. The first successful plant transformation experiments were carried out in the 1980s using the bacterium *Agrobacterium tumefaciens*. Some key challenges for the development of stable transformation in crop plants during this period include optimizing transformation efficiency, gene delivery efficiency, selection, and regeneration from transformed explants (Song et al., 2019).

The combination of conventional plant breeding and genetic transformation in crop plants is being more consistently applied to many fruits and vegetable crops, resulting in the development of several elite cultivars with different agronomic traits (Basso et al., 2020; Kumar et al., 2020). The first commercialization of a transgenic tomato, "Flavr Savr," developed by Calgene (Monsanto), was reported in 1994 using antisense RNA (asRNA) to regulate the expression of the enzyme polygalacturonase (PG) to delay ripening or extend shelf life (Kramer & Redenbaugh, 1994). Additionally, many transgenic fruits and vegetable crops, such as papaya (*Carica papaya* L.), pepper (*Capsicum annuum*), plum (*Prunus domestica*), eggplant (*Solanum melogena* L.), apple (*Malus domestica* Borkh.), melon (*Cucumis melo* L.), and pineapple (*Ananas comosus* L. Merr.), have been approved for commercial production (Fitch et al., 1992; Chen et al., 2001; Chen et al., 2003; Scorza et al., 1994; Shelton et al., 2017; USDA, 2015; Biosafety, 1999; Firoozhady & Young, 2015). However, challenges in leveraging transgenic traits remain a dilemma due to the insertion of foreign genes of interest and controversies regarding safety, health, and environmental conservation.

Genetic transformation is one of the most effective methods for improving the yield and quality of fruit and vegetable crops (Figure 2.1). Several recent review articles have focused on the characterization of genetic transformation in crops (Sedeek et al., 2019; Ashray et al., 2020; Kumar et al., 2020; Ye et al., 2022). In this chapter, we aim to provide an up-to-date summary of the methods of genetic transformation in fruit and vegetable crops. Furthermore, we will explore future opportunities and challenges of transgenic fruit and vegetable crops, including potential biosafety and health concerns associated with the use of genetic transformation tools.

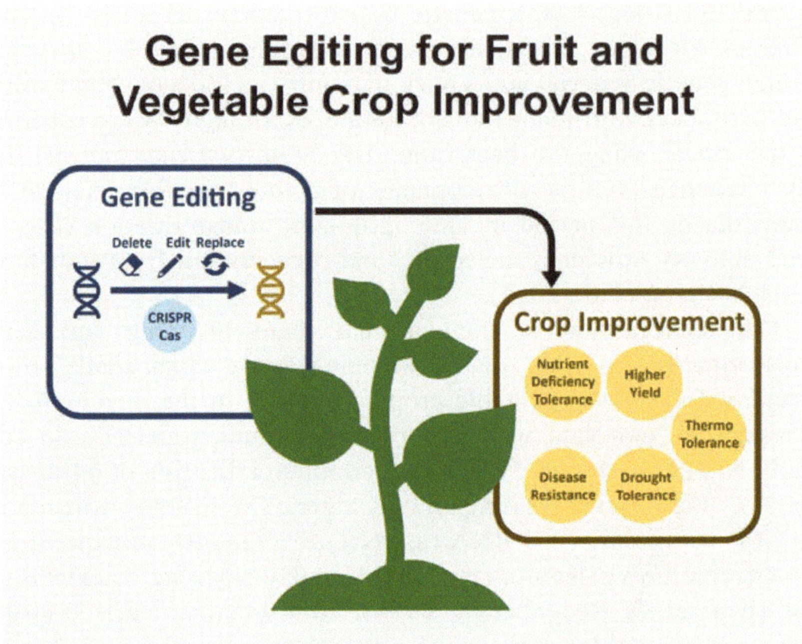

FIGURE 2.1 CRISPR/Cas9 gene editing technologies have opened up new possibilities for improving fruit and vegetable crops by allowing precise and targeted modifications to their genomes. With gene editing, fruit and vegetable crops can be modified to yield more, contain more nutrients, and resist pests and diseases. As a result of gene editing, new crops can be developed that require less water, fertilizer, and chemical inputs, which can support sustainable agriculture and address environmental concerns.

2.2 DIFFERENT METHODS ARE AVAILABLE FOR TRANSFER OF EXPRESSION CASSETTES INTO PLANT CELLS

2.2.1 *AGROBACTERIUM TUMEFACIENS*

Agrobacterium-mediated transformation was a common technique used for transgenic transformation in Plants. The transformation mechanism inserts the T-DNA region and vir region of the target gene(s) of interest into plant cells. The target sequence is designed into a binary vector with the multi-clone site of *Agrobacterium tumefaciens*. The use of *Agrobacterium tumefaciens* results in efficient and stable transformation in plant cells (Wang et al., 2018; Mahto et al., 2018). The presence of T-DNA as a part of the Ti-plasmid in *Agrobacterium tumefaciens* enables integration at the target nuclear chromosome (Bundock et al., 1995). *Agrobacterium*-mediated transformation has been successfully achieved in fruit and vegetable crops such as Tomato (*S. lycopersicum*), Carrot (*Daucus carota*), Spinach (*Spinacia oleracea*), Lettuce (*Lactuca sativa*), Potato (*Solanum tuberosum* L.), and Eggplant (*Solanum melongena*) (Klocke et al., 2010; Punja et al., 2007; Davey et al., 2007; Song et al., 2014; Allah Baksh, 2020; Irfan et al., 2022).

The benefit of *Agrobacterium tumefaciens* transformation is demonstrated by a simple, effective method, and lower cost (Ye et al., 2022). The most significant results of *Agrobacterium* transformation reported a low transgene copy number and high transgene expression (Dai et al., 2001). Despite the beneficial application of *Agrobacterium* transformation in crops, various limiting factors have been identified, such as plants' susceptibility to infection, lower yield of stable transgenic plants, longer tissue culture conditions, and a limited range of host plants (Hwang et al., 2013; Guo et al., 2011; Beaujean et al., 1998).

Although the successful transformation has been highlighted in fruits and vegetables, pre- and post-preparation of transformation were influenced by various parameters to achieve a successful *Agrobacterium* system such as pre-culture explants, bacterial density, *Agrobacterium* strains, bacteria cell density, infection time, co-cultivation duration, antibiotic selection (Madhulatha et al., 2007). Different genotypes of fruit and vegetable crops required the optimization of the protocol due to the impact of the transformation efficiency to produce sufficient numbers of transgenic plants (Harwood 2012). It is reported that vegetable variety belonging to *Brassicaceae* was successfully transformed with direct gene transfers and *Agrobacterium*-mediated transformation through protoplast by inoculating seedling explants which depends on the genotype, susceptibility of *Agrobacterium tumefaciens,* and regeneration ability (Sparrow et al., 2004). For example, *the Brassica oleracea* and cabbage (*B. oleracea var. capitata*) genotypes were the most successful in many transformations than the *Brassica rapa* genotype (Cardi et al., 2017). The improvement of the transformation protocol in the leafy vegetable *Spinacia oleracea* (Chenopodiaceae) showed a significant increase in transformation efficiency and regeneration performance. Through organogenesis or embryogenesis, it has been shown that the development of transformant regeneration procedures in Spinach depends on explant type and auxin/cytokinin interaction that regulates the plant tissue growth. Cotyledon explants with a low ratio promoted the initial shoot regeneration, while root explants transition developed at a later stage (Cardi et al., 2017; Geekiyanage et al., 2006). Low temperature and light exposure demonstrated the induction of shoot regeneration in transformant *Spinacia oleracea* (Geekiyanage et al., 2006; Chin et al., 2009). Despite the improvement of a reproducible protocol made by inoculating the *Agrobacterium*-mediated transformation from different somatic embryo regeneration including leaf, cotyledon, and hypocotyl, the effort was not potentially efficient in enhancing the transformation efficiency compared to using embryonic calluses in Parsley (Baranski, 2008; Song et al., 2007). These results suggest that distinct genotypes of crops may require several modification procedures to acquire the stability of the transformants and enhance the transformation efficiency, including the introduction of DNA into plant chromosomes, selection, and regeneration of transformed cells.

2.2.2 CRISPR/CAS9

The development of genetic editing targets the natural DNA by deleting, replacing, or inserting specific genes that have been developed to generate desired novel phenotypes (Corte et al., 2019). The fundamentals of genetic editing employ a sequence-specific nuclease to produce a DNA double-strand break (DSB) at a target site (Zhu et al., 2023). To date, three types of endonucleases have been applied in plant genome editing, including zinc finger nuclease (ZFN), transcription activator-like effector nuclease (TALENs), and the clustered regularly interspersed short palindromic repeats (CRISPR)/Cas system (Malzahn et al., 2017; Shah et al., 2018; Zhang et al., 2018; Bao et al., 2019). ZFN is a DNA-targeted modification that comprises a synthetic zinc finger (ZF) DNA domain and a cleavage domain of FokI endonuclease (Mahfouz et al., 2011). This technology has been applied to engineer the genomes of maize, rice, and Arabidopsis (Osakabe et al., 2010; Ainley et al., 2013; Shukla et al., 2009).

Transcriptor-like effector nucleases (TALENs) are programmable DNA-binding domains composed of an N-terminal translocation domain, a C-terminal nuclear localization signal, and a central tandem repeat DNA binding (DBD) (Boch & Bonas, 2010; Zhang et al., 2019). The use of engineered TALENs targets a specific binding site by creating a highly conserved repeated domain similarly encoded by plant-pathogenic bacteria

Xanthomonas to enhance gene transcription in host cells (Boch & Bonas, 2010). TALENs gene editing tools have been successfully demonstrated in crops including potato, sugarcane, soybean, and Brachypodium (Shan et al., 2013, 2015; Clasen et al., 2016; Nicolia et al., 2015; Du et al., 2016; Jung & Altpeter, 2016; Lor et al., 2014). However, the application of ZFN and TALENs has shown particular limitations in tools to facilitate advanced gene editing such as expensive and complex construction for engineering (Becker & Bosch, 2021). The clustered regularly interspersed short palindromic repeats (CRISPR)/Cas system has demonstrated a new system for gene editing tools that have been widely used to edit the genome using a synthetic single-guided RNA (sgRNA). This adaptive phage mechanism is efficiently applied compared to previous gene editing tools TALENs and ZFN. There are several advantages of using the CRISPR-Cas9 system for genome editing, such as simplicity, cost-effectiveness, the ability to target multiple sites, and efficiency in targeting. Therefore, the exploration of the CRISPR/Cas system could facilitate groundbreaking technology for fruit and vegetable crops in the future.

2.2.3 NANOPARTICLES

Nanoparticles have emerged as a promising tool for genetic transformation in a wide range of organisms, including plants (Lv et al., 2020). One of the mechanisms through which nanoparticles can induce genetic transformation is through their ability to penetrate the plant cell wall and membrane, allowing them to deliver genetic material directly to the plant cell (Wang et al., 2019; Ma et al., 2023). Additionally, nanoparticles can act as carriers for other transformation agents such as plasmid DNA or RNA, which can be complexed or coated onto the surface of the nanoparticles, enhancing their delivery and uptake by the plant cell (Torney et al., 2007; Chang et al., 2013). One of the key advantages of using nanoparticles for genetic transformation is their ability to deliver genetic material directly to the target cells in a highly efficient and specific manner (Riley & Vermerris, 2017). In addition to their ability to deliver genetic material, nanoparticles can also enhance the efficiency of gene expression and transformation through their unique physicochemical properties (Thakur et al., 2020). For example, certain types of nanoparticles, such as gold nanoparticles, have been shown to induce localized stress responses in plant cells, which can result in increased uptake of exogenous DNA and greater integration of the foreign genetic material into the host genome (Ferrari et al., 2021; Siddiqi & Husen, 2016). Overall, nanoparticles have the potential to revolutionize genetic transformation in plants, as they offer a unique set of advantages that can be harnessed to optimize the efficiency and specificity of the transformation process.

Particles that have at least one dimension measuring less than 100 nm, known as nanoparticles, can be produced using a variety of materials including polymers, metals, ceramics, carbon, and other particulate substances (Holister et al., 2003; Gonzalez-Melendi et al., 2008). These particles can be categorized based on their size, morphology, and physical, chemical, and biological characteristics, and can be classified as either magnetic or non-magnetic (Khan et al., 2019). They can also be classified as organic, inorganic, or carbon-based nanomaterials, and are known to exhibit improved properties when compared to larger bulk materials. For example, carbon-based nanomaterials such as graphene, carbon nanotubes (CNTs), and fullerenes, ceramic nanomaterials such as TiO_2, ZnO, and FeO_2, and metal-based nanomaterials such as Au, Ag, Cu, and Ni, can all be created (Zhang et al., 2019; Demirer et al., 2019). Additionally, organic nanomaterials, such as dendrimers, which are derived from symmetrical organic NPs, can also be produced.

The use of nanoparticle-mediated transformation in agriculture has shown promise for improving crop yields and increasing disease resistance. However, the transformation of nanoparticles in fruits and crops presents several challenges that need to be addressed in order to ensure their safe and effective use (McClements & Xiao, 2017). One challenge is the potential accumulation of nanoparticles in the edible portions of fruits and vegetables, which could have adverse effects on human health. A study by Wang et al. (2019) found that nanoparticles applied to tomato plants could accumulate in the fruit, with higher concentrations found in the peel compared to the pulp. There is a strong need to determine the potential risks associated with consuming these nanoparticles (Sharma et al., 2022b). Nanoparticles-mediated transformation could impact soil microbial communities, with potential downstream effects on nutrient cycling and plant health (Sharma et al., 2022b). It is reported that a need to understand these indirect effects in order to develop sustainable practices for using nanoparticle-mediated transformation in agriculture.

Despite these challenges, the use of Nanoparticle-mediated transformation in agriculture also presents several perspectives for improving crop production and sustainability. The nanoparticle-mediated transformation has shown promise for improving nutrient uptake, reducing pesticide use, and increasing plant disease resistance (Lv et al., 2020). Additionally, the potential for targeted delivery of nutrients or agrochemicals using Nanoparticle-mediated transformation could help to reduce waste and minimize environmental impacts. To realize the potential benefits of Nanoparticle-mediated transformation while addressing the associated challenges, it is important to take a holistic and systems-level approach that considers the interactions between Nanoparticle-mediated transformation, plants, and the environment. This will require further research to understand the impacts of Nanoparticle-mediated transformation on plant and soil systems, as well as the development of effective regulations and best management practices.

2.3 FUTURE PROSPECTS

Genetic transformation is one of the technological advancements in the agricultural sector aimed at creating novel traits that are more robust, sustainable, and effective in addressing food shortages. Over the past few decades, genetic transformation has led to the development of hundreds of new crop varieties, addressing quality and performance issues (Corte et al., 2019; Zhang et al., 2018). By utilizing DNA modification and assessing the genetic background of crop germplasm, genetic improvement has successfully enhanced crop quality and performance, reducing reliance on environmentally harmful chemical fertilizers and pesticides. The application of genetic transformation in fruit and vegetable crops has brought significant changes to global agriculture, although the progress and acceptance of genetically modified crops vary across regions.

Given that the development of genetic transformation and transgenic regeneration ability has been achieved in various species of fruit and vegetable crops, there is a strong need to develop protocols regarding gene delivery, culture parameters, targeting tissue, and selectable marker genes (Anjanappa & Gruissem, 2021). For instance, selectable marker genes encode an enzyme that can confer resistance of transformed cells to specific antibiotics (such as kanamycin, hygromycin, phosphinothricin, glyphosate) (Sundar & Sakthivel, 2008), and can also alter metabolized substrates in transformed cells that are not metabolized by plant cells (Aragao & Brasileiro, 2002; Anjanappa & Gruissem, 2021). Alternatively, strategies to generate selectable marker-free transformation by using the Cre/lox recombination system could help address gene flow and reduce unwanted metabolic gene expression in transgenic crops (Borrill, 2020; Dale & Ow, 1991; Sreekala et al., 2005). However, there are concerns about the induction and maintenance of tissue culture complexity, including callus formation from explants. To achieve the expected transformation efficiency, it is necessary to establish reproducible protocols before treating the callus tissue. The expression of plant growth factors such as maize baby boom (BBM), AtGRF5 transcription factor, and growth-regulating factors (GRFs) along with its co-factor GRF-Interacting Factor (GIF1) has shown success in enhancing transformation and regeneration efficiencies (Jha et al., 2018, 2020; Luo & Palmgren, 2020; Kong et al., 2020). However, these applications are primarily limited to a few plant species, such as sugar, soybean, and sunflower, and their broad applicability remains challenging in fruits and vegetable crops (Anjanappa & Gruissem, 2021).

The advancement of gene transformation has successfully created a new trait for fruits and vegetable crops. However, challenges such as complex tissue culture, reproducible protocol, pleiotropic effect, and lack of whole-genome sequence are still restricted in many crops (Irfan et al., 2023). Yet, high-quality genomes were sequenced for species such as peach, papaya, clementine, grape, apple, strawberry, sweet cherry, tomato, pepper, potato, and pear while the complete genomic information for many fruit and vegetable crops on the entire existing species variability are still lacking (Chen et al., 2019; Corte et al., 2019). Newer of all commercially important crop genomes would be valuable to facilitate a better understanding of their evolutionary history, genome structure, and gene function for future genetic modification and breeding.

2.4 CONCLUSION

Gene transformation technology has brought about a revolutionary breakthrough in the production and sustainability of horticultural systems for fruit and vegetable crops. Currently, three types of gene transformation tools have been successfully developed, including *Agrobacterium tumefaciens*,

CRISPR-Cas, and Nanoparticles. These tools offer numerous advantages for creating new crop varieties through gene editing technology. Enhancing our understanding of gene transformation tools and refining protocol strategies could simplify methods and improve their effectiveness. Despite the significant benefits of transformation tools, challenges persist, such as low stable transgene expression, off-target effects, and the emergence of new phenotypic traits. Developing a new, efficient, and promising technology or a combination of these methods could help achieve specific desirable genomic traits in fruit and vegetable crops.

KEYWORDS

- **CRISPR/Cas9**
- **fertilizer**
- **gene editing**
- **horticultural systems**
- **nanoparticles**
- **phenotypic change**
- **plant cells**
- **sustainability**

REFERENCES

Ainley, W. M., Sastry-Dent, L., Welter, M. E., Murray, M. G., Zeitler, B., Amora, R., Corbin, D. R., et al., (2013). Trait stacking via targeted genome editing. *Plant Biotechnology Journal, 11*(9), 1126–1134.

Anfoka, G., Moshe, A., Fridman, L., et al., (2016). Tomato yellow leaf curl virus infection mitigates the heat stress response of plants grown at high temperatures. *Scientific Reports, 6*, 19715.

Anjanappa, R. B., & Gruissem, W., (2021). Current progress and challenges in crop genetic transformation. *Journal of Plant Physiology, 261*, 153411.

Aragao, F. J. L., & Brasileiro, A. C. M., (2002). Positive, negative, and marker-free strategies for transgenic plant selection. *Brazilian Journal of Plant Physiology, 14*, 1–10.

Bao, A., Burritt, D. J., Chen, H., Zhou, X., Cao, D., & Tran, L. P., (2019). The CRISPR/Cas9 system and its applications in crop genome editing. *Critical Reviews in Biotechnology, 39*(3), 321–336.

Baranski, R., (2008). Genetic transformation of carrot (*Daucus carota*) and other *Apiaceae* species. *Transgenic Plant Journal, 2*, 18–38.

Basso, M. F., Arraes, F. B. M., Grossi-de-Sa, M., Moreira, V. J. V., & Alves-Ferreira, M., (2020). Insights into genetic and molecular elements for transgenic crop development. *Frontiers in Plant Science, 11*, 509.

Beaujean, A., Sangwan, R. S., Lecardonnel, A., & Sangwan-Norreel, B. S., (1998). *Agrobacterium*-mediated transformation of three economically important potato cultivars using sliced internodal explants: An efficient protocol of transformation. *Journal of Experimental Botany, 49*(326), 1589–1595.

Becker, S., & Boch, J., (2021). TALE and TALEN genome editing technologies. *Gene Genome Edit, 2,* 100007.

Biosafety Clearing-House, (1999). *Melon A and B. Country's Decision or Any Other Communication.* Available at: http://bch.cbd.int/database/record.shtml?documentid=6351 (accessed on 7 February 2024).

Boch, J., & Bonas, U., (2010). Xanthomonas AvrBs3 family-type III effectors: Discovery and function. *Annual Review of Phytopathology, 48*, 419–436.

Borrill, P., (2020). Blurring the boundaries between cereal crops and model plants. *New Phytologist, 228*, 1721.

Bundock, P., Den Dulk-Ras, A., Beijersbergen, A., & Hooykaas, P. J., (1995). Trans-kingdom T-DNA transfer from *Agrobacterium tumefaciens* to *Saccharomyces cerevisiae*. *EMBO Journal, 14*(13), 3206–3214.

Cardi, T., D'Agostino, N., & Tripodi, P., (2017). Genetic transformation and genomic resources for next-generation precise genome engineering in vegetable crops. *Frontiers in Plant Science, 8*, 241.

Chang, F. P., Kuang, L. Y., Huang, C. A., Jane, W., Hung, Y., Yu-ie, C., & Mou, C., (2013). A simple plant gene delivery system using mesoporous silica nanoparticles as carriers. *Journal of Materials Chemistry B, 1*, 5279–5287.

Chen, G., Ye, C. M., Huang, J. C., Yu, M., & Li, B. J., (2001). Cloning of the papaya ringspot virus (PRSV) replicase gene and generation of PRSV-resistant papayas through the introduction of the PRSV replicase gene. *Plant Cell Reports, 20*, 272–277.

Chen, K., Wang, Y., Zhang, R., Zhang, H., & Gao, C., (2019). CRISPR/Cas genome editing and precision plant breeding in agriculture. *Annual Review of Plant Biology, 70*, 667–697.

Chen, Z. L., Gu, H., Li, Y., Su, Y., Wu, P., Jiang, Z., Ming, X., et al., (2003). Safety assessment for genetically modified sweet pepper and tomato. *Toxicology, 188*(2, 3), 297–307.

Chin, D. P., Bao, J. H., & Mii, M., (2009). Transgenic spinach plants produced by agrobacterium-mediated method based on the low temperature-dependent high plant regeneration ability of leaf explants. *Plant Biotechnology, 26*, 243–248.

Clasen, B. M., Stoddard, T. J., Luo, S., Demorest, Z. L., Li, J., Cedrone, F., Tibebu, R., et al., (2016). Improving cold storage and processing traits in potato through targeted gene knockout. *Plant Biotechnology Journal, 14*, 169–176.

Dai, S., Zheng, P., Marmey, P., Zhang, S., Tian, W., Chen, S., Beachy, R. N., & Fauquet, C., (2001). Comparative analysis of transgenic rice plants obtained by agrobacterium-mediated transformation and particle bombardment. *Molecular Breeding, 7*(2), 25–33.

Dale, E. C., & Ow, D. W., (1990). Intra- and intermolecular site-specific recombination in plant cells mediated by bacteriophage P1 recombinase. *Gene, 91*, 79–85.

Davey, M. W., Auwerkerken, A., & Keulemans, J., (2007). Relationship of apple vitamin C and antioxidant contents to harvest date and postharvest pathogen infection. *Journal of Science of Food and Agriculture, 87*(5), 802–813.

Demirer, G. S., Zhang, H., Goh, N. S., Gonzalez-Grandio, E., & Landry, M. P., (2019). Carbon nanotube-mediated DNA delivery without transgene integration in intact plants. *Nature Protocols, 14*, 2954–2971.

Du, H., Zeng, X., Zhao, M., Cui, X., Wang, Q., Yang, H., Cheng, H., & Yu, D., (2016). Efficient targeted mutagenesis in soybean by TALENs and CRISPR/Cas9. *Journal of Biotechnology, 217*, 90–97.

Erpen-Dalla, C. L., Mahmoud, L. M., Moraes, T. S., Mou, Z., Grosser, J. W., & Dutt, M., (2019). Development of improved fruit, vegetable, and ornamental crops using the CRISPR/Cas9 genome editing technique. *Plants, 8*, 601.

Ferrari, E., Barbero, F., Busquets-Fité, M., Franz-Wachtel, M., Köhler, H. R., Puntes, V., & Kemmerling, B., (2021). Growth-promoting gold nanoparticles decrease stress responses in *Arabidopsis* seedlings. *Nanomaterials, 11*(12), 3161.

Firoozbady, E., & Young, T. R., (2015). *Pineapple Plant Named "Rose."* US Patent PP25,763 P3.

Fitch, M. M. M., Manshardt, R. M., Gonsalves, D., Slightom, J. L., & Sanford, J. C., (1990). Stable transformation of papaya via microprojectile bombardment. *Plant Cell Reports, 9*, 189–194.

Geekiyanage, S., Takase, T., Watanabe, S., Fukai, S., & Kiyosue, T., (2006). The combined effect of photoperiod, light intensity, and GA3 on adventitious shoot regeneration from cotyledons of spinach (*Spinacia oleracea* L.). *Plant Cell Reports, 25*, 431–435.

Ghanim, M., Morin, S., Zeidan, M., & Czosnek, H., (1998). Evidence for transovarial transmission of tomato yellow leaf curl virus by its vector the whitefly *Bemisia tabaci*. *Virology, 240*, 295–303.

González-Melendi, P., Fernández-Pacheco, R., Coronado, M. J., Corredor, E., Testillano, P. S., Risueño, M. C., Marquina, C., et al., (2008). Nanoparticles as smart treatment-delivery systems in plants: Assessment of different techniques of microscopy for their visualization in plant tissues. *Annals of Botany, 101*(1), 187–195.

Guo, M., Bian, X., Wu, X., & Wu, M., (2011). *Agrobacterium*-mediated genetic transformation: History and progress. In: Alvarrez, M. A., (ed.), *Genetic Transformation* (pp. 5–28). Rijeka. InTech.

Harwood, W. A., (2012). Advances and remaining challenges in the transformation of barley and wheat. *Journal of Experimental Botany, 63*(5), 1791–1798.

Hollister, P., Weener, J. W., Román, C. V., & Harper, T., (2003). *Nanoparticles. Cientifica Technology White Papers*, pp. 1–12.

Horsch, R., Fry, J., Hofmann, N. D. E., Rogers, S., & Fraley, T., (1985). A simple and general method for transferring genes into plants. *Science, 227*, 1229.

Hwang, H. H., Wu, E. T., Liu, S. Y., Chang, S. C., Tzeng, K. C., & Kado, C. I., (2013). Characterization and host range of five tumorigenic *Agrobacterium tumefaciens* strains and possible application in plant transient transformation assays. *Plant Pathology, 62*, 1384–1397.

Irfan, M., Kumar, P., Ahmad, M. F., & Siddiqui, M. W., (2023). Biotechnological interventions in reducing losses of tropical fruits and vegetables. *Current Opinion in Biotechnology, 79*, 102850.

Irfan, M., Kumar, P., Kumar, V., & Datta, A., (2022). Fruit ripening specific expression of β-D-N-acetylhexosaminidase (β-Hex) gene in tomato is transcriptionally regulated by ethylene response factor 6 (SlERF6). *Plant Science, 323*, 111380. doi.org/10.1016/j.plantsci.2022.111380; https://doi.org/10.1016/j.plantsci.2022.111380.

Jha, P., & Kumar, V., (2018). BABY BOOM (BBM): A candidate transcription factor gene in plant biotechnology. *Biotechnology Letters, 40*, 1467–1475.

Jha, P., Ochatt, S. J., & Kumar, V., (2020). WUSCHEL: A master regulator in plant growth signaling. *Plant Cell Reports, 39*, 431–444.

Jung, J. H., & Altpeter, F., (2016). TALEN mediated targeted mutagenesis of the caffeic acid O-methyltransferase in highly polyploid sugarcane improves cell wall composition for production of bioethanol. *Plant Molecular Biology, 92*(1, 2), 131–142.

Kapoor, B., Kumar, P., Sharma, R., & Irfan, M., (2023). Drought-induced plant miRNAome and phytohormone signaling cross-talk. In: Khan, M. I. R., Singh, A., & Poor, P., (eds.), *Plant Hormones in Crop Improvement* (pp. 235–247). Academic Press Elsevier. ISBN 9780323918862, https://doi.org/10.1016/B978-0-323-91886-2.00006-9.

Khan, I., Saeed, K., & Khan, I., (2019). Nanoparticles: Properties, applications, and toxicities. *Arabian Journal of Chemistry, 12*, 908–931.

Klocke, E., Nothnagel, T., & Schumann, G., (2010). Vegetables. In: Kempken, F., & Jung, C., (eds.), *Genetic Modification of Plants* (1st edn., pp. 307–326). Springer-Verlag.

Kong, J., Martin-Ortigosa, S., Finer, J., Orchard, N., Gunadi, A., Batts, L. A., Thakare, D., et al., (2020). Overexpression of the transcription factor GROWTH-REGULATING FACTOR5 improves transformation of dicot and monocot species. *Frontiers in Plant Science, 11*, 572319.

Kramer, M. G., & Redenbaugh, K., (1994). Commercialization of a tomato with an antisense polygalacturonase gene: The FLAVR SAVR™ tomato story. *Euphytica, 79*, 293–297.

Kumar, K., Gambhir, G., Dass, A., Tripathi, A. K., Singh, A., Jha, A. K., Yadava, P., Choudhary, M., & Rakshit, S., (2020). Genetically modified crops: Current status and future prospects. *Planta, 251*, 91.

Kumari, C., Sharma, M., Kumar, V., Sharma, R., Kumar, V., Sharma, P., Kumar, P., & Irfan, M., (2022). Genome editing technology for genetic amelioration of fruits and vegetables for alleviating post-harvest loss. *Bioengineering, 9*(4), 176. doi.org/10.3390/bioengineering9040176.

Lei, Y., Tang, X. B., Jayawardena, R. S., Yan, J. Y., Wang, X. D., Liu, M., Chen, T., et al., (2016). Identification and characterization of *Colletotrichum* species causing grape ripe rot in southern China. *Mycosphere, 7*, 1191.

Lobato-Gómez, M., Hewitt, S., Capell, T., et al., (2021). Transgenic and genome-edited fruits: Background, constraints, benefits, and commercial opportunities. *Horticulture Research, 8*, 166.

Lobell, D. B., & Field, C. B., (2007). Global scale climate-crop yield relationships and the impacts of recent warming. *Environmental Research Letters, 2*, 014002.

Lor, V. S., Starker, C. G., Voytas, D. F., Weiss, D., & Olszewski, N. E., (2014). Targeted mutagenesis of the tomato PROCERA gene using transcription activator-like effector nucleases. *Plant Physiology, 166*(3), 1288–1291.

Luo, G., & Palmgren, M., (2021). GRF-GIF chimeras boost plant regeneration. *Trends in Plant Science, 26*(3), 201–204.

Lv, Z., Jiang, R., Chen, J., & Chen, W., (2020). Nanoparticle-mediated gene transformation strategies for plant genetic engineering. *The Plant Journal, 104*(4), 880–891.

Ma, J., Li, Y., Chen, F., Sun, Y., Zhu, Y., & Wang, L., (2023). Bacillus mycoides PM35 in combination with titanium dioxide (TiO_2) nanoparticles enhanced morpho-physio-biochemical attributes in barley (*Hordeum vulgare* L.) under cadmium stress. *Chemosphere, 323*, 138224.

Madhulatha, P., Pandey, R., Hazarika, P., & Rajam, M. V., (2007). High transformation frequency in agrobacterium-mediated genetic transformation of tomato by using polyamines and maltose in shoot regeneration medium. *Physiology and Molecular Biology of Plants, 13*, 191–198.

Mahfouz, M. M., Li, L., Shamimuzzaman, M., Wibowo, A., Fang, X., & Zhu, J. K., (2011). De novo-engineered transcription activator-like effector (TALE) hybrid nuclease with novel DNA binding specificity creates double-strand breaks. *Proceedings of the National Academy of Sciences of the United States of America, 108*(6), 2623–2628.

Mahto, B. K., Sharma, P., Rajam, M. V., & Dhar-Ray, S., (2018). An efficient method for agrobacterium-mediated genetic transformation of chilli pepper (*Capsicum annuum* L.). *Indian Journal of Plant Physiology, 23*, 573–581.

Malzahn, A., Lowder, L., & Qi, Y., (2017). Plant genome editing with TALEN and CRISPR. *Cell Bioscience, 7*, 21.

Mattos, L. M., Moretti, C. L., Jan, S., Sargent, S. A., Lima, C. E. P., & Fontenelle, M. R., (2014). Climate changes and potential impacts on quality of fruit and vegetable crops. In: Ahmad, P., & Rasool, S., (eds.), *Emerging Technologies and Management of Crop Stress Tolerance: Biological Techniques* (Vol. 1, pp. 467–486). San Diego: Elsevier.

McClements, D. J., & Xiao, H., (2017). Is nano safe in foods? Establishing the factors impacting the gastrointestinal fate and toxicity of organic and inorganic food-grade nanoparticles. *NPJ Science of Food, 1*, 6.

McCormick, S., Niedermeyer, J., Fry, J., Barnason, A., Horsch, R., & Fraley, A., (1986). Leaf disc transformation of cultivated tomato (*L. esculentum*) using agrobacterium tumefaciens. *Plant Cell Reports, 5*, 81–84.

Moriones, E., & Navas-Castillo, J., (2000). Tomato yellow leaf curl virus, an emerging virus complex causing epidemics worldwide. *Virus Research, 71*, 123–134.

Newell, C. A., (2000). Plant transformation technology. *Molecular Biotechnology, 16*, 53–65.

Nicolia, A., Proux-Wéra, E., Åhman, I., Onkokesung, N., Andersson, M., Andreasson, E., & Zhu, L. H., (2015). Targeted gene mutation in tetraploid potato through transient TALEN expression in protoplasts. *Journal of Biotechnology, 204*, 17–24.

O'Brien, (1983). In: *Bio/Technology Miami Winter Symposium* (Vol. 1, pp. 36–38).

Osakabe, K., Osakabe, Y., & Toki, S., (2010). Site-directed mutagenesis in *Arabidopsis* using custom-designed zinc finger nucleases. *Proceedings of the National Academy of Sciences of the United States of America, 107*(26), 12034–12039.

Our World in Data, (2022). *Population Past & Future*. Retrieved from: https://ourworldindata.org/grapher/population-past-future?tab=map (accessed on 7 February 2024).

Papayiannis, L., Katis, N., Idris, A., & Brown, J., (2011). Identification of weed hosts of tomato yellow leaf curl virus in Cyprus. *Plant Disease, 95*, 120–125.

Punja, Z. K., Jayaraj, J., & Wally, O., (2007). Carrot. In: Pua, E. C., & Davies, D. R., (eds.), *Transgenic Crops* (pp. 277–294). Springer.

Riley, M. K., & Vermerris, W., (2017). Recent advances in nanomaterials for gene delivery: A review. *Nanomaterials (Basel), 7*(5), 94.

Scorza, R., Ravelonandro, M., Callahan, A. M., Cordts, J. M., Fuchs, M., Dunez, J., & Gonsalves, D., (1994). Transgenic plums (*Prunus domestica* L.) express the plum pox virus coat protein gene. *Plant Cell Reports, 14*(1), 18–22.

Sedeek, K. E. M., Mahas, A., & Mahfouz, M., (2019). Plant genome engineering for targeted improvement of crop traits. *Frontiers in Plant Science, 10*, 114.

Shah, S. A., Alkhnbashi, O. S., Behler, J., Han, W., She, Q., Hess, W. R., Garrett, R. A., & Backofen, R., (2019). Comprehensive search for accessory proteins encoded with archaeal and bacterial type III CRISPR-Cas gene cassettes reveals 39 new Cas gene families. *RNA Biology, 16*(4), 530–542.

Shan, Q., Wang, Y., Chen, K., Liang, Z., Li, J., Zhang, Y., Zhang, K., et al., (2013). Rapid and efficient gene modification in rice and *Brachypodium* using TALENs. *Molecular Plant, 6*, 1365–1368.

Shan, Q., Zhang, Y., Chen, K., Zhang, K., & Gao, C., (2015). Creation of fragrant rice by targeted knockout of the OsBADH2 gene using TALEN technology. *Plant Biotechnology Journal, 13*, 791–800.

Sharma, M., Kumar, P., Verma, V., Sharma, R., Bhargava, B., & Irfan, M., (2022a). Understanding plant stress memory response for abiotic stress resilience: Molecular insights and prospects. *Plant Physiology and Biochemistry, 179*, 10–24. doi.org/10.1016/j.plaphy.2022.03.004.

Sharma, S., Shree, B., Aditika, Sharma, A., Irfan, M., & Kumar, P., (2022b). Nanoparticle-based toxicity in perishable vegetable crops: Molecular insights, impact on human health and mitigation strategies for sustainable cultivation. *Environmental Research, 212*(A), 113168. https://doi.org/10.1016/j.envres.2022.113168.

Shelton, A. M., Hossain, M. J., Paranjape, V., Azad, A. K., Rahman, M. L., Khan, A. S. M. M. R., Prodhan, M. Z. H., et al., (2018). Bt eggplant project in Bangladesh: History, present status, and future direction. *Frontiers in Bioengineering and Biotechnology, 6*, 106.

Shivalingswami, T. M., Satpathy, S., & Banergee, M. K., (2002). Estimation of crop losses due to insect pests in vegetables. In: Babu, B. S., Varaprasad, K. S., Anitha, K., Prasada, R. D. V. J., Chakrabarthy, S. K., & Chandurkar, P. S., (eds.), *Resource Management in Plant Protection* (pp. 24–31).

Shukla, V. K., Doyon, Y., Miller, J. C., DeKelver, R. C., Moehle, E. A., Worden, S. E., Mitchell, J. C., et al., (2009). Precise genome modification in the crop species Zea mays using zinc-finger nucleases. *Nature, 459*(7245), 437–441.

Siddiqi, K. S., & Husen, A., (2017). Recent advances in plant-mediated engineered gold nanoparticles and their application in biological system. *Journal of Trace Elements in Medicine and Biology, 40*, 10–23.

Song, G. Q., Loskutov, A. V., & Sink, K. C., (2007). Highly efficient *Agrobacterium tumefaciens* mediated transformation of celery (*Apium graveolens* L.) through somatic embryogenesis. *Plant Cell, Tissue, and Organ Culture, 88*, 193–200.

Song, G. Q., Prieto, H., & Orbovic, V., (2019). *Agrobacterium*-mediated transformation of tree fruit crops: Methods, progress, and challenges. *Frontiers in Plant Science, 10*, 226.

Song, M. J., Kang, T. H., Han, C. S., & Oh, M. M., (2014). Air anions enhance lettuce growth in plant factories. *Horticulture, Environment, and Biotechnology, 55*(4), 293–298.

Sparrow, P. A. C., Townsend, T. M., Arthur, A. E., Dale, P. J., & Irwin, J. A., (2004). Genetic analysis of agrobacterium tumefaciens susceptibility in *Brassica oleracea. Theoretical and Applied Genetics, 108*(4), 644–650.

Sreekala, M. S., Thomas, S., & Groeninckx, G., (2005). Dynamic mechanical properties of oil palm fiber/phenol formaldehyde and oil palm fiber/glass hybrid phenol formaldehyde composites. *Polymer Composites, 26*(3), 388–400.

Sundar, I. K., & Sakthivel, N., (2008). Advances in selectable marker genes for plant transformation. *Journal of Plant Physiology, 165*(16), 1698–1716.

Tang, Y., Zhang, Z., Yang, Z., & Wu, J., (2023). CRISPR/Cas9 and *Agrobacterium tumefaciens* virulence proteins synergistically increase efficiency of precise genome editing via homology directed repair in plants. *Journal of Experimental Botany*, erad096.

Thakur, N., Thakur, S., Chatterjee, S., Das, J., & Sil, P. C., (2020). Nanoparticles as smart carriers for enhanced cancer immunotherapy. *Frontiers in Chemistry, 8*, 597806.

Torney, F., Trewyn, B., Lin, V. Y., & Wang, K., (2007). Mesoporous silica nanoparticles deliver DNA and chemicals into plants. *Nature Nanotechnology, 2*, 295–300.

USDA, (2015). *USDA Announces Deregulation of Non-browning Apples*. Available at: https://www.aphis.usda.gov/stakeholders/downloads/2015/SA_arctic_apples.pdf (accessed on 7 February 2024).

Wan, W. L., Kim, S. T., Castel, B., Charoennit, N., & Chae, E., (2021). Genetics of autoimmunity in plants: An evolutionary genetics perspective. *New Phytologist, 229*(3), 1215–1233.

Wang, X., Xu, J., Wang, X., Qiu, B., Cuthbertson, A. G. S., Du, C., Wu, J., & Ali, S., (2019). Isaria fumosorosea-based zero-valent iron nanoparticles affect the growth and survival of sweet potato whitefly, *Bemisia tabaci* (Gennadius). *Pest Management Science, 75*(8), 2174–2181.

Wang, Y. C., Yu, M., Shih, P. Y., Wu, H. Y., & Lai, E. M., (2018). Stable pH suppresses defense signaling and is the key to enhance agrobacterium-mediated transient expression in *Arabidopsis* seedlings. *Scientific Reports, 8*, 17071.

Ye, R., Yang, X., & Rao, Y., (2022). Genetic engineering technologies for improving crop yield and quality. *Agronomy, 12*(4), 759.

Zhang, H. X., Zhang, Y., & Yin, H., (2019). Genome editing with mRNA encoding ZFN, TALEN, and Cas9. *Molecular Therapy, 27*(4), 735–746.

Zhang, J., Wang, L., You, X., Xian, T., Wu, J., & Pang, J., (2019). Nanoparticle therapy for prostate cancer: Overview and perspectives. *Current Topics in Medicinal Chemistry, 19*, 57–73.

Zhang, Y., Massel, K., Godwin, I. D., & Gao, C., (2018). Applications and potential of genome editing in crop improvement. *Genome Biology, 19*, 210.

Zhu, H., Li, C., & Gao, C., (2020). Applications of CRISPR-Cas in agriculture and plant biotechnology. *Nature Reviews Molecular Cell Biology, 21*(11), 661–677.

CHAPTER 3

DEVELOPING BIOENGINEERED FRUITS AND VEGETABLES FOR POSTHARVEST QUALITY IMPROVEMENT USING HETEROLOGOUS, OVEREXPRESSION, AND COMBINATORIAL ENGINEERING

RAJUL TAYAL,[1] SEKHU ANSARI,[2] ANURADHA TYAGI,[3] VINAY KUMAR,[4] and MANI KANT CHOUDHARY[5]

[1]*National Institute of Plant Genome Research, New Delhi, India*

[2]*Division of Pathology, Cincinnati Children's Hospital Medical Center, Cincinnati, Ohio, USA*

[3]*Department of cBRN, Institute of Nuclear Medicine and Allied Sciences, Brig S.K. Mazumdar, Delhi, India*

[4]*Pennsylvania State University HMC, Hershey, Pennsylvania, USA*

[5]*Department of Biology, University of Arkansas, Little Rock, Arkansas, USA*

Bioengineered Fruit and Vegetables: Trends in Postharvest Quality Improvement.
Mohammad Irfan and Pankaj Kumar (Eds.)

ABSTRACT

Climate change and post-harvest losses cause crop production loss and food security. Plants are vulnerable to several biotic and abiotic stress factors due to changing climatic conditions, thus reducing agriculture production. Post-harvest agriculture produce is liable to changes that affect the quality and quantity of produce. Postharvest losses (PHLs) are one of the decisive factors in preventing loss and quality of produce. To advance the post-harvest quality characteristics of horticultural crops, attempts have been made to enhance agricultural production by the development of genetically improved and stress-resilient crops through biotechnological and transgenic approaches, including gene knockout and knockdown. Various genetic tools such as genome editing, RNA interference (RNAi), antisense technology, TILLING, heterologous expression, overexpression, and combinatorial engineering are successfully used in many fruit crops to enhance their production, shelf life, nutritional quality improvement, and resistance to abiotic and biotic stresses. Thus, transgenic techniques have the potential to establish food security, increase crop production, and reduce PHLs. Here, without sacrificing nutritional values, we present an overview of how to increase post-harvest stability qualities like shelf life, texture, and resistance to infections through transgenic approach applications by manipulating enzymes and transcriptional factors, and combinatorial engineering in horticultural crops.

3.1 INTRODUCTION

Postharvest losses (PHLs) are a major concern for the food industry. Improving the shelf life and quality of fruits and vegetables is crucial for reducing food waste and meeting the increasing demand for fresh produce worldwide (Kumari et al., 2022). Transgenic technology has emerged as a promising approach to enhancing the postharvest quality and shelf life of fruits and vegetables by introducing genes into plants to modify their characteristics (Kumari et al., 2022; Irfan et al., 2023). Several genes have been targeted for postharvest quality improvement using transgenic technology, including genes related to cell wall degradation, hormone signaling, and stress response. The FlavrSavr™ tomato, the first genetically modified (GM) fruit product, was introduced to the market in 1994 (Baranski et al., 2019). Transgenic fruits have an extended shelf life and take longer to soften due to the suppression of a gene that typically initiates pectin solubilization (Kramer et al., 1994). Overexpression of polygalacturonase-inhibiting proteins (PGIPs) has been demonstrated to reduce the softening rate during storage, leading to improved firmness and shelf life of the fruit (Kalunke et al., 2015). Transgenic tomatoes expressing a polygalacturonase inhibiting protein (PGIP) gene exhibited reduced susceptibility to *Botrytis cinerea* infection (Thornburg et al., 2003). Similarly, overexpression of the E8 gene in tomatoes delays fruit ripening and prolongs the shelf life of the fruit (Kneissl et al., 1996).

Direct DNA transfer and Agrobacterium-mediated transformation have both been used to create genetically engineered fruit and vegetable crops (Figure 3.1). Transforming plants efficiently is very context-sensitive, necessitating the refinement of strategies for effective gene delivery, regeneration, and selection from transformed explants (Song et al., 2019). Several fruit and vegetable crops whose features have been enhanced by genetic engineering (GE) and are designed for cultivation as either human food or animal feed have acquired regulatory approval for commercialization in various regions of the world. These include papaya (*Carica papaya* L.) (Fitch et al., 1992; Chen et al., 2001), plum (*Prunus domestica*) (Scorza et al., 1994), tomato (*Solanum lycopersicum*) (APHIS, 1994, 1995, 1998), pepper (*Capsicum annuum*) (Chen et al., 2003), eggplant (*Solanum melongena* L.) (Shelton et al., 2015), apple *(Malus domestica* Borkh.) (USDA, 2015), melon (*Cucumis melo* L.) (Biosafety Clearing-House, 1999), and pineapple (*Ananas comosus* L. Merr.) (Firoozbady et al., 2015). Major traits considered for transgenic fruits and vegetables were to enhance agronomic production by deliberating delayed ripening or pest disease resistance. Quality qualities, however, have been communicated more recently through the elimination of fruit browning or the addition of new visual traits like flesh color.

Bioengineering has emerged as another promising approach that uses enzymes to improve the postharvest quality of perishable horticultural crops. Enzymes such as cellulases, pectinases, and proteases have been extensively studied for their role in improving postharvest quality. Cell wall modification enzymes cellulases and pectinases lead to increased softening, which is desirable for some fruits like tomatoes (Kumar et al.,

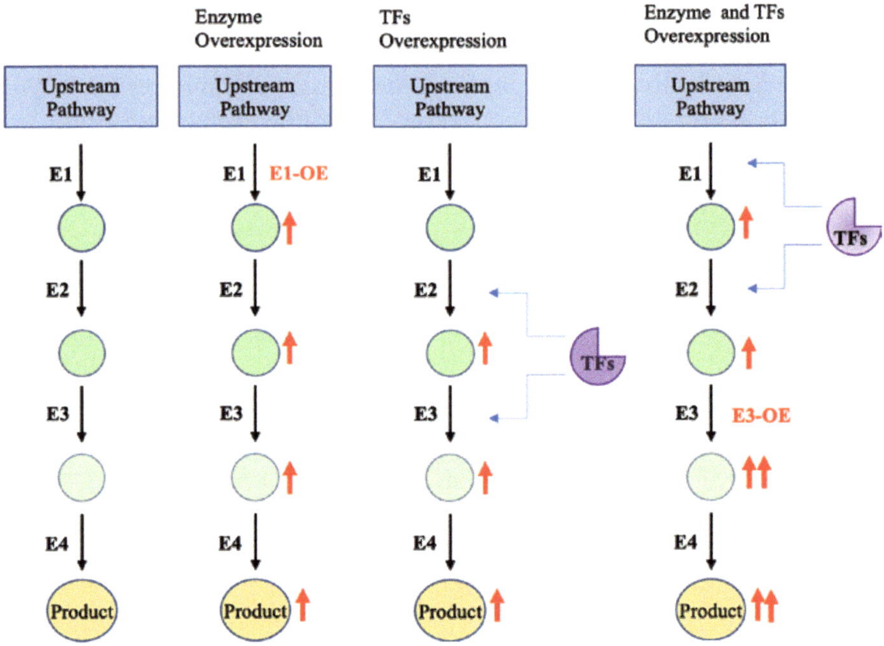

FIGURE 3.1 Strategies for synthesizing target products in plants (e: enzyme).

2016a). For some fruits, proteases can be used to reduce protein content to enhance the sweetness and flavor. Combinatorial engineering, which involves the simultaneous introduction of multiple genes into plants, has also been explored as a strategy to improve postharvest quality. For instance, a combination of genes involved in abscisic acid (ABA) signaling and cell wall modification has been shown to delay fruit softening and senescence in tomatoes, resulting in an extended shelf life (Jia et al., 2011, 2016). Another aspect of bioengineering is the use of nanotechnology and biodegradable packaging to extend shelf life and reduce food waste. Biodegradable packaging films made from natural polymers, such as starch and cellulose, have been developed and tested for their effectiveness in extending the shelf life of fruits and vegetables. These films can help maintain the quality of fruits and vegetables by reducing water loss, delaying ripening and senescence, and inhibiting the growth of microorganisms. However, the practical implementation of these approaches requires careful consideration of their efficacy, safety, and environmental impact, as well as the societal acceptance of these technologies.

The transgenic approach has resulted in the production of fruits and vegetable crops with enhanced resistance to pathogens, improved color and flavor, and extended shelf life (Meli et al., 2010; Kumar et al., 2016b; Irfan et al., 2023). Despite the potential advantages of transgenic technology in enhancing postharvest quality, there are concerns about the safety and risks associated with the introduction of GM crops into the environment. As a result, regulatory frameworks have been established to ensure the safety of GM crops for human consumption and the environment. In conclusion, transgenic technology shows significant potential for enhancing postharvest quality and prolonging the shelf life of fruits and vegetables.

3.2 OVERVIEW OF TRANSGENIC APPROACHES

The ability to manipulate plant genes using modern technology is a major step forward in agriculture and agricultural research, ushering in a new era of crop development and commercial farming (Ramkumar et al., 2022). Transgenic plant biology and other areas of plant biotechnology have been made possible by the combination of plant cells' innate totipotency and *Agrobacterium tumefaciens'* ability to infect and change plant cells. Opportunities to generate new types more quickly have arisen as a result of advancements in molecular genetics, particularly the new generation of genome editing tools (Lobato-Gómez et al., 2021). *Agrobacterium tumefaciens*, formerly known as *Bacterium tumefaciens*, is a gram-negative soilborne bacterium that was first recognized as a tumor-causing agent in plants (Smith & Townsend, 1907). Several other bacterial species have also been found to genetically change plants, not just *Agrobacterium tumefaciens*. Non-Agrobacterial but closely related species, such as *Agrobacterium rhizogenes, Rhizobium etli, Ensiferadhaerens,* and *Mesorhizobium loti*, have also been found to be capable of genetically transforming plants in the presence of T-DNA machinery. The introduction of foreign genetic material is facilitated by genetically manipulating viruses, which act as vectors. It was originally proposed that the cauliflower mosaic virus (CaMV) serve as an expression vector because it was the first virus to be investigated in detail (Szeto et al., 1997). Furthermore, Grimsley et al. demonstrated that maize streak virus (MSV) based vectors can be utilized for agro-infecting maize plants (Grimsley et al., 1987). After that, the plants were agro infected with the RNA virus beet western yellows (Leiser et al., 1992). The virus-induced gene silencing (VIGS) technique was used in the reverse genetic approach to plant functional genomics, in which viral vectors were modified to trigger gene silencing (Becker et al., 2010; Irfan et al., 2014). Furthermore, chemical or physical procedures can be used to insert foreign DNA elements. Particle bombardment and other forms of direct gene transfer do not rely on the plant-parasite connection for gene delivery as *Agrobacterium* did, hence they can be used on a wide variety of plant species. Multicopy integration and randomization of the foreign DNA element are common side effects of direct gene transfer technologies.

3.2.1 TRANSGENIC APPROACHES

The development of new and improved transgenic approaches for plant genetic transformation is imperative to producing engineered crops. Improvements in plant tissue culture methods are needed across the board to streamline processes and boost transformation efficiencies. A revolution in crop-genetic potential augmentation for a sustainable future may be at hand, thanks to advances in synthetic biology, innovative vector systems for precision genome editing, and gene integration. Therefore, the key to effectively transferring breakthroughs in plant molecular biology to crop improvement is the uniformity of plant transformation systems across species. The recent transgenic approaches used by researchers to develop various transgenic crops are discussed in succeeding subsections.

3.2.2 AGROBACTERIUM-MEDIATED PLANT GENETIC TRANSFORMATION

Both *Agrobacterium tumefaciens* and *Agrobacterium rhizogenes*, which cause plant diseases, have specialized adaptations that allow them to infect their host plants successfully. The species specificity of *A. tumefaciens* and *A. rhizogenes* is determined by the plasmids Ti (pTi) and Ri (pRi), respectively. Bacterial oncogenes (onc) that mimic the structure of plant gene cassettes can be found in the T-DNA portions of both the Ti and Ri plasmids. The Gram-negative soil bacteria known as agrobacteria can be found all over the world. The pathogenic forms of these bacteria are responsible for crown gall and hairy root disorders. When a host is present and a tumor-inducing (Ti) plasmid is present, *Agrobacterium* can either live freely in the soil or become a pathogen (Subramoni et al., 2014). *Agrobacterium* can cause drastic changes in plant life by invading through cuts or stomata (Brencic et al., 2005). The plant's virulence protein VirA responds to phenolic chemicals, the most well-known of which is acetosyringone. In turn, this protein phosphorylates and activates VirG, which then stimulates the transcription of other *vir* genes. These *vir* genes play an important role in the plant's genome by facilitating the movement and integration of transfer (T)-DNA. Between the left and right borders of the Ti plasmid is the T-DNA region, which is home to auxin, cytokinin, opiate, and other infection-related genes. Plant tumors occur as a result of these gene products being integrated into the plant, and *Agrobacterium* can utilize the resulting opines as a source of carbon and possibly nitrogen. The discovery that any other sequence Ti plasmid could substitute T-DNA between the left and right borders marked a major step forward in plant biotechnology. Any gene of interest might be placed between the two T-DNA borders by a single co-integration procedure. Rapid progress has been made in our understanding of *Agrobacterium* since the advent of molecular technology. Other advancements include the introduction of high-throughput transformation in the model plant *Arabidopsis thaliana* using the floral dip approach and the transformation of previously resistant species like rice (*Oryza sativa*) (Clough & Bent, 1998; Hiei et al., 1994).

3.2.3 PARTICLE BOMBARDMENT-MEDIATED GENETIC TRANSFORMATION OF THE NUCLEAR GENOME

Injecting foreign DNA into plant tissues or cells via particle bombardment (also known as biolistics) is a direct delivery method that utilizes high-velocity microprojectiles. Microprojectiles made of gold or tungsten are coated with the appropriate DNA and then propelled towards the plant to penetrate its cell wall. The DNA is released from the microprojectiles upon entering the cells, where it can be transiently expressed or potentially permanently integrated into the host genome (Twyman & Christou, 2004).

3.2.4 GENES TARGETED FOR IMPROVING SHELF LIFE AND POST-HARVEST MANAGEMENT IN FRUITS AND VEGETABLES

Many genes have been targeted for improving shelf life and post-harvest management in fruits and vegetables. Some examples include genes related to ethylene biosynthesis and signaling. Ethylene is a plant hormone that plays a key role in fruit ripening and senescence (Irfan et al., 2016). By manipulating the expression of genes involved in ethylene biosynthesis or signaling, it is possible to delay the onset of ripening and prolong the shelf life of fruits and vegetables. For example, transgenic tomatoes that over express the gene for ACC synthase, an enzyme involved in ethylene biosynthesis, have been shown to have delayed ripening and extended shelf life (Lucille et al., 2002). Overexpression of the *E8* gene in tomatoes was found to delay fruit ripening and extend the shelf life of the fruit by up to 40%. The *E8* gene encodes an enzyme involved in the biosynthesis of ABA, a plant hormone that regulates fruit ripening (Kneissl et al., 1996). Whereas genes involved in cell wall degradation were targeted for the softening of fruits and vegetables during ripening and senescence is largely due to the degradation of cell wall components such as pectin, cellulose, and hemicellulose (Payasi et al., 2009). By manipulating the expression of genes involved in cell wall degradation, it is possible to slow down the softening process and extend shelf life. Further, by manipulating the expression of genes involved in stress responses, it is possible to enhance the tolerance of fruits and vegetables to various stresses, such as temperature extremes, drought, and pathogen infection. In eggplants, overexpression of the spermidine synthase gene was found to increase the level of spermidine, a polyamine involved in various physiological processes, and improve postharvest quality by reducing fruit decay and maintaining firmness during storage (Peng et al., 2016).

The nutritional quality of fruits and vegetables can deteriorate during post-harvest storage due to factors such as respiration, oxidation, and enzymatic activity (Tayal et al., 2022). By manipulating the expression of genes involved in nutrient metabolism, it is possible to enhance the nutritional quality and shelf life of fruits and vegetables. Transgenic tomatoes that over express the gene for photogene synthase (Fraser et al., 2002), an enzyme involved in carotenoid biosynthesis, have been shown to have increased levels of carotenoids and extended shelf life.

3.2.5 TRANSCRIPTION FACTORS (TFS) INVOLVED IN IMPROVING POST-HARVEST QUALITY AND SHELF LIFE

Transcription factors (TFs) play a vital role in the regulation of gene expression in fruits and vegetables, influencing their postharvest quality and shelf life (Irfan et al., 2014, 2016). The expression of certain TFs is known to be induced during fruit ripening and senescence, which can affect the quality attributes of fruits and vegetables (Ghosh et al., 2013; Kumar et al., 2016). A large number of studies have reported the role of various TFs in the regulation of the key enzyme pathways for delayed fruit ripening and shelf life over the decades. The expression of certain TFs is known to be induced during fruit ripening and senescence, which can affect the quality attributes of fruits and vegetables. The ERF (ethylene response factor) TF family has been shown to play a critical role in the regulation of fruit ripening and senescence (Irfan et al., 2022). The manipulation of ERF genes using biotechnological approaches, such as overexpression or silencing, has been shown to influence fruit ripening and quality. Overexpression of ERF TFs, such as AtERF1 and AtERF2, has been shown to accelerate fruit ripening in tomatoes and Arabidopsis while silencing of ERF TFs results in delayed ripening and extended shelf life (Gu et al., 2017). One of the well-characterized metabolic pathways is anthocyanin biosynthesis. Maize transcription factor LC and C1 in the fruit of genetically altered tomato plants raise flavonoid levels in tomatoes (Bovy et al., 2002). Similarly, overexpression of TFs Del and Ros1 induces the accumulation of high levels of anthocyanins in tomatoes (Butelli et al., 2008). Several other events reported to regulate anthocyanin biosynthesis through numerous crucial TFs MYB, bHLH, and WD40 (Dixon et al., 2013; Zhang et al., 2014; Yuan & Grotewold, 2015). Similarly, other TFs, such as NAC (NAM, ATAF1/2, and CUC2) and WRKY, have also been implicated in the regulation of fruit and vegetable ripening, senescence, and stress responses. For instance, the expression of a tomato NAC gene, *SlNAC1*, has been shown to increase during fruit ripening and senescence, leading to increased expression of genes involved in cell wall modification and aroma production. In contrast, the silencing of *SlNAC1* resulted in delayed fruit ripening and extended shelf life (Ma et al., 2021). Additionally, transcription factor *NAC1* alters the ripening of tomato fruit ethylene-dependent and ABA-dependent pathways (Ma et al., 2014).

TFs have also been shown to regulate the response of fruits and vegetables to various stresses, such as pathogen infection, temperature, and drought. The TFs *MYB* and *WRKY* have been implicated in the defense response of fruits and vegetables against pathogen infection, while the TFs *DREB* and *CBF* have been shown to regulate the response of crops to abiotic stresses, such as drought and cold. It has been demonstrated that the *MYB12* transcription factor from Arabidopsis thaliana controls the expression of essential enzyme genes involved in flavonoid production, increasing flavonoid accumulation (Zhang et al., 2015). *MYB75* is a key regulator of fruit quality which includes the accumulation of anthocyanins through the transcriptional regulation of downstream genes involved in a large number of metabolic pathways (Jian et al., 2019). Li et al. (2021) showed that transcription factor *MdWRKY32* is involved in starch-sugar metabolism by binding to the *MdBam5* promoter in apples during postharvest storage. It has been observed that overexpression of lycopene β-cyclase (*LCYb*) not only accumulates the β-carotene level but also displays a wide range of prolonged shelf life by delayed softening (Diretto et al., 2020). The overexpression of *MdERF1* and suppression of *MdACO1*, which are TFs involved in ethylene biosynthesis and signaling, respectively, has been shown to delay apple fruit ripening and reduce softening, resulting in an extended shelf life (Li et al., 2016). The overexpression of a transcription factor, *MaMADS*, in banana fruit, is necessary for fruit ripening and reduces the incidence of post-harvest diseases, resulting in an extended shelf life (Elitzur et al., 2016). The overexpression of a transcription factor, *SlMADS1*, in tomato fruits has been shown to delay fruit ripening and reduce ethylene production, resulting in an extended shelf life (Vrebalov et al., 2009). Overall, the manipulation of TFs presents a promising approach for improving the postharvest quality and shelf life of fruits and vegetables.

3.3 OVEREXPRESSION OF ENZYMES

Fresh fruits and vegetables are an essential part of our daily diet, providing us with vitamins, minerals, and other essential nutrients. However, postharvest losses due to spoilage and decay remain a significant problem worldwide (Kumari et al., 2022; Irfan et al., 2023). Enzymes play a crucial role in various physiological processes, including fruit ripening and senescence (Ghosh et al., 2013; Irfan et al., 2021). To increase the shelf life of fruits and vegetables, scientists have developed various strategies to regulate the activity of enzymes involved in ripening and senescence. The over-expression of enzymes in plants for the shelf life of fruits and vegetables is a promising approach that has gained significant attention in recent years. However, there are still challenges to overcome, such as the potential impact of these genetic modifications on the nutritional quality and safety of fruits and vegetables (Irfan & Datta, 2017). Strategies for overexpression of an enzyme or a series of enzymes in a complex metabolic pathway to ensure increased metabolic flow. To provide an adequate supply of precursors and increase metabolic flow along the target pathway, boosting the expression of upstream genes encoding rate-limiting enzymes or numerous critical enzymes. Over the decades, a large number of techniques have been used to develop transgenics, including the transformation of key enzymes or multiple enzymes using binary vectors or particle bombardment methods. Although it has been challenging to transform plants with multiple genes or large constructs, multigene vector systems have been created utilizing a variety of strategies to make it easier to stack and deliver varied numbers of genes into plants. In this approach, a single transformation event leads to the generation of transgenic plants carrying multiple transgenes. Presently, these strategies include restriction digestion-based cloning, Gateway recombination, and overlapping sequence-based Gibson and Golden Gate assembly (Chen et al., 2006; Wakasa et al., 2006; Gibson et al., 2009; Engler et al., 2008). Recently, multigene expression systems through a single T-DNA region or polycistronic genes using self-cleavage peptide 2A and plastid transformation have gained popularity due to ease of manipulation or without the risk of epigenetic effects (Dafny-Yelin & Tzfira, 2007; Bock, 2013; Farre et al., 2014). The multigene expression system offers a significant advantage over other techniques due to the assembly of numerous target genes in a single cassette for T-DNA of binary vectors or vectors for particle bombardment or other approaches (Dafny-Yelin et al., 2007). Boehm & Bock (2019) discuss the benefits and drawbacks of a multigene vector system. Recently, a highly efficient multigene assembly and transformation system has been developed, i.e., TGSII (Zhu et al., 2017). This system used Cre/loxP-mediated recombination in which multigene assembly can be introduced to obtain marker-free transgenic plants. TGSII is used successfully to engineer complex metabolic pathways. In recent reports, 10 anthocyanin biosynthesis genes, and

4 and 6 ketocarotenoid biosynthetic genes have been used for metabolite engineering for the biofortification of crops (Zhu et al., 2017, 2018). Many fruits and vegetables naturally contain anthocyanins, a strong antioxidant, and its accumulation in plants has been associated with increased shelf life in various fruits. Tomato fruits have low levels of anthocyanins due to the mutation of the key regulatory enzyme chalcone isomerase of anthocyanins biosynthetic pathway (Muir et al., 2001). An increase in anthocyanin production in tomatoes and other fruits could increase their shelf-life and decrease post-harvest loss. The overexpression of a gene encoding a cell wall-loosening protein, endo-beta-1,4-glucanase, in strawberry fruits has been shown to improve fruit softening and maintain quality during storage (Jia et al., 2011). The overexpression of a gene encoding a chitinase enzyme, which breaks down fungal cell walls, in mango fruits has been shown to reduce the incidence of post-harvest fungal diseases and extend shelf life (Prasanna et al., 2007).

3.4 COMBINATORIAL ENGINEERING

Combinatorial engineering involves the simultaneous manipulation of multiple genes to achieve a desired phenotype. In the case of fruits and vegetables, combinatorial engineering has been used to improve post-harvest quality through transgenic approaches. To maximize the synthesis of the target metabolites, combining the overexpression of the transcription factor with several pathway-related key enzymes can also be used for metabolite engineering. It has been reported numerous structural genes related to the key enzyme of anthocyanin biosynthesis, decorating genes, and transporters, along with the transcription factor collectively as MBW (MYB-bHLH-WD40) complex modulate the anthocyanin biosynthesis (Li et al., 2014). However, these combinatorial approaches are not successful in all cases, especially in cereals, for instance, the biosynthesis route for anthocyanins in rice endosperm is functionally deficient. Because of this, the co-expression of structural genes along with bHLH and MYB in the rice endosperm resulted in the production of intermediate flavonoid products, not the target anthocyanins (Shin et al., 2006; Ogo et al., 2013). Later, these approaches have been used to develop carotene-enriched Rice, anthocyanin-rich Purple Tomatoes, and Purple Endosperm as well as astaxanthin-enriched aSTARice (Paine et al., 2005; Butelli et al., 2008; Zhu et al., 2017, 2018). Another thoroughly researched pathway is the production of carotenoids. *Agrobacterium*-mediated multigene vector transformation to direct synthesis target products in crops using multiple essential enzyme genes from plants, algae, and microbes (Huang et al., 2013; Pierce et al., 2015; Zhu et al., 2008, 2018; Ha et al., 2019; Tian et al., 2019). The transcription factor genes along with the essential structural genes have been used for the biofortification of flavonoids and anthocyanins in tomatoes. One example of combinatorial engineering in fruits is the modification of ethylene signaling pathways through the overexpression or suppression of multiple TFs involved in the regulation of fruit ripening. For instance, the simultaneous overexpression of the TFs *RIN, NOR,* and *CNR* in tomato fruit has been shown to delay fruit ripening, resulting in an extended shelf life (Wang et al., 2020). Combinatorial engineering through transgenic approaches holds promise for improving post-harvest quality in fruits and vegetables. Further research is needed to explore the potential of combinatorial engineering and multiple gene delivery techniques for enhancing postharvest quality and reducing food waste.

3.5 FUTURE PROSPECTS

In the future, the use of transgenic approaches for post-harvest quality improvement is likely to increase due to the growing demand for high-quality and nutritious fruits and vegetables. Researchers are continuing to explore the potential of transgenic crops for enhancing traits such as shelf life, nutritional content, flavor, and appearance. However, several challenges still need to be addressed, such as the regulatory approval process for GM crops, potential environmental impacts, and public perception and acceptance. To address these challenges, researchers are exploring new genome editing technologies such as CRISPR-Cas9, which could enable precise and targeted genetic modifications without the incorporation of foreign genes. In addition, researchers are investigating alternative approaches, such as RNA interference (RNAi) and epigenetic modifications, which could provide alternative strategies for manipulating gene expression. Overall, the use of transgenic approaches for post-harvest quality improvement holds great promise for addressing the challenges of global food security and meeting the demands of a growing population. However, continued research and innovation will be necessary to ensure that these technologies are developed and implemented responsibly and sustainably.

3.6 CONCLUSION

Fruits with desirable agronomic or qualitative qualities are easier to generate with the use of GE than they would be through conventional breeding alone, which suffers from a lack of acceptable germplasm, lengthy breeding cycles, and the necessity for many rounds of back-crossing. GE allows for the rapid, often direct, introduction of desirable traits into superior crop varieties. Over the course of their 25 years on the market, genetically altered fruits have increased farmers' profits by decreasing the prevalence of pests and diseases and boosting the quality of the final product. The decreased usage of pesticides has further positive effects on farmers, shoppers, and the natural world. The promise of additional commercial fruit varieties with resistance to biotic and abiotic stresses, improved flavor and nutrient content, and modified flowering and ripening times is a major motivator for the research and development of new molecular breeding technologies like trans-grafting and genome editing. In particular, the advancement of CRISPR/Cas9 and other nuclease-based genome editing approaches provides a means to introduce precise modifications at preselected genomic regions without leaving any genetic fingerprints or off-targets.

KEYWORDS

- **combinatorial engineering**
- **enzymes**
- **genes**
- **horticulture crops**
- **post-harvest**
- **post-harvest quality**
- **shelf life**
- **transcription factors**
- **transgenics**

REFERENCES

APHIS, (1994). *Interpretative Ruling on DNA Plant Technology Corporation Petition for Determination of Non-regulated Status of Delayed-Ripening Tomato Line 1345-4.* Docket No. 94-092-2.

APHIS, (1995). *Interpretative Ruling on Agritope, Inc. Petition for Determination of Regulatory Status of Cherry Tomatoes with a S-Adenosylmethionine Hydrolase Gene.* Docket No. 95-097-1.

APHIS, (1995). *Interpretative Ruling on Zeneca Plant Science and Petoseed Company Inc. Petition for Release from Regulation for Modified Lines B, Da- and F Derived from T7 Varieties of Processing Tomatoes.* Docket No. 95–016–1.

APHIS, (1998). *Interpretive Ruling on Monsanto Petition of Nonregulated Status for Insect Resistant Tomato Line 5345.* Docket number: 97-114-1.

Baranski, R., Klimek-Chodacka, M., & Lukasiewicz, A., (2019). Approved genetically modified (GM) horticultural plants: A 25-year perspective. *Folia Hortic., 31,* 3–49.

Becker, A., & Lange, M., (2010). VIGS-genomics goes functional. *Trends Plant Sci., 15*(1), 1–4.

Biosafety Clearing-House, (1999). *Melon A and B. Country's Decision or Any Other Communication*Available at: http://bch.cbd.int/database/record.shtml?documentid=6351 (accessed on 7 February 2024).

Bock, R., (2013). Strategies for metabolic pathway engineering with multiple transgenes. *Plant Mol. Biol., 83,* 21–31.

Bovy, A., De Vos, R., Kemper, M., Schijlen, E., Almenar, P. M., Muir, S., Collins, G., Robinson, S., Verhoeyen, M., Hughes, S., et al., (2002). High-flavonol tomatoes resulting from the heterologous expression of the maize transcription factor genes LC and C1. *Plant Cell, 14,* 2509–2526.

Brencic, A., Angert, E. R., & Winans, S. C., (2005). Unwounded plants elicit agrobacterium Vir gene induction and T-DNA transfer: Transformed plant cells produce opines yet are tumor free. *Mol. Microbiol., 57*(6), 1522–1531.

Bruening, G., & Lyons, M. J., (2000). The case of the FLAVR SAVR tomato. *Calif. Agric., 54,* 6, 7.

Butelli, E., Titta, L., Giorgio, M., Mock, H. P., Matros, A., Peterek, S., Schijlen, E. G., Hall, R. D., Bovy, A. G., Luo, J., et al., (2008). Enrichment of tomato fruit with health-promoting anthocyanins by expression of select transcription factors. *Nat. Biotechnol. 26,* 1301–1308.

Chen, G., Ye, C. M., Huang, J. C., Yu, M., & Li, B. J., (2001). Cloning of the papaya ringspot virus (PRSV) replicase gene and generation of PRSV-resistant papayas through the introduction of the PRSV replicase gene. *Plant Cell Rep., 20,* 272–277.

Chen, Q. J., Zhou, H. M., Chen, J., & Wang, X. C., (2006). A Gateway-based platform for multigene plant transformation. *Plant Mol. Biol., 62,* 927–936.

Chen, Z. L., et al., (2003). Safety assessment for genetically modified sweet pepper and tomato. *Toxicology, 188,* 297–307.

Clough, S. J., & Bent, A. F., (1998). Floral dip: A simplified method for *Agrobacterium*-mediated transformation of *Arabidopsis thaliana. Plant J., 16*(6), 735–743.

Dafny-Yelin, M., & Tzfira, T., (2007). Delivery of multiple transgenes to plant cells. *Plant Physiol. 145*, 1118–1128.

Dafny-Yelin, M., Chung, S. M., Frankman, E. L., & Tzfira, T., (2007). pSAT RNA interference vectors: A modular series for multiple gene down-regulation in plants. *Plant Physiol., 145*, 1272–1281.

Diretto, G., Frusciante, S., Fabbri, C., Schauer, N., Busta, L., Wang, Z., Matas, A. J., et al., (2020). Manipulation of β-carotene levels in tomato fruits results in increased ABA content and extended shelf life. *Plant Biotechnol. J., 18*(5), 1185–1199.

Dixon, R. A., Liu, C., & Jun, J. H., (2013). Metabolic engineering of anthocyanins and condensed tannins in plants. *Curr. Opin. Biotechnol., 24*, 329–335.

Elitzur, T., Yakir, E., Quansah, L., Zhangjun, F., Vrebalov, J., Khayat, E., Giovannoni, J. J., & Friedman, H., (2016). Banana MaMADS Transcription factors are necessary for fruit ripening and molecular tools to promote shelf-life and food security. *Plant Physiol., 171*(1), 380–391. doi: 10.1104/pp.15.01866. Epub 2016 Mar 8. PMID: 26956665; PMCID: PMC4854685.

Engler, C., Kandzia, R., & Marillonnet, S., (2008). A one pot, one step, precision cloning method with high throughput capability. *PLoS One, 3*, e3647.

Farre, G., Blancquaert, D., Capell, T., Van, D. S. D., Christou, P., & Zhu, C., (2014). Engineering complex metabolic pathways in plants. *Annu. Rev. Plant Biol., 65*, 187–223.

Firoozbady, E., & Young, T. R., (2015). *Pineapple Plant Named "Rose."* US Patent PP25,763 P3.

Fitch, M., Manshardt, R., Gonsalves, D., Slightom, J., & Sanford, J., (1992). Virus resistant papaya plants derived from tissues bombarded with the coat protein gene of papaya ringspot virus. *Nat. Biotechnol., 10*, 1466–1472.

Fraser, P. D., Romer, S., Shipton, C. A., Mills, P. B., Kiano, J. W., Misawa, N., Drake, R. G., Schuch, W., & Bramley, P. M., (2002). Evaluation of transgenic tomato plants expressing an additional phytoene synthase in a fruit-specific manner. *Proc. Natl. Acad. Sci. U S A., 99*(2), 1092–1097. doi: 10.1073/pnas.241374598. PMID: 11805345; PMCID: PMC117435.

Ghosh, S., Singh, U. K., Meli, V. S., Kumar, V., Kumar, A., Irfan, M., Chakraborty, N., et al., (2013). Induction of senescence and identification of differentially expressed genes in tomato in response to monoterpene. *PLoS One, 8*(9), e76029. https://doi.org/10.1371/journal.pone.0076029.

Gibson, D. G., Benders, G. A., Axelrod, K. C., Zaveri, J., Algire, M. A., Moodie, M., Montague, M. G., et al., (2008). One-step assembly in yeast of 25 overlapping DNA fragments to form a complete synthetic *Mycoplasma genitalium* genome. *Proc. Natl. Acad. Sci. U S A, 105*, 20404–20409.

Grimsley, N., Hohn, T., & Davies, J., (1987). *Agrobacterium*-mediated delivery of infectious maize streak virus into maize plants. *Nature, 325*, 177–179.

Gu, C., Guo, Z. H., Hao, P. P., et al., (2017). Multiple regulatory roles of AP2/ERF transcription factor in angiosperm. *Bot Stud., 58*, 6. https://doi.org/10.1186/s40529-016-0159-1.

Ha, S. H., Kim, J. K., Jeong, Y. S., You, M. K., Lim, S. H., & Kim, J. K., (2019). Stepwise pathway engineering to the biosynthesis of zeaxanthin, astaxanthin and capsanthin in rice endosperm. *Metab. Eng., 52*, 178–189.

Hiei, Y., Ohta, S., Komari, T., & Kumashiro, T., (1994). Analysis of the boundaries of the T-DNA. *Plant J., 6*(2), 271–282.

Huang, J. C., Zhong, Y. J., Liu, J., Sandmann, G., & Chen, F., (2013). Metabolic engineering of tomato for high-yield production of astaxanthin. *Metab. Eng., 17*, 59–67.

Irfan, M., & Datta, A., (2017). Improving food nutritional quality and productivity through genetic engineering. *International Journal of Cell Science & Molecular Biology, 2*(1), 555576. https://doi.org/10.19080/IJCSMB.2017.02.555576.

Irfan, M., Ghosh, S., Kumar, V., Chakraborty, N., Chakraborty, S., & Datta, A., (2014). Insights into transcriptional regulation of β-DN-acetylhexosaminidase, an N-glycan-processing enzyme involved in ripening-associated fruit softening. *Journal of Experimental Botany, 65*(20), 5835–5848.

Irfan, M., Ghosh, S., Meli, V. S., Kumar, A., Kumar, V., Chakraborty, N., Chakraborty, S., & Datta, A., (2016). Fruit ripening regulation of α-mannosidase expression by the MADS box transcription factor RIPENING INHIBITOR and ethylene. *Frontier Plant Science, 7*, 10.

Irfan, M., Kumar, P., Ahmad, I., & Datta, A., (2021). Unraveling the role of tomato Bcl-2-associated athanogene (BAG) proteins during abiotic stress response and fruit ripening. *Scientific Reports, 11*(1), 1–15.

Irfan, M., Kumar, P., Ahmad, M. F., & Siddiqui, M. W., (2023). Biotechnological interventions in reducing losses of tropical fruits and vegetables. *Current Opinion in Biotechnology, 79,* 102850. doi.org/10.1016/j.copbio.2022.102850.

Irfan, M., Kumar, P., Kumar, V., & Datta, A., (2022). Fruit ripening specific expression of β-D-N-acetylhexosaminidase (β-Hex) gene in tomato is transcriptionally regulated by ethylene response factor SlERF.E4. *Plant Science, 323*, 111380.

Jia, H. F., Chai, Y. M., Li, C. L., Lu, D., Luo, J. J., Qin, L., & Shen, Y. Y., (2011). Abscisic acid plays an important role in the regulation of strawberry fruit ripening. *Plant Physiol., 157*(1), 188–199. doi: 10.1104/pp.111.177311. Epub 2011 Jul 6. PMID: 21734113; PMCID: PMC3165869.

Jia, H., Jiu, S., Zhang, C., Wang, C., Tariq, P., Liu, Z., Wang, B., Cui, L., & Fang, J., (2016). Abscisic acid and sucrose regulate tomato and strawberry fruit ripening through the abscisic acid-stress-ripening transcription factor. *Plant Biotechnol. J., 14*(10), 2045–2065. doi: 10.1111/pbi.12563. Epub 2016 May 4. PMID: 27005823; PMCID: PMC5043491.

Jian, W., Cao, H., Yuan, S., Liu, Y., Lu, J., Lu, W., Li, N., Wang, J., Zou, J., Tang, N., et al., (2019). SlMYB75, an MYB-type transcription factor, promotes anthocyanin accumulation and enhances volatile aroma production in tomato fruits. *Hortic. Res., 6*, 22.

Julia, V., Irvin, L. P., Antonio, J. M. A., McQuinn, R., MiYoung, C., Mervin, P., Jocelyn, R., et al., (2009). Fleshy fruit expansion and ripening are regulated by the tomato SHATTERPROOF gene TAGL1. *The Plant Cell, 21*(10), 3041–3062. https://doi.org/10.1105/tpc.109.066936.

Kalunke, R. M., Tundo, S., Benedetti, M., Cervone, F., De Lorenzo, G., & D'Ovidio, R., (2015). An update on polygalacturonase-inhibiting protein (PGIP), a leucine-rich repeat protein that protects crop plants against pathogens. *Front Plant Sci., 6*, 146. doi: 10.3389/fpls.2015.00146. PMID: 25852708; PMCID: PMC4367531.

Kneissl, M. L., & Deikman, J., (1996). The tomato E8 gene influences ethylene biosynthesis in fruit but not in flowers. *Plant Physiol., 112*(2), 537–547. doi: 10.1104/pp.112.2.537. PMID: 12226407; PMCID: PMC157976.

Kramer, M. G., & Redenbaugh, K., (1994). Commercialization of a tomato with an antisense polygalacturonase gene: The FLAVR SAVR™ tomato story. *Euphytica, 79*, 293–297.

Kumar, V., Irfan, M., Ghosh, S., Chakraborty, N., Chakraborty, S., & Datta, A., (2016a). Fruit ripening mutants reveal cell metabolism and redox state during ripening. *Protoplasma, 253*, 581–594. https://doi.org/10.1007/s00709-015-0836-z.

Kumar, V., Irfan, M., Ghosh, S., Chakraborty, N., Chakraborty, S., & Datta, A., (2016b). Fruit ripening mutants reveal cell metabolism and redox state during ripening. *Protoplasma, 253*, 581–594. https://doi.org/10.1007/s00709-015-0836-z.

Kumari, C., Sharma, M., Kumar, V., Sharma, R., Kumar, V., Sharma, P., Kumar, P., & Irfan, M., (2022). Genome editing technology for genetic amelioration of fruits and vegetables for alleviating post-harvest loss. *Bioengineering, 9*(4), 176.

Leiser, R. M., Ziegler-Graff, V., Reutenauer, A., Herrbach, E., Lemaire, O., Guilley, H., Richards, K., & Jonard, G., (1992). Agroinfection as an alternative to insects for infecting plants with beet western yellows luteovirus. *Proc. Natl. Acad. Sci., 89*(19), 9136–9140.

Li, A., Chen, J., Lin, Q., Zhao, Y., Duan, Y., Wai, S. C., Song, C., & Bi, J., (2021). Transcription factor MdWRKY32 participates in starch-sugar metabolism by binding to the MdBam5 promoter in apples during postharvest storage. *Journal of Agricultural and Food Chemistry, 69*(49), 14906–14914. https://doi.org/10.1021/acs.jafc.1c03343.

Li, S., (2014). Transcriptional control of flavonoid biosynthesis: Fine-tuning of the MYB-bHLH-WD40 (MBW) complex. *Plant Signal Behav., 9*(1), e27522. doi: 10.4161/psb.27522. Epub 2014 Jan 6. PMID: 24393776; PMCID: PMC4091223.

Li, T., Jiang, Z., Zhang, L., Tan, D., Wei, Y., Yuan, H., Li, T., & Wang, A., (2016). Apple (*Malus domestica)* MdERF2 negatively affects ethylene biosynthesis during fruit ripening by suppressing MdACS1 transcription. *Plant J., 88*(5), 735–748. doi: 10.1111/tpj.13289. Epub 2016 Sep 17. PMID: 27476697.

Liu, J., Xu, B., Hu, L., Li, M., Su, W., Wu, J., Yang, J., & Jin, Z., (2009). Involvement of a banana MADS-box transcription factor gene in ethylene-induced fruit ripening. *Plant Cell Rep., 28*(1), 103–111. doi: 10.1007/s00299-008-0613-y. Epub 2008 Sep 24. PMID: 18820933.

Lobato-Gómez, M., Hewitt, S., Capell, T., Christou, P., Dhingra, A., & Girón-Calva, P. S., (2021). Transgenic and genome-edited fruits: Background, constraints, benefits, and commercial opportunities. *Hortic Res., 8*(1), 166.

Lucille, A., & Don, G., (2002). Ethylene biosynthesis and action in tomato: A model for climacteric fruit ripening, *Journal of Experimental Botany, 53*(377), 2039–2055. https://doi.org/10.1093/jxb/erf072.

Ma, N., Feng, H., Meng, X., et al., (2014). Overexpression of tomato *SlNAC1*transcription factor alters fruit pigmentation and softening. *BMC Plant Biol.,* 14, 351. https://doi.org/10.1186/s12870-014-0351-y.

Muir, S. R., Collins, G. J., Robinson, S., Hughes, S., Bovy, A., Ric De Vos, C. H., Van Tunen, A. J., & Verhoeyen, M. E., (2001). Overexpression of petunia chalcone isomerase in tomato results in fruit containing increased levels of flavonols. *Nat. Biotechnol., 19*, 470–474.

Ogo, Y., Ozawa, K., Ishimaru, T., Murayama, T., & Takaiwa, F., (2013). Transgenic rice seed synthesizing diverse flavonoids at high levels: A new platform for flavonoid production with associated health benefits. *Plant Biotechnol. J., 11*, 734–746.

Paine, J. A., Shipton, C. A., Chaggar, S., Howells, R. M., Kennedy, M. J., Vernon, G., Wright, S. Y., Hinchliffe, E., Adams, J. L., Silverstone, A. L., et al., (2005). Improving the nutritional value of golden rice through increased pro-vitamin A content. *Nat. Biotechnol., 23*, 482–487.

Payasi, A., Mishra, N. N., Chaves, A. L., & Singh, R., (2009). Biochemistry of fruit softening: An overview. *Physiol. Mol. Biol. Plants, 15*(2), 103–113. doi: 10.1007/s12298-009-0012-z. Epub 2009 Jun 28. PMID: 23572919; PMCID: PMC3550369.

Peng, H., Yang, T., Whitaker, B., et al., (2016). Characterization of spermidine hydroxycinnamoyl transferases from eggplant (*Solanum melongena* L.) and its wild relative *Solanum richardii* Dunal. *Hortic. Res.,* 3, 16062. https://doi.org/10.1038/hortres.2016.62.

Pierce, E. C., LaFayette, P. R., Ortega, M. A., Joyce, B. L., Kopsell, D. A., & Parrott, W. A., (2015). Ketocarotenoid production in soybean seeds through metabolic engineering. *PLoS One, 10*, e0138196.

Prasanna, V., Prabha, T. N., & Tharanathan, R. N., (2007). Fruit ripening phenomena: An overview. *Crit. Rev. Food Sci. Nutr., 47*(1), 1–19. doi: 10.1080/10408390600976841. PMID: 17364693.

Ramkumar, T. R., Lenka, S. K., Arya, S. S., & Bansal, K. C., (2020). A short history and perspectives on plant genetic transformation. *Methods Mol. Biol., 2124*, 39–68.

Scorza, R., et al., (1994). Transgenic plums (*Prunus domestica* L.) express the plum pox virus coat protein gene. *Plant Cell Rep., 14*, 18–22.

Shelton, A. M., et al., (2017). *Bt Eggplant: A Genetically Engineered 'Minor' Crop Comes of Age in Bangladesh and the Philippines*. ISB News Rep. VTechWorks.

Shin, Y. M., Park, H. J., Yim, S. D., Baek, N. I., Lee, C. H., An, G., & Woo, Y. M., (2006). Transgenic rice lines expressing maize C1 and R-S regulatory genes produce various flavonoids in the endosperm. *Plant Biotechnol. J., 4*, 303–315.

Smith, E. F., & Townsend, C. O., (1907). A plant-tumor of bacterial origin. *Science, 25*(643), 671–673.

Song, G., Prieto, H., & Orbovic, V., (2019). *Agrobacterium*-mediated transformation of tree fruit crops: Methods, progress, and challenges. *Front. Plant Sci., 10*, 226.

Subramoni, S., Nathoo, N., Klimov, E., & Yuan, Z. C., (2014). *Agrobacterium tumefaciens* responses to plant-derived signaling molecules. *Front Plant Sci., 5*, 322.

Szeto, W. W., Hamer, D. H., Carlson, P. S., & Thomas, C. A., (1977). Cloning of cauliflower mosaic virus (CLMV) DNA in *Escherichia coli. Science, 196*(4286), 210–212.

Tayal, R., Kumar, V., & Irfan, M., (2022). Harnessing the power of hydrogen sulphide (H_2S) for improving fruit quality traits. *Plant Biology, 24*(4), 594–601.

Thornburg, R. W., Carter, C., Powell, A. L., Mittler, R., Rizhsky, R., & Horner, H. T., (2003). A major function of the tobacco floral nectary is defense against microbial attack. *Plant Syst. Evol., 238,* 211–218. doi: 10.1007/s00606-003- 0282-9.

Tian, Y. S., Wang, B., Peng, R. H., Xu, J., Li, T., Fu, X. Y., Xiong, A. S., et al., (2019). Enhancing carotenoid biosynthesis in rice endosperm by metabolic engineering. *Plant Biotechnol. J., 17*, 849–851.

Tieman, D. M., Harriman, R. W., Ramamohan, G., & Handa, A. K., (1992). An antisense pectin methylesterase gene alters pectin chemistry and soluble solids in tomato fruit. *Plant Cell, 4*(6), 667–679. doi: 10.1105/tpc.4.6.667. PMID: 12297658; PMCID: PMC160163.

Twyman, R. M., & Christou, P., (2004). Plant transformation technology: Particle bombardment. In: Christou, P., & Klee, H., (eds.), *Handbook of Plant Biotechnology*.

USDA, (2015). *USDA Announces Deregulation of Non-browning Apples*. Available at: https://www.aphis.usda.gov/stakeholders/downloads/2015/SA_arctic_apples.pdf (accessed on 7 February 2024).

Wakasa, Y., Yasuda, H., & Takaiwa, F., (2006). High accumulation of bioactive peptide in transgenic rice seeds by expression of introduced multiple genes. *Plant Biotechnol. J., 4*, 499–510.

Wang, R., Lammers, M., Tikunov, Y., Bovy, A. G., Angenent, G. C., & De Maagd, R. A., (2020). The *rin, nor* and *Cnr* spontaneous mutations inhibit tomato fruit ripening in additive and epistatic manners. *Plant Sci., 294*, 110436. doi: 10.1016/j.plantsci.2020.110436. Epub 2020 Feb 4. PMID: 32234221.

Yuan, L., & Grotewold, E., (2015). Metabolic engineering to enhance the value of plants as green factories. *Metab. Eng., 27*, 83–91.

Zhang, Y., Butelli, E., & Martin, C., (2014). Engineering anthocyanin biosynthesis in plants. *Curr. Opin. Plant Biol., 19*, 81–90.

Zhang, Y., Butelli, E., Alseekh, S., Tohge, T., Rallapalli, G., Luo, J., Kawar, P. G., Hill, L., Santino, A., Fernie, A. R., et al., (2015). Multi- level engineering facilitates the production of phenylpropanoid compounds in tomato. *Nat. Commun., 6*, 8635.

Zhu, C., Naqvi, S., Breitenbach, J., Sandmann, G., Christou, P., & Capell, T., (2008). Combinatorial genetic transformation generates a library of metabolic phenotypes for the carotenoid pathway in maize. *Proc. Natl. Acad. Sci. U S A., 105*(47), 18232–18237. doi: 10.1073/pnas.0809737105. Epub 2008 Nov 14. PMID: 19011084; PMCID: PMC2587607.

Zhu, L., & Qian, Q., (2017). Development of "purple endosperm rice" by engineering anthocyanin biosynthesis in endosperm: Significant breakthrough in transgene stacking system, new progress in rice biofortification (in Chinese). *Chin. Bull. Bot., 52*, 539–542.

Zhu, Q., Yu, S., Zeng, D., Liu, H., Wang, H., Yang, Z., Xie, X., Shen, R., Tan, J., Li, H., et al., (2017). Development of "purple endosperm rice" by engineering anthocyanin biosynthesis in the endosperm with a high-efficiency transgene stacking system. *Mol. Plant, 10*, 918–929.

Zhu, Q., Zeng, D., Yu, S., Cui, C., Li, J., Li, H., Chen, J., Zhang, R., Zhao, X., Chen, L., et al., (2018). From golden rice to aSTARice: Bioengineering astaxanthin biosynthesis in rice endosperm. *Mol. Plant, 11*, 1440–1448.

CHAPTER 4

ENHANCING POSTHARVEST STABILITY AND SHELF LIFE OF PERISHABLE FRUITS AND VEGETABLES USING MOLECULAR APPROACHES

JITENDER KUMAR, KAJAL CHOUDHARY, and SHELJA

Department of Plant Science, Central University of Himachal Pradesh, India

ABSTRACT

Post-harvest losses, intensified by factors such as rapid perishability and environmental conditions, result in substantial waste, particularly in developing nations. Traditional post-harvest techniques focus on physical and chemical treatments, including heat treatments, edible coatings, irradiation, and chemical agents, to control infections and extend shelf life. However, these methods have limitations in terms of efficacy and safety concerns. The chapter emphasizes the emerging role of molecular approaches, particularly genome editing techniques such as zinc finger nucleases (ZFNs), transcription activator-like effector nucleases (TALENs), and the CRISPR/Cas9 system, in crop improvement. These techniques offer precise and efficient means of enhancing crop traits, including postharvest stability and shelf life, without the introduction of foreign genes. By elucidating the mechanisms underlying postharvest processes and employing targeted genetic modifications, this chapter underscores the potential of

Bioengineered Fruit and Vegetables: Trends in Postharvest Quality Improvement.
Mohammad Irfan and Pankaj Kumar (Eds.)

molecular approaches to revolutionize postharvest management strategies and address global food security challenges.

4.1 INTRODUCTION

All fruits, as well as plants, are part of horticultural crops, with the majority of them holding the highest economic position due to their significant contribution to global crop output. Fruits and vegetables contain essential components for the human diet, such as minerals, vitamins, fibers, carbohydrates, proteins, and organic acids. However, these horticultural crops face challenges in post-harvest storage due to their rapid perishability, respiration, and transpiration, which result in increased ripening-related softness. This leads to post-harvest losses or waste, which vary depending on plant types, climate zones, and handling practices in different countries. These losses affect the quality, nutrition, viability as seeds, and selling price of the crops in the food chain from harvest to consumption. Globally, the annual post-harvest loss is estimated to be over 1.3 billion tons, with developing nations facing more significant difficulties. For example, India experiences the annual waste of 30–40% of the fruits and vegetables produced (Gosh et al., 2011).

Fruits and vegetables are lost after harvest due to their perishable nature. The delay in biochemical processes is associated with the need for essential measures to reduce post-harvest losses, such as ethylene production, softening, pigment alteration, respiratory rate, and weight loss. The reduction of post-harvest damages depends on various factors. Post-harvest processes, including product selection, cleaning, washing, grading, disinfection, storage, drying, packing, temperature maintenance, relative humidity, and coatings, aim to enhance the shelf life of the products (Tiwari et al., 2022). Edible coatings are beneficial in postharvest management techniques as they help preserve the freshness of fruits and vegetables. By using edible coatings made of proteins, lipids, and carbohydrates, physical changes during storage can be prevented. The primary purpose of edible coatings is to act as a barrier against respiration and transpiration. Natural biomolecules, known as edible coatings, are used to improve the appearance and durability of fruits and vegetables (Prasad et al., 2015).

The primary method for determining how genes work in plants and for crop genetic improvement is the study of gene mutants. Numerous research conducted in this area over the past several years has aided in the identification of mutants and enabled the creation of mutant libraries that correspond to the hundreds of genes in the plants under study. This research, however, employed physical, chemical, and biological mutagenesis methods, which can lead to random mutagenesis, resulting in unwanted and undesirable changes, leading to large-scale mutant screening, which is an expensive and time-consuming process. Zinc finger nucleases (ZFN), the first-ever target-specific reagent, were utilized in 2002 and marked the beginning of the usage of programmable sequence-specific nucleases (SSNs) for genome modification. As a dimer, two ZFN DNA binding (DBD) domains are linked to the Fok 1 monomer, which will bind to the site with a spacer region of 5–6 base pairs, activating Fok 1 and enabling DNA double-strand breaks (DSBs) to occur. ZFN DBD domain recognizes three base pairs at the binding location (Ma et al., 2012). Numerous studies using crops including tobacco, *Arabidopsis*, and maize revealed the usage of ZFN. Then came TALENs, which was initially utilized in plants after being developed from the bacteria *Xanthomonas.* Compared to ZFN, which recognizes three nucleotides, TALENs recognize only one, making it more specific and desired. It required intricate tandem repeat domains, similar to those seen in TAL proteins, despite being simple to utilize (Weeks et al., 2016). Clustered regularly interspaced short palindromic repeats (CRISPR)-associated protein, a new genome modification technique, was recently produced from type 2 CRISPR adaptive immunity system discovered in bacteria *Streptococcus pyogenes*, which protects bacteria and archaea from invading phages (Cas9). By base pairing to the target DNA site utilizing engineered single guide RNA (sgRNA), they can more easily cause site-directed specific alterations than first-generation editing tools like ZFN and TALENs. Although the existence of such a CRISPR in bacteria was discovered in 1987, its purpose remained unknown until 2005. It wasn't until 2007 that it was discovered that CRISPR collaborates with the Cas protein to offer immunity to the bacteria.

4.2 CONTROL OF POST-HARVEST INFECTIONS AND POST-HARVEST LOSSES OF CROPS, FRUITS, AND VEGETABLES

4.2.1 PHYSICAL TREATMENTS

4.2.1.1 HEAT TREATMENT

For harvested fresh fruit and vegetables, heat treatment is utilized. Hot water rinse, hot dry air, saturated water vapor heat, and hot water dip with brushing are the heat remedies (Schirra et al., 2000). These heat treatments have positive effects on plant growth in three ways. It alters the biological processes, like reducing chilling injury, prolongs ripening processes by inactivating degradative enzymes with heat (Lurie, 1998); eliminating harmful insect germs, and delay the start of fungal attack (Schirra et al., 2000). The length of a heat treatment might range from one hour and up to many weeks (up to 4 days). The heat treatments are used on carrots, potatoes, strawberries, and tomatoes, to maintain the color of kiwi fruits, lettuce, and green beans; to stop the ripening of other melons, and to lengthen the shelf life of grapes, plums, etc. (Falik, 2004). *Penicillium* sp., which causes blue mold on grapefruit, is managed by soaking grapefruit in boiling water for two minutes at 50°C (Schirra et al., 2000).

4.2.1.2 EDIBLE COATINGS

A thin covering of material known as edible coatings protects the food against moisture, air, and solute movement. It can be used as a continuous layer or as a complete food covering. Edible coatings act as a moisture barrier and reduce moisture loss while being stored. The color, texture, and volatile components of fresh fruits as well as vegetables are preserved by edible coatings. Additionally, it safeguards against mechanical harm and preserves structural integrity. The hydrophobic groups increase the coating's functionality and make up edible coatings (Gol et al., 2013). The different edible coatings, such as chitosan, polyvinyl acetate, *Aloe vera,* etc., had demonstrated favorable characteristics on fresh products, including excellent shield capabilities, no lasting flavor or odor, and effective antibacterial activity (Dhall, 2013). The list of edible coatings that are utilized for various fruits and vegetables is presented in Table 4.1 and Figure 4.1.

4.2.1.3 IRRADIATION

Food that has been irradiated is exposed to radiant energy from e-beams and rays (high-energy electrons), which permeate objects and dissolve molecules, even DNA in life forms. To increase the fresh product's shelf

TABLE 4.1 Various Fruits and Vegetables Covered with Edible Coatings

Sl. No.	Coating Film	Function	Fruit/ Vegetable	References
1.	Chitosan, Twin 80	Reduce the generation of ethylene and postpone ripening as demonstrated by a high titrable acidity concentration.	Pear	Li & Yu (2001)
2.	Aloe vera gel	Increase the keeping standards and shelf life.	Mushroom	Mohebbi et al. (2012)
3.	Chitosan/aloe vera coatings	Less fungus and bacterial contamination; increased the shelf life to five days roughly.	Blue berry	Vieira et al. (2016)
4.	Soy protein isolate (SPI)/hydroxypropyl methylcellulose (HPMC)/ olive oil coatings.	Postharvest life was prolonged by coating by up to 15 days.	Pear fruit	Dave, Ramana Rao, & Nandane (2017)
5.	Chitosan/zein/α-tocopherol film.	Less weight loss.	Mushroom	Zhang et al. (2020)
6.	Cellulose and glycerol	Slow down physiological processes.	Carrot	Li et al. (1998)
7.	A blend of sodium carboxy methyl cellulose, mono- and diglycerides, and sucrose fatty acid esters.	Delay ripening, lessen chlorophyll loss and weight loss.	Mango	Motlagh et al. (1988)

life, the source of irradiation is ionizing radiation from cesium-137, cobalt-60, or artificial electron beams (Roberts, 2019). By inhibiting cellular proliferation, irradiation helps to resolve pest and food safety problems.

4.2.2 CHEMICAL TREATMENTS

Chlorine-based treatments for antimicrobial and anti-browning agents include hydrogen peroxide (H_2O_2), electrolyzed water, and peroxyacetic

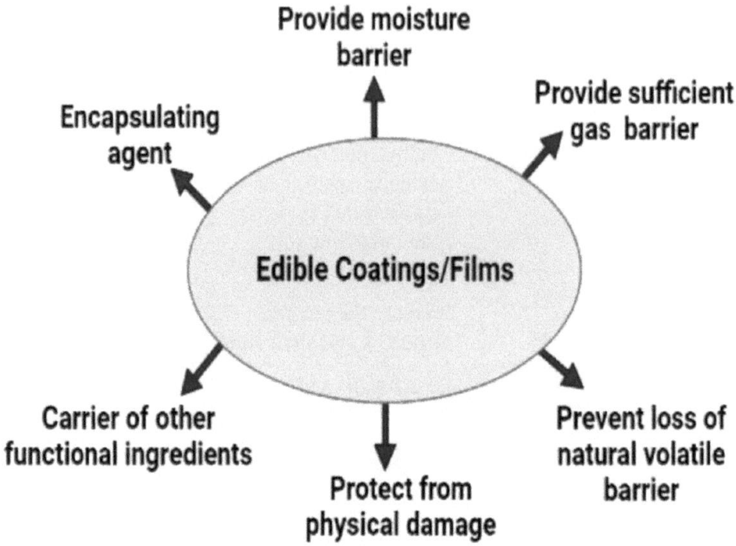

FIGURE 4.1 Various functions of edible coatings/films.

acid (PAA) as examples of chemical-based agents (Artes et al., 2009). Nitric oxide (NO), a free radical gas known for its high reactivity, acts as a versatile signaling agent crucial for physiological processes in plants, such as fruit maturation and senescence (Wendehenne et al., 2004). Table grapes are commonly fumigated with sulfur dioxide (SO_2) to prevent decay during storage (Palou et al., 2010). The efficacy of SO_2 technology in inhibiting postharvest deterioration in various fruits like litchi, fig, banana, lemon, and apple has also been studied (Sivakumar et al., 2010) (Table 4.2 and Figure 4.2).

4.3 GENOME EDITING

Increased production, robust tolerance to abiotic and biotic stressors, and improved nutritional values are ongoing goals of crop improvement. Breeding methods have been extensively used to select desired qualities within species that hybridize to achieve these goals. In addition to naturally occurring genetic variations, artificially induced mutations

TABLE 4.2 List of Some Fruits and Vegetables Showing Anti-Browning Properties

Sl. No.	Fruits/ Vegetables	Bioactive Compound	Biological Activity	References
1.	Tomato	Carotenoid (lycopene)	Antioxidant, antimicrobial	Martinez et al. (2019)
2.	Strawberry	Phenolic compounds (procyanidin B type, gallic acid, flavonols, epicatechin, and catechin).	Anti-browning, antioxidant	Dias et al. (2020)
3.	Potato	Phenolic compounds (chlorogenic acid and caffeic acid).	Antioxidant	Venturi et al. (2019)
4.	Mango	Phenolic compounds (protocatechuic, gallic acid, and mangiferin).	Anti-browning	Jirasuteeruk & Theerakulkait (2019)
5.	Ginger	Ginger extract	Anti-browning	Weerawardana et al. (2020)

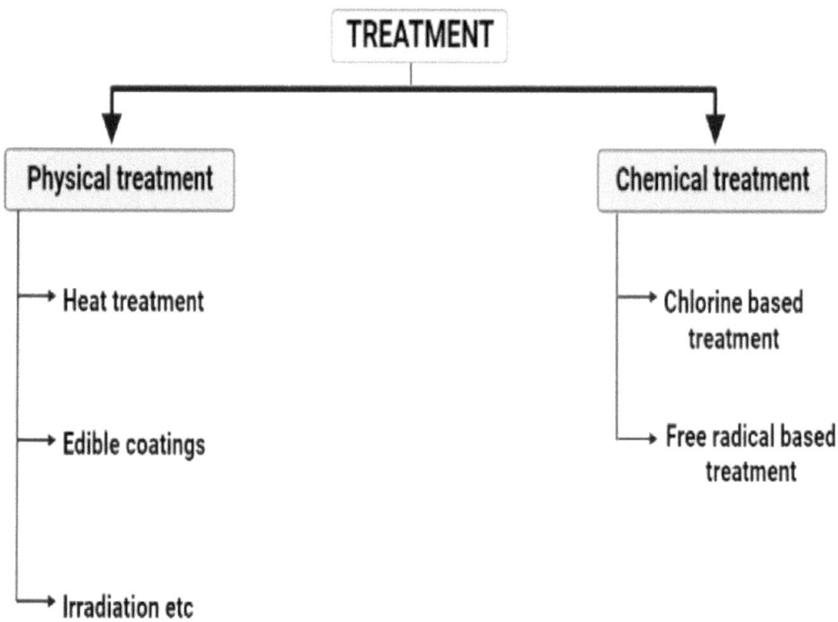

FIGURE 4.2 Various treatments of fruits and vegetables against postharvest problems.

through chemical or radioactive irradiation have significantly increased diversity in selected traits. The development of new commercial varieties is hindered by the laborious and time-consuming nature of traditional methods, as well as the unpredictability of major mutations and their rarity in producing desired traits (Pacher & Puchta, 2017). Since 1996, genetically modified (GM) crop varieties have been created by incorporating genetic components or genes with specific functions to express desired traits. However, concerns regarding the safety of GM crops, particularly those created by introducing foreign genes, require substantial time and financial resources to comply with rigorous and costly safety assessments.

Tools for genome editing have become a new breeding technology with the potential to address these issues. They are successfully used to alter crop genomes in a range of plants without leaving any signs of the introduction of foreign genes. They have significantly impacted crop development in a short amount of time, outperforming conventional breeding in terms of precision and efficiency when it comes to genetic alteration. Because of this, many crops with altered genomes have been developed and they are multiplying faster than those created using any additional methods, such as artificial and natural mutant screening, traditional breeding, and GM technology (Park et al., 2019). Several technologies for genome modification, including CRISPR (Jinek et al., 2012), ZFNs (Kim et al., 1996), and TALENs (Christain et al., 2010) are extensively employed to improve the quality of a variety of crops.

4.3.1 ZFN

Since a decade ago, scientists have been investigating the possibility of connecting the non-specific Fok1 endonuclease domain to specific DBD domains in order to develop techniques for precise DNA recognition and restriction. Eventually, ZF domains were chosen because they are the most prevalent DBD motif (Kim et al., 1996). Following the discovery of the functioning Cys2-His2 ZF domain's working principles, ZFN uses chimerically engineered nucleases (Palpant et al., 2013). Each Cys2-His2 ZF domain has 30 amino acid residues and is folded into $\beta\beta\alpha$ configuration (Petolino, 2015). Three tandem nucleotides in the DNA can be recognized by each ZF protein. The artificial ZF Cys2-His2 domain is located at the N-terminal area of the generalized ZFN monomer, and a non-specific FokI DNA cleavage domain is located at the C-terminal region. For ZFN to function enzymatically, FokI domain dimerization is essential (Kim et al., 1996). ZFNs enable targeted gene editing by inducing DNA double-strand breaks (DSBs) that allow homologous recombination (HR) to replace the gene. Each ZFN has a DNA-binding domain that recognizes a distinct 6 bp hexamer in DNA sequence and a DNA-cleaving domain that is made up of the Fok1 nuclease domain. ZFP is created by merging these domains. A highly specialized genomic scissor is created when the DNA-cleaving and DNA-binding domains are coupled (Gupta et al., 2012). To create a ZFN, ZFP is fused with the FokI-cleavage domain (N), methylase domain (M), transcription repressor domain (R), and transcription activator domain (A) (Alwin et al., 2005). By proving the ability to change specific genomic locations, ZFN has revolutionized the area of genome editing and made way for both fundamental and applied research. With regard to high specificity, effectiveness, and minimum nontarget consequences, ZFNs are superior to other techniques, and ongoing research is concentrated on enhancing ZFN design and delivery and broadening the range of crops in which they can be used (Figure 4.3).

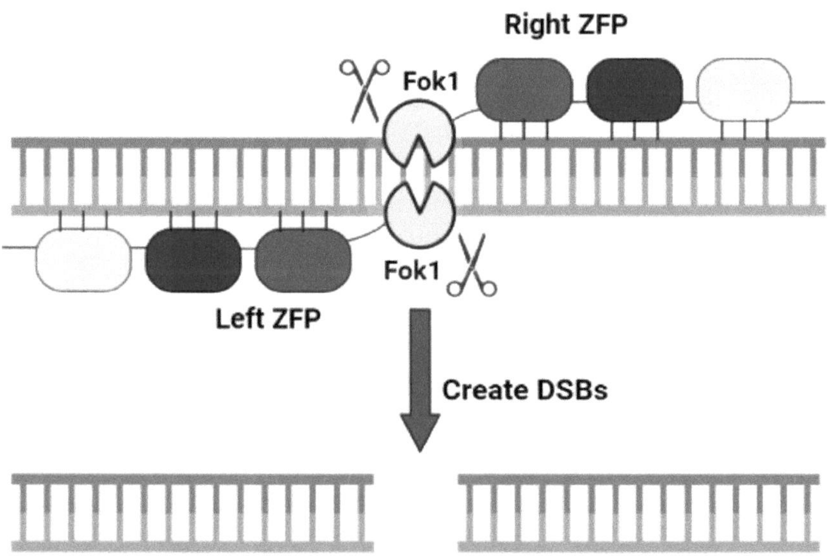

FIGURE 4.3 Mechanism of ZFN.

Source: Adapted from Hillary and Ceasar, 2021

4.3.2 TALEN

Effective genome editing techniques for plants now include Transcription Activator-Like Effector Nucleases (TALENs), which have replaced ZFNs (Joung et al., 2013). ZFNs and TALENs are both designed nucleases; however, in this case, the non-specific DNA endonuclease FokI is fused with a Transcription Activator-Like Effector (TALE) DBD protein to create TALENs (Miller et al., 2011). *Xanthomonas* bacteria, which release TALEs to change gene transcription in host plants, contain TALE proteins (Boch et al., 2010). An N-terminal translocation signal, a C-terminal acidic transcription-activation domain, and a central DNA-binding domain are elements of a typical TALE protein (Bogdanove et al., 2010). The Fok1 nuclease domain of the TALEN can attach to the other Fok1 nuclease domain of the TALEN bound to the opposite DNA strand, causing a double-strand break (DSB) in the target site between the TALENs. TALENs are also chimeric proteins with two primary subunits. TALEs have up to 30 copies of highly conserved 33–34 amino acid sequences in their DNA-binding domains, with the exception of positions 12th and 13th. Positions 12th and 13th are known as the repeat variable diresidues (RVD), and they have a strong association with the recognition of particular nucleotides. Since each repeat can distinguish between a single nucleotide, any DNA sequence can have additional binding sites created. The DNA-binding domain's repeat units bind to a particular nucleotide, and the RVDs inside the repeats have a high association with a particular nucleotide (A, G, C, or T) (Deng et al., 2012). The FokI domain works as a dimer, and a hybrid nuclease can be created using the FokI endonucleases non-specific DNA cleavage domain (Mani et al., 2005). Even though TALEN technology is superior and more effective than ZFN, developing engineered TALE repetitions is challenging since they require several identical repeat sequences (Joung et al., 2013) (Figure 4.4).

4.3.3 CRISPR/CAS9

The CRISPR/CRISPR-associated (Cas) protein system, of which the most well-known variant is CRISPR/Cas9 (based on the Cas9 protein), is another revolutionary genome editing technique that has recently gained

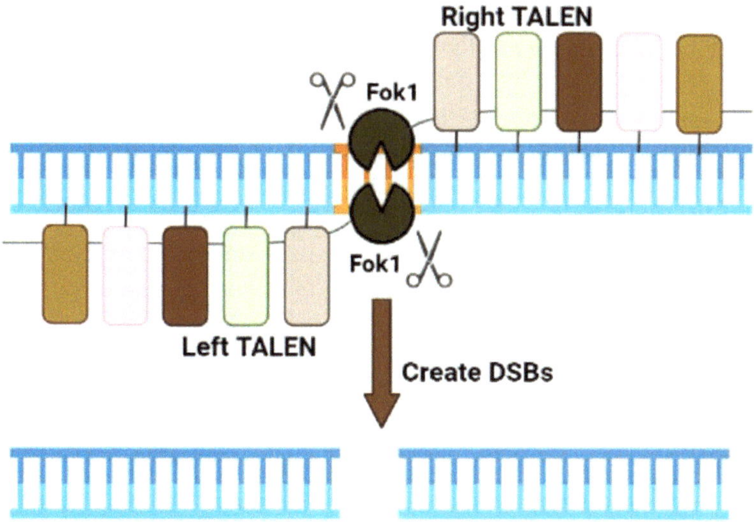

FIGURE 4.4 Mechanism of TALEN.

widespread popularity. Target site recognition by the CRISPR/Cas9 system is performed by the complementary sequence-based interaction between the guide (noncoding) RNA and DNA of the target site and the guide. When Cas9 endonuclease is used, the RNA and Cas protein combination possess the nuclease activity to precisely cleave double-stranded DNA (Cong et al., 2013). The Cas9 protein, two non-coding CRISPR RNAs (crRNAs), a trans-activating crRNA (tracrRNA), and a precursor crRNA make up the three primary parts of the type II CRISPR system (Bhaya et al., 2011). The foreign DNA sequence is incorporated into the CRISPR cluster by naturally occurring CRISPR systems (Sander et al., 2014). The foreign DNA-containing CRISPR cluster generates a crRNA that is roughly 40 nt long and contains a PAM area, which is complementary to the foreign DNA location. Pre-crRNA is the newly transcribed RNA produced by spacers. It consists of two regions: the protospacer adjacent motif (PAM) and the spacer region. When combined with Cas9 and bound to any sequence to initiate cleavage, pre-crRNA is broken down into smaller crRNA by the action of endoribonuclease. In eukaryotes, Cas nuclease and a guide RNA known as gRNA or sgRNA can be used to modify a genome. This sgRNA, which is a chimera of crRNA and would be complementary to the target site and tracrRNA, is created by the fusion of crRNA and tracer RNA. As a result, it acts as a connection between Cas9 and crRNA (Patidae et al., 2016). The Cas9 endonuclease, which is utilized to cut the target sequence, and a sgRNA, which specifies the significance of the Cas9, and directs Cas9 to target DNA, are the two elements needed for CRISPR/Cas9 action.

In the modified system, Cas9 can be made to identify a particular genomic sequence by simply adding a 20-bp spacer region to the sgRNA. Because of this, CRISPR/Cas9 is considerably simpler to employ than the older methods ZFNs and TALENs, and most importantly, it is much more successful in causing precise changes. It is simple to create numerous sgRNAs having various target regions for parallel multiplexed editing. CRISPR is technically straightforward and effective, and non-specialized laboratories can employ it frequently, which is a real leap in genome editing (Tian et al., 2017) (Figure 4.5 and Table 4.3).

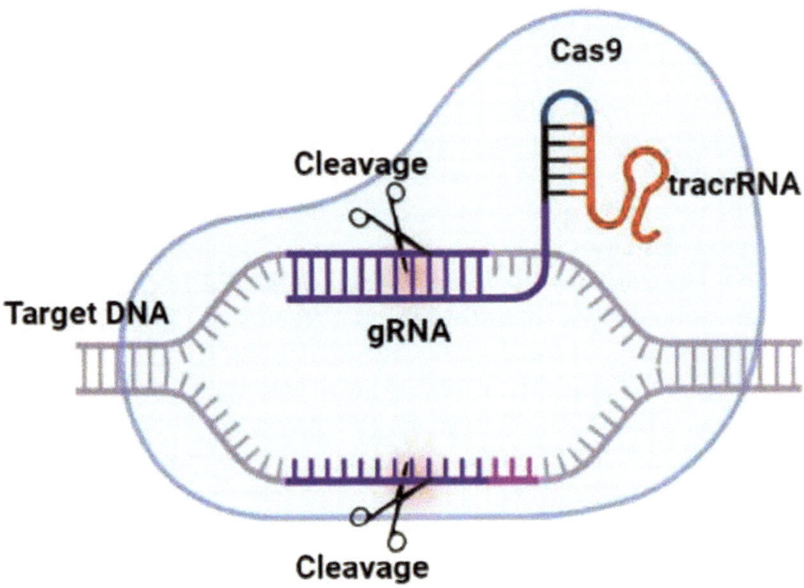

FIGURE 4.5 Mechanism of genome editing tools CRISPR/Cas9.

TABLE 4.3 Use of Genome Modification to Enhance the Quality and Longevity of Horticulture Crops After Harvest

Sl. No.	Crop	Genome Editing Tool	Target Gene	Function of Target Gene	References
1.	Potato	TALEN (SDN1)	Vinv	Converts the sucrose formed when starch undergoes hydrolysis, releasing one glucose and one fructose molecule.	Clasen et al. (2017)
2.	Tomato	CRISPR/Cas9 (SDN1)	SBP-CNR and NAC-NOR	Ripening genes transcription factor.	Gao et al. (2019)
3.	White button mushroom.	CRISPR/Cas9 (SDN1)	StPPO2	Facilitates phenolic oxidation into quinines.	Waltz (2016)
4.	Potato	CRISPR/Cas9	StGBSS	Producing amylase in the starch biosynthetic pathway.	Kusano et al. (2018)
5.	Tomato	CRISPR/Cas9	ALC	Inhibits ethylene synthesis.	Lu et al. (2018)
6.	Petunia	CRISPR/Cas9 (SDN1)	PhACO	Enables the ethylene biosynthesis pathway to catalyze the conversion of aminocyclopropane-1-carboxylic acid to ethylene.	Xu et al. (2020)
7.	Apple	CRISPR/Cas9	IdnDH	An essential enzyme in the biosynthetic route for TA.	Oskabe et al. (2018)
8.	Citrus	CRISPR/Cas9	CsLOB1	Citrus bacterial canker disease susceptibility gene.	Tripathi et al. (2020)
9.	Chili pepper	CRISPR/Cas9 (SDN1)	CaERF28	The gene for anthracnose disease susceptibility.	Mishra et al. (2021)

4.4 CONCLUSION

The market has a high demand for fruits and vegetables because of their nutritional value. Vegetables and fruits have a limited shelf life due to their perishable nature. The use of edible coatings and molecular techniques such as genome editing can extend the lifespan of both vegetables and fruits. Edible coatings have been utilized for many years as a method of food preservation for fruits and vegetables. These coatings help delay ripening and reduce water loss in produce. By employing edible coatings, there is potential to reduce packaging waste. Coating produce with edible materials results in a longer shelf life and helps preserve more of their nutritional content. Genome editing tools are increasingly used in crop improvement, including extending shelf life. Major methods for genome editing in plants include the CRISPR/Cas9 system, TALENs, and ZFNs.

KEYWORDS

- **coating**
- **CRISPR/Cas9**
- **edible coatings**
- **fruits**
- **genome**
- **molecular techniques**
- **nutritional qualities**
- **perishable**
- **postharvest**
- **shelf life**
- **vegetables**

REFERENCES

Alwin, S., Gere, M. B., Guhl, E., Effertz, K., Barbas, III. C. F., Segal, D. J., & Cathomen, T., (2005). Custom zinc-finger nucleases for use in human cells. *Molecular Therapy, 12*(4), 610–617.

Artes, F., Gomez, P., Aguayo, E., Escalona, V., & Artes-Hernandez, F., (2009). Sustainable sanitation techniques for keeping quality and safety of fresh-cut plant commodities. *Postharvest Biology and Technology, 51*(3), 287–296.

Bhaya, D., Davison, M., & Barrangou, R., (2011). CRISPR-Cas systems in bacteria and archaea: Versatile small RNAs for adaptive defense and regulation. *Annual Review of Genetics, 45*(1), 273–297.

Boch, J., & Bonas, U., (2010). Xanthomonas AvrBs3 family-type III effectors: Discovery and function. *Annual Review of Phytopathology, 48*, 419–436.

Bogdanove, A. J., Schornack, S., & Lahaye, T., (2010). TAL effectors: Finding plant genes for disease and defense. *Current Opinion in Plant Biology*, *13*(4), 394–401.

Cermak, T., Doyle, E. L., Christian, M., Wang, L., Zhang, Y., Schmidt, C., & Voytas, D. F., (2011). Efficient design and assembly of custom TALEN and other TAL effector-based constructs for DNA targeting. *Nucleic Acids Research*, *39*(12), e82–e82.

Chen, X., Lu, X., Shu, N., Wang, S., Wang, J., Wang, D., & Ye, W., (2017). Targeted mutagenesis in cotton (*Gossypium hirsutum* L.) using the CRISPR/Cas9 system. *Scientific Reports*, *7*(1), 1–7.

Christian, M., Cermak, T., Doyle, E. L., Schmidt, C., Zhang, F., Hummel, A., & Voytas, D. F., (2010). Targeting DNA double-strand breaks with TAL effector nucleases. *Genetics*, *186*(2), 757–761.

Clasen, B. M., Stoddard, T. J., Luo, S., Demorest, Z. L., Li, J., Cedrone, F., & Zhang, F., (2016). Improving cold storage and processing traits in potato through targeted gene knockout. *Plant Biotechnology Journal*, *14*(1), 169–176.

Cong, L., Ran, F. A., Cox, D., Lin, S., Barretto, R., Habib, N., & Zhang, F., (2013). Multiplex genome engineering using CRISPR/Cas systems. *Science*, *339*(6121), 819–823.

Dave, R. K., Ramana, R. T. V., & Nandane, A. S., (2017). Improvement of post-harvest quality of pear fruit with optimized composite edible coating formulations. *Journal of Food Science and Technology*, *54*(12), 3917–3927.

Deng, D., Yan, C., Pan, X., Mahfouz, M., Wang, J., Zhu, J. K., & Yan, N., (2012). Structural basis for sequence-specific recognition of DNA by TAL effectors. *Science*, *335*(6069), 720–723.

Dhall, R. K., (2013). Advances in edible coatings for fresh fruits and vegetables: A review. *Critical Reviews in Food Science and Nutrition*, *53*(5), 435–450.

Dias, C., Fonseca, A. M., Amaro, A. L., Vilas-Boas, A. A., Oliveira, A., Santos, S. A., & Pintado, M., (2020). Natural-based antioxidant extracts as potential mitigators of fruit browning. *Antioxidants*, *9*(8), 715.

Fallik, E., (2004). Prestorage hot water treatments (immersion, rinsing and brushing). *Postharvest Biology and Technology*, *32*(2), 125–134.

Gao, Y., Zhu, N., Zhu, X., Wu, M., Jiang, C. Z., Grierson, D., & Qu, G., (2019). Diversity and redundancy of the ripening regulatory networks revealed by the fruitENCODE and the new CRISPR/Cas9 CNR and NOR mutants. *Horticulture Research*, *6*.

Ghosh, S., Meli, V. S., Kumar, A., Thakur, A., Chakraborty, N., Chakraborty, S., & Datta, A., (2011). The N-glycan processing enzymes α-mannosidase and β-D-N-acetylhexosaminidase are involved in ripening-associated softening in the non-climacteric fruits of capsicum. *Journal of Experimental Botany*, *62*(2), 571–582.

Gol, N. B., Patel, P. R., & Rao, T. R., (2013). Improvement of quality and shelf-life of strawberries with edible coatings enriched with chitosan. *Postharvest Biology and Technology*, *85*, 185–195.

Grissa, I., Vergnaud, G., & Pourcel, C., (2007). The CRISPRdb database and tools to display CRISPRs and to generate dictionaries of spacers and repeats. *BMC Bioinformatics*, *8*(1), 1–10.

Gupta, A., Christensen, R. G., Rayla, A. L., Lakshmanan, A., Stormo, G. D., & Wolfe, S. A., (2012). An optimized two-finger archive for ZFN-mediated gene targeting. *Nature Methods*, *9*(6), 588–590.

Gupta, P., Chikala, S., & Reddy, K. T., (2017). Storage and shelf life of horticultural crops. *Advances in Horticultural Crop Management and Value Addition*, 283.

Jinek, M., Chylinski, K., Fonfara, I., Hauer, M., Doudna, J. A., & Charpentier, E., (2012). A programmable dual-RNA–guided DNA endonuclease in adaptive bacterial immunity. *Science*, *337*(6096), 816–821.

Jirasuteeruk, C., & Theerakulkait, C., (2019). Ultrasound-assisted extraction of phenolic compounds from mango (*Mangifera indica* cv. Chok Anan) peel and its inhibitory effect on enzymatic browning of potato puree. *Food Technology and Biotechnology*, *57*(3), 350.

Joung, J. K., & Sander, J. D., (2013). TALENs: A widely applicable technology for targeted genome editing. *Nature Reviews Molecular Cell Biology*, *14*(1), 49–55.

Jung, J. H., & Altpeter, F., (2016). TALEN mediated targeted mutagenesis of the caffeic acid O-methyltransferase in highly polyploid sugarcane improves cell wall composition for production of bioethanol. *Plant Molecular Biology*, *92*(1), 131–142.

Kim, Y. G., Cha, J., & Chandrasegaran, S., (1996). Hybrid restriction enzymes: Zinc finger fusions to Fok I cleavage domain. *Proceedings of the National Academy of Sciences*, *93*(3), 1156–1160.

Kumari, C., Sharma, M., Kumar, V., Sharma, R., Kumar, V., Sharma, P., & Irfan, M., (2022). Genome editing technology for genetic amelioration of fruits and vegetables for alleviating post-harvest loss. *Bioengineering*, *9*(4), 176.

Kusano, H., Ohnuma, M., Mutsuro-Aoki, H., Asahi, T., Ichinosawa, D., Onodera, H., & Shimada, H., (2018). Establishment of a modified CRISPR/Cas9 system with increased mutagenesis frequency using the translational enhancer dMac3 and multiple guide RNAs in potato. *Scientific Reports*, *8*(1), 1–9.

Li, H., & Yu, T., (2001). Effect of chitosan on incidence of brown rot, quality, and physiological attributes of postharvest peach fruit. *Journal of the Science of Food and Agriculture*, *81*(2), 269–274.

Li, P., & Barth, M. M., (1998). Impact of edible coatings on nutritional and physiological changes in lightly-processed carrots. *Postharvest Biology and Technology*, *14*(1), 51–60.

Lloyd, A., Plaisier, C. L., Carroll, D., & Drews, G. N., (2005). Targeted mutagenesis using zinc-finger nucleases in *Arabidopsis. Proceedings of the National Academy of Sciences*, *102*(6), 2232–2237.

Lü, P., Yu, S., Zhu, N., Chen, Y. R., Zhou, B., Pan, Y., & Zhong, S., (2018). Genome encode analyses reveal the basis of convergent evolution of fleshy fruit ripening. *Nature Plants*, *4*(10), 784–791.

Lurie, S., (1998). Postharvest heat treatments. *Postharvest Biology and Technology*, *14*(3), 257–269.

Ma, S., Zhang, S., Wang, F., Liu, Y., Liu, Y., Xu, H., & Xia, Q., (2012). Highly efficient and specific genome editing in silkworm using custom TALENs. *PLoS ONE, 7*(9), e45035. doi: 10.1371/journal.pone.0045035.

Mali, P., Yang, L., Esvelt, K. M., Aach, J., Guell, M., DiCarlo, J. E., & Church, G. M., (2013). RNA-guided human genome engineering via Cas9. *Science*, *339*(6121), 823–826.

Mani, M., Smith, J., Kandavelou, K., Berg, J. M., & Chandrasegaran, S., (2005). Binding of two zinc finger nuclease monomers to two specific sites is required for effective double-strand DNA cleavage. *Biochemical and Biophysical Research Communications*, *334*(4), 1191–1197.

Martínez-Hernández, G. B., Castillejo, N., & Artés-Hernández, F., (2019). Effect of fresh–cut apples fortification with lycopene microspheres, revalorized from tomato by-products, during shelf life. *Postharvest Biology and Technology*, *156*, 110925.

Miller, J. C., Tan, S., Qiao, G., Barlow, K. A., Wang, J., Xia, D. F., & Rebar, E. J., (2011). A TALE nuclease architecture for efficient genome editing. *Nature Biotechnology*, *29*(2), 143–148.

Mishra, R., Mohanty, J. N., Mahanty, B., & Joshi, R. K., (2021). A single transcript CRISPR/Cas9 mediated mutagenesis of CaERF28 confers anthracnose resistance in chilli pepper (*Capsicum annuum* L.). *Planta*, *254*(1), 1–17.

Mohebbi, M., Ansarifar, E., Hasanpour, N., & Amiryousefi, M. R., (2012). Suitability of aloe vera and gum tragacanth as edible coatings for extending the shelf life of button mushroom. *Food and Bioprocess Technology*, *5*(8), 3193–3202.

Motlagh, F. H., & Quantick, P. C., (1988). Effect of permeable coatings on the storage life of fruits. I. Pro-long treatment of limes (*Citrus aurantifolia* cv. Persian). *International Journal of Food Science & Technology*, *23*(1), 99–105.

Osakabe, Y., Liang, Z., Ren, C., Nishitani, C., Osakabe, K., Wada, M., & Nagamangala, K. C., (2018). CRISPR–Cas9-mediated genome editing in apple and grapevine. *Nature Protocols*, *13*(12), 2844–2863.

Palou, L., Serrano, M., Martínez-Romero, D., & Valero, D., (2010). New approaches for postharvest quality retention of table grapes. *Fresh Produce, 4*(1), 103–110.

Palpant, N. J., & Dudzinski, D., (2013). Zinc finger nucleases: Looking toward translation. *Gene Therapy*, *20*(2), 121–127.

Park, S. C., Joung, Y. H., Kim, K. M., Kim, J. K., & Koh, H. J., (2019). Gene-edited crops: Present status and their future. *Korean Society of Breeding Science*, *51*(3), 175–183.

Patidar, O. P., Gautam, C., Tantuway, G., Kumar, S., Yadav, A., Meena, D. S., & Nagar, A., (2016). RNA-guided genome editing tool CRISPR-Cas9: Its applications and achievements in model and crop plants. *Journal of Pure and Applied Microbiology*, *10*(4), 3035–3042.

Petolino, J. F., (2015). Genome editing in plants via designed zinc finger nucleases. *In Vitro Cellular & Developmental Biology-Plant*, *51*(1), 1–8.

Prasad, N., & Batra, E., (2015). Edible coating (the future of packaging): Cheapest and alternative source to extend the post-harvest changes: A review. *Asian J. Biochem. Pharm. Res.*, *3*(5), 2231–2560.

Ramirez, C. L., Foley, J. E., Wright, D. A., Müller-Lerch, F., Rahman, S. H., Cornu, T. I., & Joung, J. K., (2008). Unexpected failure rates for modular assembly of engineered zinc fingers. *Nature Methods*, *5*(5), 374, 375.

Roberts, P. B., (2014). Food irradiation is safe: Half a century of studies. *Radiation Physics and Chemistry*, *105*, 78–82.

Sander, J. D., & Joung, J. K., (2014). CRISPR-Cas systems for editing, regulating, and targeting genomes. *Nature Biotechnology*, *32*(4), 347–355.

Schirra, M., D'hallewin, G., Ben-Yehoshua, S., & Fallik, E., (2000). Host–pathogen interactions modulated by heat treatment. *Postharvest Biology and Technology*, *21*(1), 71–85.

Shan, Q., Wang, Y., Chen, K., Liang, Z., Li, J., Zhang, Y., & Gao, C., (2013). Rapid and efficient gene modification in rice and *Brachypodium* using TALENs. *Molecular Plant*, *6*(4), 1365–1368.

Sivakumar, D., Terry, L. A., & Korsten, L., (2010). An overview on litchi fruit quality and alternative postharvest treatments to replace sulfur dioxide fumigation. *Food Rev. Int., 26*, 162–188.

Tian, S., Jiang, L., Gao, Q., Zhang, J., Zong, M., Zhang, H., & Xu, Y., (2017). Efficient CRISPR/Cas9-based gene knockout in watermelon. *Plant Cell Reports*, *36*(3), 399–406.

Tiwari, V. K., Verma, V. C., Khushboo, A., Kumar, K., Tsewang, T., Verma, A., & Acharya, S., (2022). Edible coating for postharvest management of fruits and vegetables. *Pharma Innovation Journal*, *11*, 970–978.

Tripathi, L., Ntui, V. O., & Tripathi, J. N., (2020). CRISPR/Cas9-based genome editing of banana for disease resistance. *Current Opinion in Plant Biology*, *56*, 118–126.

V. Edwin Hillary, S. Antony Ceasar. Chapter Seven - Genome engineering in insects for the control of vector borne diseases, Progress in Molecular Biology and Translational Science, Academic Press, Volume 179, 2021, Pages 197–223

Venturi, F., Bartolini, S., Sanmartin, C., Orlando, M., Taglieri, I., Macaluso, M., & Mensuali, A., (2019). Potato peels as a source of novel green extracts suitable as antioxidant additives for fresh-cut fruits. *Applied Sciences*, *9*(12), 2431.

Vieira, J. M., Flores-López, M. L., De Rodríguez, D. J., Sousa, M. C., Vicente, A. A., & Martins, J. T., (2016). Effect of chitosan–Aloe vera coating on postharvest quality of blueberry (*Vaccinium corymbosum)* fruit. *Postharvest Biology and Technology*, *116*, 88–97.

Waltz, E., (2016). Gene-edited CRISPR mushroom escapes US regulation. *Nature*, *532*(7599), 293.

Weeks, D. P., Spalding, M. H., & Yang, B., (2016). Use of designer nucleases for targeted gene and genome editing in plants. *Plant Biotechnology Journal*, *14*(2), 483–495.

Weerawardana, M. B. S., Thiripuranathar, G., & Paranagama, P. A., (2020). Natural anti-browning agents against polyphenol oxidase activity in *Annona muricata* and *Musa acuminata. J. Chem., 2020,* 1904798.

Wendehenne, D., Durner, J., & Klessig, D. F., (2004). Nitric oxide: A new player in plant signaling and defense responses. *Current Opinion in Plant Biology*, *7*(4), 449–455.

Xu, J., Kang, B. C., Naing, A. H., Bae, S. J., Kim, J. S., Kim, H., & Kim, C. K., (2020). CRISPR/Cas9-mediated editing of 1-aminocyclopropane-1-carboxylate oxidase1 enhances *Petunia* flower longevity. *Plant Biotechnology Journal*, *18*(1), 287–297.

Zhang, L., Liu, Z., Sun, Y., Wang, X., & Li, L., (2020). Combined antioxidant and sensory effects of active chitosan/zein film containing α-tocopherol on *Agaricus bisporus. Food Packaging and Shelf Life*, *24*, 100470.

CHAPTER 5

RNA INTERFERENCE: APPLICATIONS IN POSTHARVEST QUALITY IMPROVEMENT OF FRUITS AND VEGETABLES

ANUPAMA SINGH, SURENDER KUMAR, AAYUSHEE THAKUR, ANCHAL TANDON, PARUL TOMAR, and RIDDHI RATHORE

Department of Biotechnology, Dr. Yashwant Singh Parmar University of Horticulture and Forestry, Nauni, Solan, Himachal Pradesh, India

ABSTRACT

Post-harvest losses of crop products have become a significant burden to modern agricultural practices, decreasing their economic importance and posing a serious threat to global food security. Various techniques aimed at enhancing nutritional quality and minimizing such losses have been developed, most of which are not very effective. Therefore, it is essential to develop new crop varieties with longer shelf life and consistent nutritional quality. Conventional crop improvement methods do not apply to all crops due to a lack of genetic diversity for post-harvest traits and hybridization incompatibility. Alternatively, genetic engineering (GE) techniques offer opportunities to alter genetic makeup to minimize post-harvest losses and enhance nutritional quality. In particular, RNA interference (RNAi) that mediates gene silencing at the transcriptional and post-transcriptional levels in eukaryotic organisms can be used to control post-harvest damage of crop yields, such as by delaying fruit ripening, reducing browning, and eliminating toxic compounds. The small RNAs associated with RNAi

Bioengineered Fruit and Vegetables: Trends in Postharvest Quality Improvement.
Mohammad Irfan and Pankaj Kumar (Eds.)

also mediate crosstalk between plants and their associated microorganisms, including pathogens and pests, helping to generate the disease resistance response in plant tissues, including fruits and grains. Furthermore, transformation-independent RNAi using SIGS (spray-induced gene silencing) and similar techniques make RNAi more feasible for nearly all crop species, including those that are otherwise recalcitrant to genetic transformation. This chapter describes the mechanism of gene silencing and its applications in mitigating post-harvest losses and enhancing the quality of crops.

5.1 INTRODUCTION

Global food security is one of the major challenges in the 21st century due to climate change and the increasing population. Crop yields are significantly impacted by various biotic and abiotic stresses in the fields and post-harvest, leading to reduced shelf life and making them unsuitable for consumption. Postharvest losses (PHLs) result in direct physical and nutritional quality losses, causing significant economic damage to crops worldwide. This, in turn, severely affects food availability for millions of people, contributing to a rise in malnutrition rates in many middle-class and poor countries.

Plant foods such as cereals, fruits, vegetables, tubers, and legumes, along with dairy products, constitute approximately 92% of total post-harvest food losses in developing countries (Kader et al., 2012). The major causes of post-harvest crop yield losses are harvesting at incorrect stages, inefficient harvesting techniques, poor food supply chains (transport, storage, processing, packaging, and distribution), and consumption problems (Aulakh, Regmi, Fulton, & Alexander, 2013; Etefa, Forsido, & Kebede, 2022). Major factors associated with such post-harvest losses include short shelf life, physical, physiological, and mechanical injuries, and poor hygienic conditions (Kumar & Kalita, 2017; Yahaya & Mardiyya, 2019). These factors increase the susceptibility of grains, fruits, and vegetables to pests, pathogens, birds, rodents, and natural calamities (Elik et al., 2019).

In addition, other environmental factors such as temperature, relative humidity, and aeration during storage contribute significantly to such economic losses. The high relative humidity and temperature stimulate microbial growth and also increase the respiration rate of stored fruits and vegetables, which enhances their decay process (Yahaya, 2005). The low temperature resists microbial attack; however, it may induce chilling injury at an extreme level. Therefore, appropriate cultural operations, harvesting, storage transportation, and pre- and post-harvest treatments are required to minimize the PHLs (Yahaya & Mardiyya, 2019).

Generally, ripening at the harvest stage determines the crop product quality and shelf life, as unripened fruits are more prone to physical illness with possible implications on nutritional qualities, such as high acidity and low sugar, whereas over-ripe ones have a short shelf life and high susceptibility to physiological injuries (Elik et al., 2019). Hence, ripening is a destructive and essential stage in developing fruits and vegetables for postharvest study (Sharma & Kumari, 2020). The ripening of climacteric fruits (such as bananas, melons, peaches, apples, avocados, etc.) and vegetables (such as tomatoes, etc.) is caused by ethylene, a gaseous phytohormone that is responsible for their short durability (Almansa et al., 2016; Qiao, Zhang, Wang, & Shen, 2021). In the case of non-climacteric fruits and vegetables, ABA is the responsible phytohormone (Chai, Jia, Li, Dong, & Shen, 2011.

Maintaining the freshness of agricultural commodities for a longer duration is another major challenge in modern crop improvement programs apart from developing high-yielding varieties for optimizing the quality of crops. The new cultivars possessing good yield durability and nutritional qualities are traditionally obtained by selection of late ripening genotypes, hybridization, mutation breeding, and polyploidy. Some successful examples are mango cultivars 'Keith' derived by selection that are more resistant to fungal infections (Prusky, 2011), a polyploid variety of grape, Sunny Rog (Sharma & Kumari, 2020), and melon obtained by TILLING, with a delayed ripening trait (Dahmani-Mardas et al., 2010). However, the use of these traditional approaches for the management of post-harvest losses is limited by the availability of genetic variability in germplasm and the lack of sexually compatible parents with useful post-harvest traits.

Transgenic crops obtained by the introduction of desirable DNA fragments or genes from different sources using genetic engineering (GE) tools provide a better alternative for developing new crop varieties with high adaptability in contemporary climate scenarios, tolerance to postharvest diseases, securing nutritional status, simultaneous ripening for proper harvesting, and acquiring extended shelf life. The information about mechanisms associated with the postharvest attributes in most of the cultivated crops has been obtained in detail at the physiological, biochemical, and molecular levels, which can be useful in advanced genetic studies and also for manipulation of the corresponding characters to create improved varieties (Rai, Rai, Singh, & Singh, 2019). Genetic manipulation for down-regulating ethylene production in plants is one of the most used transgenic approaches to enhance the storage durability of crop yields (Sharma & Kumari, 2020). However, strict regulations for the cultivation of genetically modified (GM) crops and unfavorable public perception of the use of transgenic crops in many countries hinder the successful utilization of this approach. Alternatively, an RNAi-based transgenic approach in which no transgenic protein formation takes place seems to be more acceptable than traditional ones that produce transgenic protein products. However, they still have the same risks that need to be thoroughly analyzed to avoid any off-target effects (Rajam, 2020).

5.2 RNA INTERFERENCE (RNAI)

RNA interference (RNAi) is the most prevalent post-transcriptional gene regulatory mechanism that mediates the silencing of the corresponding genes at the RNA levels. It was first observed by Jorgensen (Napoli, Lemieux, & Jorgensen, 1990) in petunia plants containing a transgenic copy of the *chs* gene encoding Chalcone synthase, in order to overexpress the flower pigment. Surprisingly, he obtained white or variegated flowers instead of purple ones, which subsequently led to unraveling the silencing of the endogenous *chs* gene by its homologous transgene copy named co-suppression (Van, Mur, Beld, Mol, & Stuitje, 1990). A similar phenomenon named Quelling was later discovered by Romano & Macino (1992) in the filamentous fungus *Neurospora crassa* containing a transgenic copy of the homologous endogenous *albino-1* gene, which encoded the carotenoids responsible for orange pigmentation. Both these phenomena were concluded to be due to the decay of the mRNA products encoded by endogenous genes and their homologous transgenic copies. RNAi is an evolutionary, innate genomic function regulatory process having peculiar roles in controlling the gene activities in plants, fungi (except for *Saccharomyces cerevisiae*), and other eukaryotic organisms (Hirai & Kodama, 2008; Rajput et al., 2021). The molecular mechanism of RNAi was first deduced by Fire & Mello (Fire, Xu, Montgomery, Kostas,

Driver, & Mello, 1998) in the case of *C. elegans*. They introduced single-stranded RNA copies of sense and antisense strands from a gene encoding muscle protein into *C. elegans* and observed no significant change in the worm movements. However, the introduction of dsRNA containing both sense and antisense RNA (asRNA) strands of the same gene led to unusual twitching movements in worms. Consequently, the term RNAi was coined to explain this phenomenon (Rajput et al., 2021).

RNAi is a homology-dependent and sequence-specific gene silencing process governed by double-stranded RNA (dsRNA) (Hirai & Kodama, 2008) that plays a crucial role in regulating the plant immune response against viruses, transposons, and transgenes. It also regulates the function of genes associated with plant developmental processes and stress responses (Mamta & Rajam, 2018; Mujtaba et al., 2021). Recent studies revealed the importance of RNAi in transcriptional-level gene silencing, especially in plants, most probably by DNA methylation being induced by siRNAs (small interfering RNAs) in a process called RNA-directed DNA methylation (RdDM), which is catalyzed by RNA-dependent RNA polymerase (RdRPs) (Erdmann & Picard, 2020; Matzke & Birchler, 2005). However, these RdRPs are absent in *Drosophila melanogaster* and mammals; instead, the PIWI-interacting RNAs (piRNAs), a distinct small class of RNAs, mediate RNA degradation in the cytoplasm and DNA or histone methylation in the nucleus (Ishizu, Nagao, & Siomi, 2011; Khurana & Theurkauf, 2010; Senti & Brennecke, 2010). These piRNAs are derived from single-stranded RNA precursors and are based on a 'ping-pong' cycle for cleavage of the complementary RNAs (Holoch & Moazed, 2015).

RNAi helps in maintaining genome stability, regulating transposon movement, epigenetic changes, etc., by controlling the cellular processes at the transcriptional and translational levels (Rajput et al., 2021). Several studies reported the transport of small RNAs (sRNAs) between different organisms to cause silencing of the specific gene functions in interacting organisms, this process is known as cross-kingdom RNAi. In detail, small RNAs can traffic between a pathogen pest, or symbiont and their corresponding hosts, and have significant effects on their interactions, such as suppression of host immunity by pathogen small RNAs to achieve successful infection for both plants and animals and inhibition of pathogen virulence by plant sRNAs. These RNAi mechanisms play crucial roles in developing efficient disease control measures. Furthermore, unlike insects where only the uptake of dsRNAs of usually longer than 50 bp occurs and not small RNAs, both dsRNAs and small RNAs are easily taken up by fungi and plants. It could indicate the presence of different uptake mechanisms in these organisms (Cagliari et al., 2019; Wang& Jin, 2017).

Nowadays, the direct treatments of crops and their post-harvested products with sRNAs also lead to suppression of plant diseases and induce tolerance for abiotic stresses. This strategy is known as spray-induced gene silencing (SIGS). It reduces the excessive use of chemicals in crop management and also circumvents the need for GM plants against biotic stresses. SIGS is an easy, environmentally sustainable, and friendly RNAi-based approach that can be easily used for controlling various plant disorders caused by biotic and abiotic stresses before and after harvesting crops (Wang & Jin, 2017). RNAi is generally used as an important functional genomics tool for functional annotation of many genes in eukaryotes such as plants and animals. Furthermore, it is an important crop improvement tool for developing stress-tolerant crop varieties, improving nutritional values, slowing down ripening, inducing male sterility, photoperiod manipulation, modifying plant architecture, modulating secondary metabolite production, and eliminating undesirable qualities such as allergens and toxins from crop products (Figure 5.1), and post-harvest quality management (Rajam, 2020). RNAi is applicable at any physiological stage of the crop plants and even after their harvesting. This chapter explains the different aspects of RNAi including types and mechanisms of actions and is also highlighting the applications of RNAi in post-harvest quality management of crop yields.

5.2.1 MECHANISMS OF RNA INTERFERENCE (RNAI)

RNA interference (RNAi) utilizes non-coding small RNAs, along with AGO (conserved argonaute) and PIWI proteins for post-transcriptional and/or transcriptional gene inactivation (Holoch & Moazed, 2015). Small RNAs associated with RNAi include micro RNAs (miRNAs) and siRNAs, which differ in their origins, post-transcriptional processing sites, and machinery, though they share some components of the gene silencing process (Cagliari et al., 2019). These small RNAs form a complex with AGO proteins through the interaction of their 5′ nucleotide possessing monophosphate with the middle (MID) domain and the

3′ nucleotide with the PIWI-AGO-ZWILLE (PAZ) domain of the AGO proteins (Ma, Ye, & Patel, 2004; Wang et al., 2009). These associations are unique and determined by a specific arrangement of the MID and PAZ domains and their bindings with distinct small RNA termini. The associated small RNAs provide specificity to the nuclease activity of AGO proteins by complementary binding to the target mRNAs to cause mRNA degradation, translational repression, or transcriptional silencing. There is an RNase H-like domain in AGO proteins that triggers mRNA cleavage (Holoch & Moazed, 2015). RNAi generally utilizes two ribonucleases called Dicer and RNA-induced silencing complex (RISC) containing AGO proteins for maturation and inhibitory actions of the small RNAs. Dicer is an RNase III enzyme belonging to class 3, which consists of 4 domains including a PAZ (Piwi/Argonaute/Zwille) domain, an N-terminal helicase domain, a dual RNase III domain, and a dsRNA binding domain (Rajput et al., 2021). According to the proposed Dicer catalysis model, a multidomain Dicer enzyme contains two RNase III domains that dimerize to form an active center of which each of the two domains is responsible for the degradation of the single strand from dsRNA during siRNA biogenesis (Zhang, Kolb, Jaskiewicz, Westhof, & Filipowicz, 2004). The RISC complex possesses Argonaute (AGO), an RNase core enzyme that in combination with other proteins participates in the final stage of the RNAi pathway involving gene silencing. Within an RISC complex, the AGO proteins help in the recognition of the guide strand termini, target mRNA cleavage independently (using its nuclease activity), or associate with other proteins to induce silencing (Rajput et al., 2021; Wilson & Doudna, 2013).

The biosynthesis of miRNA and siRNA mainly differs only in the beginning steps as the miRNAs have a cellular origin whereas siRNA is evolved either endogenously through Dicer-mediated cleavage of dsRNA or derived externally via transgenes, transposons, or viruses. Both types of small RNAs share similarities in their sizes and mode of action, i.e., sequence-specific inhibitory functions (Rajput et al., 2021). In the RNAi pathway, there are mainly four basic steps:

- Formation of small RNA (non-coding);
- Loading of small RNA into the silencing apparatus (RISC);
- Activation of loaded silencing complex; and
- Targeted mRNA degradation.

5.2.1.1 SMALL INTERFERING RNA (SIRNA)-MEDIATED RNAI

siRNAs are small RNAs that can originate either exogenously, introduced directly into the cytoplasm from outside the cells (from viruses or artificial supply), or endogenously (Cagliari et al., 2019). These RNAi pathways are mediated by long dsRNA with perfect complementarity to the target gene and can also be induced by external intruders (viruses, transposons, or transgenes) that recruit Dicer-related enzymes (Mello & Conte, 2004). The Dicer enzyme processes the long dsRNAs in the cytoplasm into short siRNA duplexes (21–26 nt long) with a 3′OH end containing 2 nt overhangs and a 5′ phosphorylated end (Holoch & Moazed, 2015; Rajput et al., 2021). These short RNA duplexes are loaded into the siRNA-induced silencing complex (siRISC), leading to cleavage of their sense strand (identical to the mRNA to be silenced or degraded), while the complementary antisense siRNA strands associate with siRISC. The antisense siRNA in the siRISC hybridizes specifically with the complementary mRNA. AGO proteins, along with other proteins in the siRISC, induce degradation of the targeted mRNA or repress its translation (post-transcriptional gene silencing (PTGS)) (Figure 5.1) (Rajput et al., 2021). Additionally, siRNAs also play a role in transcriptional-level gene silencing by inducing epigenetic modifications. Interestingly, there have been reports of Dicer-independent siRNA biogenesis in *Arabidopsis*, *Neurospora, C. elegans,* and *Schizosaccharomyces pombe* (Aoki et al., 2007; Rajput et al., 2021.

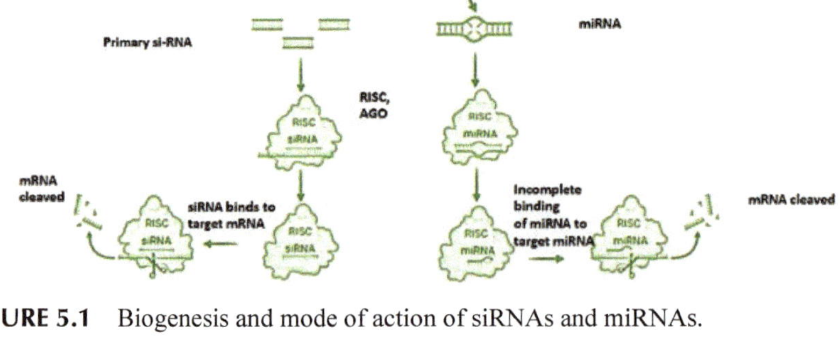

FIGURE 5.1 Biogenesis and mode of action of siRNAs and miRNAs.

5.2.1.2 MICRO-RNA-MEDIATED RNAI

The miRNAs are distinct types of small RNAs (21–23 nt long) having endogenous origin. They are transcribed from miRNA (MIR) genes in the nucleus using RNA polymerase II and produce about 1,000 nt long pri-miRNA (primary miRNA) transcript that produces an imperfect hairpin structure (folded-back and stem-looped) due to the complementarity within the same sequence. The pri-miRNA transcripts are further processed by Drosha, a class (II) RNAse III enzyme, leading to the formation of a short stem-loop precursor known as pre-miRNA. This pre-miRNA is transported into the cytosol via Exportin 5 – RanGTP complex, where it is processed by Dicer to form mature miRNA. This functional miRNA is a single-stranded RNA that is complex with RISC, which possesses AGO and other proteins, and forms a miRNA-induced silencing complex (miRISC). Such miRNA participates in base pairing with target mRNA and guides AGO protein to undergo cleavage of the target mRNA if there is a perfect or complete base pairing between miRNA and mRNA. However, if it is imperfect, i.e., incomplete base pairing, then miRISC blocks the mRNA translation without its degradation (Figure 5.2) (Guo, Liu, Smith, Liang, & Wang, 2016; Rajput et al., 2021). Moreover, amiRNA is a recently developed new RNAi vector containing a hairpin RNA structure similar to that of miRNAs to induce RNAi (Hirai & Kodama, 2008).

5.2.2 APPLICATIONS OF RNAI IN POST-HARVEST QUALITY MANAGEMENTS OF CROPS

Nowadays, a significant loss of plant yields is caused by different biotic and abiotic stresses at both pre- and post-harvest stages that have a major detrimental effect on the overall economic value of crop production. In particular, several post-harvest disorders due to biotic and abiotic factors develop during processing, transportation, and storage, accounting for up to 50% of crop yield loss in developing countries (Sharma, Singh, & Singh, 2009). Currently, these post-harvest diseases are mostly controlled through chemical treatments, which have limited efficiency and also have adverse impacts on human health and the environment. Alternatively, RNAi is one of the most effective, sustainable, and eco-friendly approaches to prevent crop yield losses from pests and pathogens both before and after crop harvesting (Wang, Thomas, & Jin, 2017).

RNAi is applicable with and without transgenic plant generation. However, the applications of RNAi-based transgenic approach are restricted by various challenges associated with their approval due to proclaimed public concerns over their effects on the environment and human health. Alternatively, exogenous treatments of given noncoding small RNAs, which could be dsRNAs, siRNAs, and hairpin RNAs (hpRNAs), are assumed to be safe and regarded as more eco-friendly, sustainable, and publicly acceptable than genetic transformation (Das & Sherif, 2020). There are various techniques for the exogenous supply of dsRNAs or siRNAs, as shown in Figure 5.2. Among them, SIGS (spray induced gene silencing) allows the application of dsRNA/siRNA to the selected plant tissues, which is getting much attention because of its high feasibility and low cost. It employs the formulation of dsRNA or siRNA, such as pesticidal dsRNA dissolved in water for spray onto the plant tissues. From treated plant tissue (say leaf surface), small RNAs or dsRNAs move either directly into the target pests or pathogens or indirectly through plant cells as mentioned above in the case of cross-kingdom RNAi (Cagliari et al., 2019; Das & Sherif, 2020). However, despite being a promising technique, some issues with RNAi such as its efficiency, dsRNA degradation, environmental risk assessments, and resistance evolution need to be addressed to make an efficient use of RNAi in pest management (Cagliari et al., 2019).

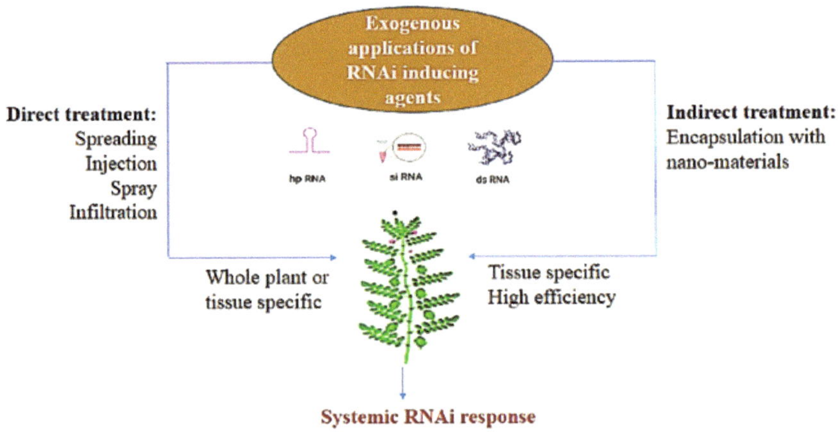

FIGURE 5.2 Different ways of exogenous application of non-coding small RNAs into plants.

A study involving the application of dsRNA on detached leaves of *Hordeum vulgare* revealed the translocation of dsRNA through the vascular system to distal parts of the plants where it subsequently matured into siRNAs by plant Dicer enzyme. It resulted in growth inhibition of the *Fusarium graminearum* in both local and distantly located untreated leaves (Koch et al., 2016). The dsRNA molecules used in this study were present in the mesophyll cells, stomata, trichomes, phloem, xylem parenchymal cells, and companion cells (Cagliari et al., 2019). Furthermore, several recent studies indicated the uptake of RNAs from the environment by plants and fungi (including plant pathogens). Indeed, Wang et al. (2016) reinforced such findings by showing the uptake of the dsRNAs and siRNAs by *B. cinerea*, a fungal pathogen, from the environment. In detail, fruits, vegetables, and flowers were treated with *B. cinerea* DCL1/2-targeting dsRNAs or sRNAs, resulting in a significant inhibition of gray mold disease progression. Similarly, the application of dsRNAs targeting *Fusarium graminearum* cytochrome P450 lanosterol C-14α-demethylase (CYP51) genes on barley leaves led to a remarkable reduction in disease symptoms (Koch et al., 2016). Spray-induced gene silencing (SIGS) is an emerging and highly effective plant disease management strategy that can substantially decrease post-harvest yield losses (Killiny, Hajeri, Tiwari, Gowda, & Stelinski, 2014; Pridgeon et al., 2014; San & Scott, 2016). The small RNAs associated with RNAi can systematically translocate within a plant host from one tissue to another, reaching non-sprayed distal parts of the plants, demonstrating the potential and versatility of this approach from seedling to harvesting (Koch et al., 2016; Wang, Thomas, & Jin, 2017). The increasing attention towards RNA-based biopesticides over chemical-based control measures is due to their target specificity and efficiency against biotic stresses (Fletcher, Reeves, Hoang, & Mitter, 2020; Wang, Thomas, & Jin, 2017).

5.2.2.1 ENCAPSULATION TECHNOLOGY FOR IMPROVING THE EFFICIENCY OF RNAI-BASED POST-HARVEST CROP MANAGEMENT

Encapsulation provides protection and stability to the dsRNAs and prevents them from degradation by RNAses enzymes or pH degradation during transportation into the target cells. It can be achieved using engineered microorganisms or synthetic micro/nanoparticles, proteins, biopolymers, clays, lipids, and capsid proteins, which also facilitate the intake of dsRNAs by target cells, thereby enhancing the efficiency of RNAi (Hernández-Soto, & Chacón-Cerdas, 2021). In particular, nano-encapsulation of the dsRNA or small RNAs using biopolymers including polysaccharides (like hyaluronic acid, chitosan, etc.), proteins, and lipids, increases their target specificity. Furthermore, these nanomaterials serve as a barrier to gas diffusion and water loss and protect dsRNA. Naked dsRNA and siRNA treatments against plant pathogenic fungi were found to remain effective for up to 10 days after spraying (Chariou, Ortega-Rivera, & Steinmetz, 2020; Koch et al., 2016; Wang, Weiberg, Lin, Thomma, Huang, & Jin, 2016).

Likewise, RNAi-based antiviral immunity was increased over 20 days while applying dsRNAs complex with layered double hydroxide (LDH) clay nanosheets (called BioClay) as it had provided protection to the dsRNAs from RNases or sunlight-induced degradation or wiping up away from tissue surface with water (Mitter et al., 2017). Encapsulation using BioClay extended the resistance to the target virus from days to weeks by giving stability to the dsRNAs and facilitating their slower release (Fletcher, Reeves, Hoang, & Mitter, 2020). Therefore, the potential applications of SIGS against plant diseases are becoming more effective with the help of nanoparticle technology.

RNAi-based post-harvest crop management is enlisted in Table 5.1. Two major areas in which RNAi is used for post-harvest management in horticultural crops are discussed in subsections.

5.2.2.2.1 Prolongation of Shelf Life

The post-harvest yield loss due to deterioration and spoilage caused during harvesting, transport, and storage is the most critical factor affecting crop yield economic value significantly. Therefore, the longer shelf life of vegetables and fruits is much more desirable. It can be achieved through delayed fruit ripening (Batra, Sane, Trivedi, Sane, & Nath, 2010). The ripening of fruits and vegetables is mainly controlled by ethylene, therefore by manipulating the genes associated with the ethylene biosynthetic pathway, the ethylene production can be reduced, and the ripening process can be delayed from many days to months (Figure 5.3) (Batra, Sane, Trivedi, Sane, & Nath, 2010). Reduction in ethylene synthesis can be achieved by silencing the expression of ethylene biosynthetic genes at the transcriptional (mainly through epigenetic modifications) and post-transcriptional levels (though mRNA degradation or translational repression).

In tomatoes (climacteric fruits), the down-regulation of *1-aminocyclopropane-1-carboxylate oxidase* (*ACCO*), a gene of the ethylene biosynthesis pathway, using RNAi reduced ethylene level in the ripening fruits and extended their shelf life beyond 120 days while retaining similar amounts of the amino acids, total soluble sugar, total soluble solids, and titratable acid, as for control plants (Xiong, Yao, Peng, Li, Han, & Fan, 2005). Furthermore, silencing of multiple genes simultaneously such as three homologs of the ACC synthase gene in tomato using chimeric hpRNA has also been proven more effective for enhancing the longevity of the tomato fruits further by 45 days and reducing post-harvest losses (Gupta, Pal, & Rajam, 2013). The miRNAs including miR156 and miR172, have been identified that regulate the expression of a fruit-ripening gene COLORLESS *NEVER RIPE* (CNR), which is associated with the development and ripening process of the tomato fruit (Karlova et al., 2013; Molesini, Pii, & Pandolfini, 2012). The Flavr-Savr variety of tomatoes was the world's first transgenic crop variety that was developed by Calgene Inc. and approved for commercial cultivation on May 21, 1994, in the USA. It was developed through *Agrobacterium*-mediated genetic modification using an antisense construct for silencing the endogenous *polygalacturonase* (PG) gene. PG is a hydrolytic enzyme responsible for cell wall degradation, especially by degrading the pectin layer present in

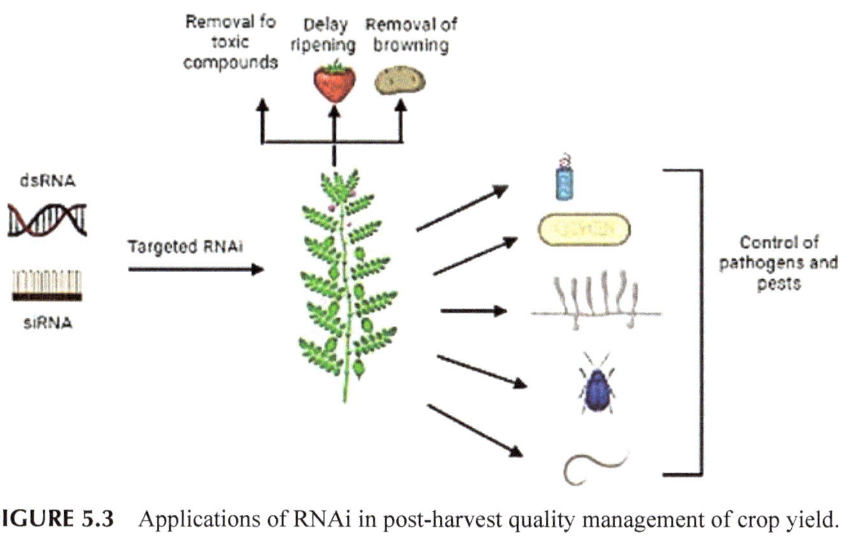

FIGURE 5.3 Applications of RNAi in post-harvest quality management of crop yield.

the middle lamella of ripening tomato fruit resulting in softening of the ripened fruits. asRNA-based suppression of PG caused delaying in this natural ripening process and softening of tomato fruits. As a consequence, it increases the shelf life of tomato fruits, which take much more time to ripen (Krieger et al., 2008; https://advocatetanmoy.com/2021/02/06/biotechnology-of-food-fda-and-new-tomato/?print=print).

In another study, the expression of the tomato ACC oxidase gene was silenced using an hpRNAi transformation vector containing a partial coding region of ACO1 cloned from the lowland tomato cultivar (MT1) and 35S promoter. It reduced ethylene production in the resulting transgenic lines, which in turn enhanced the shelf life of tomato fruits from 10 to 32 days compared to the wild-type fruits (Behboodian, Mohd, Ismail, & Zainal, 2012). Polyphenol oxidase (PPO) is an enzyme that induces the oxidation of phenolic compounds into quinones and causes browning in potatoes and apples. Potatoes and apples with reduced PPO contents were developed by using either antisense or hpRNA (Bachem et al., 1994; Coetzer, Corsini, Love, Pavek, & Tumer, 2001; Murata, Nishimura, Murai, Haruta, Homma, & Itoh, 2001). The silencing of four members of the PPO gene family individually or in combination using artificial miRNA (amiRNA) in potatoes and apples led to the development of their low-browning varieties (Chi et al., 2014; Kumar, Mundiyara, & Jitarwal, 2017). Arctic Apples were the world's first GM apples developed by Okanagan Specialty Fruits Inc., Canada, using a sense-posttranscriptional silencing approach to eliminate browning. They used a chimeric PPO gene targeted to disrupt the expression of endogenous PPO to avoid browning in fruit flesh (Stowe & Dhingra, 2021). In detail, a gene construct named PGAS containing DNA fragments from each of the four genes (*PPO2, GPO3, APO5, and pSR7*), in a Ti-plasmid vector GEN-03 containing cauliflower mosaic virus (CaMV) 35S promoter and nopaline synthase terminator (NOS), was used for *Agrobacterium*-based genetic transformation of apple (Baker & Horin, n.d.). Two modified apple varieties Granny Smith and Golden Delicious, with the trade names Arctic® Golden Delicious apple and the Arctic® Granny Smith apple, were devoid of the browning trait. They were approved for human consumption in the US and Canada in 2017 (Kumar, Mundiyara, & Jitarwal, 2017). These developments of potatoes and apples not only reduced browning during food processing and consumption but also reduced waste derived as a consequence of browning (Guo, Liu, Smith, Liang, & Wang, 2016). Three different GM events using potato varieties,

Russet Burbank, Ranger Russet, and Atlantic by researchers at J.R. Simplot Company, led to the development of Simplot's Innate™ Russian Burbank, which contains a resistance gene from *Solanum venturii* against late blight (*Phytophthora infestans*) and RNAi cassettes for silencing the expression of four different proteins that are mentioned in Table 5.1 (https://www.canada.ca/en/health-canada/services/food-nutrition/genetically-modified-foods-other-novel-foods/approved-products/novel-food-information-simplot-innate-potato-gen2-w8-x17-y9.html).

TABLE 5.1 Characteristics of Four Genes Present in the T-DNA of RNAi Expression Cassette pSIM1278 for the Development of the Simplot's Innate™ Russian BurBank

Gene	Trait	Benefits
Asn1: asparagine synthetase-1	Decreases free asparagine	Reduced acrylamide potential
R1: water dikinase	Decreases reducing sugars	Reduced acrylamide potential
PhL: phosphorylase L	Decreases reducing sugars	Reduced acrylamide potential
Ppo5: polyphenol oxidase 5	Decreases polyphenol oxidase	Reduced black spot

It was observed that defense response against *P. infestans* and reduction in the discoloration (browning and black spots from harvesting and transport-associated damages) were due to suppression of the expression of polyphenol oxidase-5. Additionally, the repression of the expression of asparagine synthetase (resulting in reduced free asparagine), water dikinase (leading to decreased reducing sugars), and phosphorylase-L (resulting in decreased reducing sugars) contributed to the lowering of acrylamide formation during baking or frying at high temperatures. Moreover, the decline in the formation of invertase allowed the storage of potatoes at colder temperatures (Kumar, Mundiyara, & Jitarwal, 2017). Kiwi fruit lines with RNAi-*ACCO* knockdown were developed to prevent the softening of fruits at the desired stage (Atkinson et al., 2011). Similarly, strawberry fruits, which are soft with a short shelf life due to the rapid loss of their firm texture, had these traits controlled by silencing the function of the PL gene using asRNA technology (Jiménez-Bermudez et al., 2002), and the *FaPYR1* gene using a tobacco rattle virus-based gene silencing method (Chai, Jia, Li, Dong, & Shen, 2011). Furthermore, Qiao et al. (2021) utilized an RNAi cassette to silence the expression of the ABA receptor *FcPYL8* gene, interrupting the ABA signaling and down regulating ethylene biosynthesis, which led to the delayed ripening of fig (*Ficus carica*) fruit (Qiao, Zhang, Wang, & Shen, 2021).

5.2.2.2.2 Elimination of Toxic Compounds From Edible Crop

Cassava generally carries cyanogenic glycosides (linamarin) in its leaves and roots at potentially toxic levels to protect itself from herbivory and theft (Siritunga & Sayre, 2004). Consumption by humans results in the formation of a toxic compound by enzymes and gut flora of the human intestine. This issue was resolved with the development of transgenic cassava in which the formation of cyanogenic glycoside in leaves and roots was selectively repressed by using an antisense construct against the CYP79D1/D2 gene (Tilahun, Bezie, Kerisew, & Taye, 2021) (Table 5.2). Furthermore, there are several reports of using dsRNA specific to β2-tubulin mRNA of *Fusarium asiaticum* to induce broad-spectrum resistance against different fungal pathogens, such as *Magnaporthe oryzae,*

TABLE 5.2 Examples of RNAi Applications in Post-Harvest Crop Management

Crop	Trait	Gene Silenced	References
Papaya	Delay ripening	ACC oxidase	López-Gómez et al. (2009)
Apple	Enhanced shelf life	ACC synthase, ACO ACC oxidase	Dandekar et al. (2004)
Strawberry	Delayed fruit ripening	FaPYR1	Chai, Jia, Li, Dong, & Shen (2011)
Charentais type cantaloupe	Delayed fruit ripening	ACC oxidase	Ayub et al. (1996); Li, Yao, Yang, & Li (2006)
Banana	Enhanced shelf life	MaMADS1/S2	Elitzur et al. (2016)
Tomato	Delayed ripening	Three homologs of the ACC synthase (ACS) gene	Gupta, Pal, & Rajam (2013)
	Enhanced shelf life	α-mannosidase (α-Man) and β-D-N-acetyl hexosaminidase (β-Hex)	Meli et al. (2010)
Cotton	Downregulated gossypol synthesis	cdn1-Cl	Martin, Liu, Benedict, Stipanovic, & Magill (2003)
Coffee	Decaffeination	CaMXMT1	Pathak& Gogoi (2016)

Colletotrichum truncatum, B. cinerea, and other *Fusarium* spp. These dsRNAs influenced the mycelium growth, asexual reproduction, and virulence of these pathogens and provided effective control of postharvest spoilage in different crops (De Oliveira, Silva, Cipriano, Gomes, & Egea, 2021; Gu et al., 2019; Qiu, Huang, Xu, Bi, Chen, & Zhou, 2012).

5.3 CONCLUSION

RNAi is the most used GE approach for enhancing the nutritional features and perishables, imparting resistance to pests, pathogens, and other stresses. It has become a new generation of RNA-based pesticides that outweigh the demerits of chemical-based ones and offer a sustainable and environmentally friendly means of reducing crop yield losses caused during harvesting and by post-harvest damages.

KEYWORDS

- **crop improvement**
- **edible crop**
- **post-harvest losses**
- **post-harvest quality**
- **RNA interference**
- **shelf life**
- **small RNAs**
- **spray-inducedgenesilencing**
- **toxic compounds**

REFERENCES

Almansa, S., Hernandez, F., Legua, P., Nicolas-Almansa, M., & Amoros, A., (2016). Physico-chemical and physiological changes during fruit development and on-tree ripening of two Spanish jujube cultivars (*Ziziphus jujuba* Mill.). *J. Sci. Food Agric., 96,* 4098–4105.

Aoki, K., Moriguchi, H., Yoshioka, T., Okawa, K., & Tabara, H., (2007). *In vitro* analyses of the production and activity of secondary small interfering RNAs in *C. elegans*. *EMBO J., 26*(24), 5007–5019.

Atkinson, R. G., Gunaseelan, K., Wang, M. Y., Luo, L., Wang, T., Norling, C. L., Johnston, S. L., et al., (2011). Dissecting the role of climacteric ethylene in kiwifruit (*Actinidia chinensis*) ripening using a 1-aminocyclopropane-1-carboxylic acid oxidase knockdown line. *J. Exp. Bot.*, *62*(11), 3821–3835.

Aulakh, J., Regmi, A., Fulton, J. R., & Alexander, C. E., (2013). *Estimating Post-Harvest Food Losses: Developing a Consistent Global Estimation Framework.* Agricultural & Applied Economics Association's 2013 AAEA & CAES Joint Annual Meeting, Washington, DC.

Ayub, R., Guis, M., Amor, M. B., Gillot, L., Roustan, J. P., Latché, A., Bouzayen, M., & Pech, J. C., (1996). Expression of ACC oxidase antisense gene inhibits ripening of cantaloupe melon fruits. *Nat. Biotechnol.*, *14*, 862–866.

Bachem, C. W., Speckmann, G. J., Van, D. L. P. C., Verheggen, F., Hunt, M. D., Steffens, J. C., & Zabeau, M., (1994). Antisense expression of polyphenol oxidase genes inhibits enzymatic browning in potato tubers. *Biotechnol.*, *12*(11), 1101–1105.

Baker, A., & Horin, L. Arctic Apples: A Fresh New Take on Genetic Engineering. https://sitn.hms.harvard.edu/flash/2018/arctic-apples-fresh-new-take-genetic-engineering/ (accessed on 7 February 2024).

Batra, A., Sane, V. A., Trivedi, P. K., Sane, A. P., & Nath, P., (2010). Suppression of ACC oxidase expression in tomato using heterologous gene from banana prolongs shelf-life both on vine and post-harvest. *Curr. Sci.*, 1243–1250.

Behboodian, B., Mohd, A. Z., Ismail, I., & Zainal, Z., (2012). Postharvest analysis of lowland transgenic tomato fruits harboring hpRNAi-ACO1 construct. *Sci. World J.*, *2012*, 439870. doi: 10.1100/2012/439870.

Biotechnology of Food, FDA, and New Tomato. https://advocatetanmoy.com/2021/02/06/biotechnology-of-food-fda-and-new-tomato/?print=print (accessed on 7 February 2024).

Cagliari, D., Dias, N. P., Galdeano, D. M., Dos Santos, E. Á., Smagghe, G., & Zotti, M. J., (2019). Management of pest insects and plant diseases by non-transformative RNAi. *Front. Plant Sci.*, *10*, 1319.

Chai, Y. M., Jia, H. F., Li, C. L., Dong, Q. H., & Shen, Y. Y., (2011). FaPYR1 is involved in strawberry fruit ripening. *J. Exp. Bot.*, *62*(14), 5079–5089.

Chariou, P. L., Ortega-Rivera, O. A., & Steinmetz, N. F., (2020). Nanocarriers for the delivery of medical, veterinary, and agricultural active ingredients. *ACS Nano*, *14*(3), 2678–2701.

Chi, M., Bhagwat, B., Lane, W. D., Tang, G., Su, Y., Sun, R., Oomah, B. D., et al., (2014). Reduced polyphenol oxidase gene expression and enzymatic browning in potato (*Solanum tuberosum* L.) with artificial microRNAs. *BMC Plant Biol.*, *14*(1), 1–18.

Coetzer, C., Corsini, D., Love, S., Pavek, J., & Tumer, N., (2001). Control of enzymatic browning in potato (*Solanum tuberosum* L.) by sense and antisense RNA from tomato polyphenol oxidase. *J. Agric. Food Chem.*, *49*(2), 652–657.

Dahmani-Mardas, F., Troadec, C., Boualem, A., Leveˆque, S., Alsadon, A. A., Aldoss, A. A., Dogimont, C., & Bendahmane, A., (2010). Engineering melon plants with improved fruit shelf life using the TILLING approach. *PloS One*, *5*(12), e15776.

Dandekar, A. M., Teo, G., Defilippi, B. G., Uratsu, S. L., Passey, A. J., Kader, A. A., Stow, J. R., Colgan, R. J., & James, D. J., (2004). Effect of down-regulation of ethylene biosynthesis on fruit flavor complex in apple fruit. *Transgenic Res.*, *13*(4), 373–384.

Das, P. R., & Sherif, S. M., (2020). Application of exogenous dsRNAs-induced RNAi in agriculture: Challenges and triumphs. *Front. Plant Sci., 11*, 946.

De Oliveira, F. J. G., Silva, G. D. C., Cipriano, L., Gomes, M., & Egea, M. B., (2021). Control of postharvest fungal diseases in fruits using external application of RNAi. *J. Food Sci., 86*(8), 3341–3348.

Elik, A., Yanik, D. K., Istanbullu, Y., Guzelsoy, N. A., Yavuz, A., & Gogus, F., (2019). Strategies to reduce post-harvest losses for fruits and vegetables. *Strategies, 5*(3), 29–39.

Elitzur, T., Yakir, E., Quansah, L., Zhangjun, F., Vrebalov, J., Khayat, E., Giovannoni, J. J., & Friedman, H., (2016). Banana MaMADS transcription factors are necessary for fruit ripening and molecular tools to promote shelf-life and food security. *Plant Physiol., 171*(1), 380–391.

Erdmann, R. M., & Picard, C. L., (2020). RNA-directed DNA methylation. *PLoS Genet., 16*(10), e1009034.

Etefa, O. F., Forsido, S. F., & Kebede, M. T., (2022). Postharvest loss, causes, and handling practices of fruits and vegetables in Ethiopia: Scoping review. *J. Hortic. Res., 30*(1), 1–10.

Fire, A., Xu, S., Montgomery, M. K., Kostas, S. A., Driver, S. E., & Mello, C. C., (1998). Potent and specific genetic interference by double-stranded RNA in *Caenorhabditis elegans*. *Nat., 391*(6669), 806–811.

Fletcher, S. J., Reeves, P. T., Hoang, B. T., & Mitter, N., (2020). A perspective on RNAi-based biopesticides. *Front. Plant Sci., 11*, 51.

Gu, K. X., Song, X. S., Xiao, X. M., Duan, X. X., Wang, J. X., Duan, Y. B., Hou, Y. P., & Zhou, M. G., (2019). A β2-tubulin dsRNA derived from *Fusarium asiaticum* confers plant resistance to multiple phytopathogens and reduces fungicide resistance. *Pestic. Biochem. Physiol., 153*, 36–46.

Guo, Q., Liu, Q., Smith, N. A., Liang, G., & Wang, M. B., (2016). RNA silencing in plants: Mechanisms, technologies, and applications in horticultural crops. *Curr. Genomics, 17*(6), 476–489.

Gupta, A., Pal, R. K., & Rajam, M. V., (2013). Delayed ripening and improved fruit processing quality in tomato by RNAi-mediated silencing of three homologs of 1-aminopropane-1-carboxylate synthase gene. *J. Plant Physiol., 170*(11), 987–995.

Hernández-Soto, A., & Chacón-Cerdas, R., (2021). RNAi crop protection advances. *Int. J. Mol. Sci., 22*(22), 12148.

Hirai, S., & Kodama, H., (2008). RNAi vectors for manipulation of gene expression in higher plants. *The Open Plant Sci. J., 2*(1), 21–30.

Holoch, D., & Moazed, D., (2015). RNA-mediated epigenetic regulation of gene expression. *Nat. Rev. Genet., 16*(2), 71–84.

Ishizu, H., Nagao, A., & Siomi, H., (2011). Gatekeepers for Piwi–piRNA complexes to enter the nucleus. *Curr. Opin. Genet. Dev., 21*(4), 484–490.

Jiménez-Bermudez, S., Redondo-Nevado, J., Munoz-Blanco, J., Caballero, J. L., López-Aranda, J. M., Valpuesta, V., Pliego-Alfaro, F., et al., (2002). Manipulation of strawberry fruit softening by antisense expression of a pectate lyase gene. *Plant Physiol., 128*(2), 751–759.

Kader, A. A., Kitinoja, L., Hussein, A. M., Abdin, O., Jabarin, A., & Sidahmed, A. E., (2012). *Role of Agro-industry in Reducing Food Losses in the Middle East and North*

Africa. Agro industry and Infrastructure, FAO, Regional Office for the Near East, Cairo, Egypt.

Karlova, R., Van, H. J. C., Maliepaard, C., Van De, G. H., Bovy, A. G., Lammers, M., Angenent, G. C., & De Maagd, R. A., (2013). Identification of microRNA targets in tomato fruit development using high-throughput sequencing and degradome analysis. *J. Exp. Bot.*, *64*(7), 1863–1878.

Khurana, J. S., & Theurkauf, W., (2010). piRNAs, transposon silencing, and *Drosophila* germline development. *J. Cell Biol.*, *191*(5), 905–913.

Kikulwe, E. M., Okurut, S., Ajambo, S., Nowakunda, K., Stoian, D., & Naziri, D., (2018). Postharvest losses and their determinants: A challenge to creating a sustainable cooking banana value chain in Uganda. *Sustainability*, *10*(7), 2381.

Killiny, N., Hajeri, S., Tiwari, S., Gowda, S., & Stelinski, L. L., (2014). Double-stranded RNA uptake through topical application mediates silencing of five CYP4 genes and suppresses insecticide resistance in *Diaphorina citri*. *PloS One*, *9*(10), e110536.

Koch, A., Biedenkopf, D., Furch, A., Weber, L., Rossbach, O., Abdellatef, E., Linicus, L., et al., (2016). An RNAi-based control of *Fusarium graminearum* infections through spraying of long dsRNAs involves a plant passage and is controlled by the fungal silencing machinery. *PLoS Pathogens*, *12*(10), e1005901.

Krieger, E. K., Allen, E., Gilbertson, L. A., Roberts, J. K., Hiatt, W., & Sanders, R. A., (2008). The Flavr Savr tomato, an early example of RNAi technology. *Hort. Sci.*, *43*(3), 962–964.

Kumar, D., & Kalita, P., (2017). Reducing postharvest losses during storage of grain crops to strengthen food security in developing countries. *Foods*, *6*(1), 8.

Kumar, P., Mundiyara, R., & Jitarwal, R. C., (2017). RNA interference: New approach of gene silencing in plants. *Int. J. Pure Appli. Biosci.*, *5*(3), 421–425.

Li, Z., Yao, L., Yang, Y., & Li, A., (2006). Transgenic approach to improve quality traits of melon fruit. *Scientia Horticulturae*, *108*(3), 268–277.

López-Gómez, R., Cabrera-Ponce, J. L., Saucedo-Arias, L. J., Carreto-Montoya, L., Villanueva-Arce, R., Díaz-Perez, J. C., Gómez-Lim, M. A., & Herrera-Estrella, L., (2009). Ripening in papaya fruit is altered by ACC oxidase cosuppression. *Transgenic Res.*, *18*(1), 89–97.

Ma, J. B., Ye, K., & Patel, D. J., (2004). Structural basis for overhang-specific small interfering RNA recognition by the PAZ domain. *Nat.*, *429*(6989), 318–322.

Mamta, B., & Rajam, M. V., (2018). RNA interference: A promising approach for crop improvement. In: *Biotechnologies of Crop Improvement* (Vol. 2, pp. 41–65) Springer, Cham.

Martin, G. S., Liu, J., Benedict, C. R., Stipanovic, R. D., & Magill, C. W., (2003). Reduced levels of cadinane sesquiterpenoids in cotton plants expressing antisense (+)-δ-cadinene synthase. *Phytochem.*, *62*(1), 31–38.

Matzke, M. A., & Birchler, J. A., (2005). RNAi-mediated pathways in the nucleus. *Nat. Rev. Genet.*, *6*(1), 24–35.

Meli, V. S., Ghosh, S., Prabha, T. N., Chakraborty, N., Chakraborty, S., & Datta, A., (2010). Enhancement of fruit shelf life by suppressing N-glycan processing enzymes. *Proc. Natl. Acad. Sci. U. S. A.*, *107*(6), 2413–2418.

Mello, C. C., & Conte, D., (2004). Revealing the world of RNA interference. *Nat.*, *431*(7006), 338–342.

Mitter, N., Worrall, E. A., Robinson, K. E., Li, P., Jain, R. G., Taochy, C., & Fletcher, S. J., (2017). Clay nanosheets for topical delivery of RNAi for sustained protection against plant viruses. *Nat. Plants*, *3*(2), 1–10.

Molesini, B., Pii, Y., & Pandolfini, T., (2012). Fruit improvement using intragenesis and artificial microRNA. *Trends Biotechnol.*, *30*(2), 80–88.

Mujtaba, M., Wang, D., Carvalho, L. B., Oliveira, J. L., Espirito, S. P. A. D., Sharif, R., & Jogaiah, S., (2021). Nanocarrier-mediated delivery of miRNA, RNAi, and CRISPR-Cas for plant protection: Current trends and future directions. *ACS Agric. Sci. Technol.*, *1*(5), 417–435.

Murata, M., Nishimura, M., Murai, N., Haruta, M., Homma, S., & Itoh, Y., (2001). A transgenic apple callus showing reduced polyphenol oxidase activity and lower browning potential. *Biosci. Biotechnol. Biochem.*, *65*(2), 383–388.

Napoli, C., Lemieux, C., & Jorgensen, R., (1990). Introduction of a chimeric chalcone synthase gene into petunia results in reversible co-suppression of homologous genes in trans. *The Plant Cell*, *2*(4), 279–289.

Novel Food Information-Simplot Innate® Potato Events Gen2-W8, Gen2-X17, and Gen2-Y9. Completed safety assessments of novel foods including genetically modified (GM) foods. Government of Canada. https://www.canada.ca/en/health-canada/services/food-nutrition/genetically-modified-foods-other-novel-foods/approved-products/novel-food-information-simplot-innate-potato-gen2-w8-x17-y9.html (accessed on 7 February 2024).

Pathak, K., & Gogoi, B., (2016). RNA interference (RNAi): Application in crop improvement: A review. *Agric. Rev.*, *37*(3), 245–249.

Pridgeon, J. W., Zhao, L., Becnel, J. J., Strickman, D. A., Clark, G. G., & Linthicum, K. J., (2014). Topically applied AaeIAP1 double-stranded RNA kills female adults of *Aedes aegypti*. *J. Med. Entomol.*, *45*(3), 414–420.

Prusky, D., (2011). Reduction of the incidence of postharvest quality losses, and future prospects. *Food Secur.*, *3*(4), 463–474.

Qiao, H., Zhang, H., Wang, Z., & Shen, Y., (2021). Fig fruit ripening is regulated by the interaction between ethylene and abscisic acid. *J. Integr. Plant Biol.*, *63*(3), 553–569.

Qiu, J., Huang, T., Xu, J., Bi, C., Chen, C., & Zhou, M., (2012). βTubulins in *Gibberella zeae:* Their characterization and contribution to carbendazim resistance. *Pest Manage. Sci.*, *68*(8), 1191–1198.

Rai, A. C., Rai, A., Singh, M., & Singh, A. K., (2019). Biotechnological applications in post-harvest management of vegetable crops. *Applications in Post-Harvest Management of Vegetable Crops* (1st edn.). Taylor & Francis group.

Rajam, M. V., (2020). RNA silencing technology: A boon for crop improvement. *J. Biosci.*, *45*(1), 1–5.

Rajput, M., Choudhary, K., Kumar, M., Vivekanand, V., Chawade, A., Ortiz, R., & Pareek, N., (2021). RNA interference and CRISPR/Cas gene editing for crop improvement: Paradigm shift towards sustainable agriculture. *Plants*, *10*(9), 1914.

Romano, N., & Macino, G., (1992). Quelling: Transient inactivation of gene expression in *Neurospora crassa* by transformation with homologous sequences. *Mol. Microbiol.*, *6*(22), 3343–3353.

San, M. K., & Scott, J. G., (2016). The next generation of insecticides: DsRNA is stable as a foliar-applied insecticide. *Pest Manage. Sci.*, *72*(4), 801–809.

Senti, K. A., & Brennecke, J., (2010). The piRNA pathway: A fly's perspective on the guardian of the genome. *Trends Genet.*, *26*(12), 499–509.

Sharma, B., & Kumari, A., (2020). Role of post-harvest physiology in evolution of transgenics crops. *Plant Breeding – Current and Future Views* (pp. 1–21). IntechOpen.

Sharma, R. R., Singh, D., & Singh, R., (2009). Biological control of postharvest diseases of fruits and vegetables by microbial antagonists: A review. *Biol. Control*, *50*(3), 205–221.

Siritunga, D., & Sayre, R., (2004). Engineering cyanogen synthesis and turnover in cassava (*Manihot esculenta*). *Plant Mol. Biol.*, *56*(4), 661–669.

Stowe, E., & Dhingra, A., (2021). Development of the Arctic® apple. *Plant Breed. Rev.*, *44*, 273–296.

Tilahun, T., Bezie, Y., Kerisew, B., & Taye, M., (2021). The application of antisense technology for crop improvement: A review. *Cogent Food Agric.*, *7*(1), 1910157.

Van, D. K. A. R., Mur, L. A., Beld, M., Mol, J. N. M., & Stuitje, A. R., (1990). Flavonoid genes in petunia: Addition of a limited number of gene copies may lead to a suppression of gene expression. *Plant Cell*, *2*, 291–299.

Wang, M., & Jin, H., (2017). Spray-induced gene silencing: A powerful innovative strategy for crop protection. *Trends Microbiol.*, *25*(1), 4–6.

Wang, M., Thomas, N., & Jin, H., (2017). Cross-kingdom RNA trafficking and environmental RNAi for powerful innovative pre-and post-harvest plant protection. *Curr. Opin. Plant Biol.*, *38*, 133–141.

Wang, M., Weiberg, A., Lin, F. M., Thomma, B. P., Huang, H. D., & Jin, H., (2016). Bidirectional cross-kingdom RNAi and fungal uptake of external RNAs confer plant protection. *Nat. Plants*, *2*(10), 1–10.

Wang, Y., Juranek, S., Li, H., Sheng, G., Wardle, G. S., Tuschl, T., & Patel, D. J., (2009). Nucleation, propagation, and cleavage of target RNAs in ago silencing complexes. *Nat.*, *461*(7265), 754–761.

Wilson, R. C., & Doudna, J. A., (2013). Molecular mechanisms of RNA interference. *Annu. Rev. Biophys.*, *42*, 217–239.

Xiong, A. S., Yao, Q. H., Peng, R. H., Li, X., Han, P. L., & Fan, H. Q., (2005). Different effects on *ACC oxidase* gene silencing triggered by RNA interference in transgenic tomato. *Plant Cell Rep.*, *23*(9), 639–646.

Yahaya, S. M., & Mardiyya, A. Y., (2019). Review of post-harvest losses of fruits and vegetables. *Biomed. J. Sci. Tech. Res.*, *13*(4), 10192–10200.

Yahaya, S. M., (2005). *Contribution of Harvest to Pathogenic and Nonpathogenic Losses of Vegetables in Kano State, Nigeria*. MSC Thesis, Bayero University Kano, Nigeria.

Zhang, H., Kolb, F. A., Jaskiewicz, L., Westhof, E., & Filipowicz, W., (2004). Single processing center models for human Dicer and bacterial RNase III. *Cell*, *118*(1), 57–68.

CHAPTER 6

EMERGING TRENDS OF GENOME EDITING TECHNOLOGY FOR IMPROVEMENT OF POSTHARVEST QUALITY OF HORTICULTURAL CROPS

BHUVNESH KAPOOR,[1] VIKRANT,[2] and PANKAJ KUMAR[2]

[1]*University Institute of Biotechnology (UIBT), Chandigarh University, Gharuan, Mohali, Punjab, India*

[2]*Department of Biotechnology, Dr. Yashwant Singh Parmar University of Horticulture and Forestry, Nauni, Solan, Himachal Pradesh, India*

ABSTRACT

Horticultural crops are an integral part of the human diet and are important to ensure global food security as well as economics. For decades, horticulture production has heavily relied upon different breeding methods to develop new varieties, the use of chemical fertilizers and pesticides for disease and nutrient management, along with other engineering practices. In recent decades, multiple factors such as changes in global climatic conditions with increasing anthropogenic activities, the emergence of new diseases, and post-harvest losses are presenting threatening challenges to the current state of the global horticulture system. As a part of various horticultural crop improvement programs, a lot of emphasis has been paid to yield and quality attributes. Meanwhile, recent developments in genome editing technologies have been widely adopted in horticultural

Bioengineered Fruit and Vegetables: Trends in Postharvest Quality Improvement.
Mohammad Irfan and Pankaj Kumar (Eds.)

crops all over the world. Advancements of genome editing technologies comprise zinc finger nucleases (ZNFs), transcription activator-like effector nucleases (TALENs), clustered regularly interspaced palindromic repeats (CRISPR)/CRISPR-associated 9 (Cas9). The CRISPR-Cas system offers novel and effective solutions for horticulture crop improvements. This chapter aims to summarize the important aspects of genome editing tools in the improvement of post-harvest quality of horticultural crops with recent scientific progress.

6.1 INTRODUCTION

Horticultural crops serve a wide range of purposes for humans, including special dietary, medicinal, and aesthetic needs. Since their domestication, horticultural crops have been selectively bred to enhance their unique quality characteristics. These quality attributes play a crucial role in determining the nutritional and economic values of horticultural crops. Despite the vast diversity among horticultural crops, breeding objectives often overlap, focusing on enhancing nutrient content, bioavailability, shelf life, and disease resistance in consumable vegetables and fruits. Historically, both spontaneous and induced mutations have been utilized to create genetic resources with diverse properties for breeding purposes. Scientists are now exploring research strategies to induce precise mutations in specific areas, given the rarity and unpredictability of such changes. While these technologies have shown some success in improving crop quality, they are known to be time-consuming, labor-intensive, and subject to controversy due to concerns about GMO safety. The quality attributes of horticultural crops are influenced by factors such as genetic profiles, geo-climatic conditions, soil properties, and post-harvest handling. Post-harvest, perishable crops continue to ripen due to active respiration, highlighting the importance of preserving and selecting the best genotypes using advanced biotechnological methods to maintain or enhance nutritional quality and flavors from harvest to consumption. Crop breeders are constantly exploring alternative and combined techniques to ensure optimal nutritional distribution, storage suitability, and quality attributes like color, texture, size, and flavor, aiming for increased market value in targeted horticultural crops.

Therefore, in comparison to the levels of accuracy and efficiency of conventional breeding methods, genome editing technological tools such as mega-nucleases (MNs), ZFNs, TALENs, and especially CRISPR-Cas have been proven more reliable and adaptive in the improvement of horticultural crops (Khan et al., 2021; Karmakar et al., 2022). CRISPR has proved to be a very adaptable technique with more practicality, targeting ability, cost-effectiveness, and precision when compared to ZFNs and TALENs, which are more difficult, time-consuming, and inefficient in the application (Ahmad et al., 2021; Mushtaq et al., 2021). These techniques involve the interplay of promoter and transcription factors (TFs) in defining the sequence-specific DNA binding (DBD). In the case of MNs, ZFNs, and TALENs, for example, double-strand breaks are generated by nucleases with protein-DNA interaction, whereas CRISPR uses RNA-dependent DNA base pairing. Originally, CRISPR-Cas9 represents an active immune system of bacteria *Streptococcus pyogenes* against the viruses that cleave the foreign DNA in RNA sequences dependent way. The CRISPR-Cas9 system is composed of a Cas9 endonuclease that is directed by an RNA molecule called single guide RNA (sgRNA) to a protospacer region representing a specific genomic site forming a double-strand break (DSBs) (Jiang & Doudna, 2015; Jiang & Doudna, 2017). The resulting DSBs further activate the DNA repair mechanisms either by non-homologous end joining (NHEJ) or homologous recombination (HR). Usually, when DNA repair occurs at coding regions, it leads to the formation of insertions or deletions (InDel) mutations that further result in gene knockouts. On the other hand, if a similarity sequence is available in the vicinity of the DSBs, gene editing by HR is simple. Hence, using different DNA repair modes, CRISPR-Cas9 can achieve many such site-specific alterations. This technology enables modern crop improvement programs to easily design site-specific genomic sites to re-engineer the gene architecture. However, the only concern with the existing CRISPR modules is the Cas9 off-target cleavage that can be prevented with the use of off-targeting tools such as GT-scan (O'Brien & Bailey, 2014), CRISPR-Plant (Xie et al., 2014), CRISPR direct (Naito et al., 2015) effectively. Until recently, several genome-edited crops with increased economic value have been introduced not just by multinational corporations but also by smaller laboratories in universities, implying minimal entry barriers in terms of people and finance. Combining this fast-evolving genome-editing technique with traditional breeding will considerably expand the potential and commercialization of genome-edited crops. Among horticultural crops, vegetables are the preferred group for genome editing comprising more than 72% of studies followed by floricultural and medicinal plants. Meanwhile, tomato is the crop that received much attention among other horticultural crops accounting for ~42% of genome-editing studies followed by potato with ~13% (Xu et al., 2019). With a wide range of possible applications in fundamental research, disease diagnosis, and crop improvement, genome editing technology, specifically CRISPR, is developing at a rapid rate.

6.2 IMPROVEMENT IN NUTRIENT AND QUALITY PROFILES OF HORTICULTURAL CROPS

Horticultural crops are a source of a substantial amount of nutrients such as vitamins, essential amino acids, organic acids, antioxidants, minerals, dietary fibers, and a plethora of metabolites having therapeutic values. The use of genome editing technologies in crop improvement has increased substantially during the last decade. This is in part because technological advances in genome editing have enabled improved targeting as well as a variety of simple commercial applications made possible by the regulatory framework for genome editing crops. Horticultural crops' productivity, ripening, appearance, flavor, and storage quality have all been improved using genome editing technologies. For example, in tomatoes, the important genes and TFs associated with the ripening process have been mapped and manipulated using genome editing technology such as genes ripening inhibitor (RIN), non-ripening (NOR), colorless non-ripening (CNR) and fruitful (FUL); the DNA demethylase *SlDML2*; and the flowering repressor *SP5G* (Kumar et al., 2016). Nowadays, genome editing technologies, particularly CRISPR/Cas, are widely utilized to alter the nutritional profiles and quality features of horticulture crops, especially in vegetable crops like tomatoes (*Solanum lycopersicum*) (Tashkandi et al., 2018; Tran et al., 2021), potato (*Solanum tuberosum*) (Tuncel & Qi, 2022), grape (*Vitis vinifera*) (Ren et al., 2016), cabbage (*Brassica oleracea* var. capitata) (Stajič et al., 2019; Jeong et al., 2019), cucumber (*Cucumis sativus*) (Hu et al., 2017), apple (*Malus domestica*) (Nishitani et al., 2016), *Petunia* spp. (Zhang et al., 2016), and *Dendrobium officinale* (Kui et al., 2017) (Table 6.1).

TABLE 6.1 List of Genome-Edited Crops

Crop Species	Editing Tool	Gene Target	Trait Target	References
Solanum lycopersicum	CRISPR-Cas9	RIN	Fruit ripening	Ito et al. (2015)
Solanum lycopersicum	CRISPR-Cas9	ANT1 promoter	Purple fruit color	Čermák et al. (2015)
Vitis vinifera	CRISPR-Cas9	MLO-7	Fire blight	Malnoy et al. (2016)
Malus domestica		DIPM-1, DIPM-2, and DIPM-4		
Vitis vinifera	CRISPR-Cas9	L-idonate dehydrogenase gene IdnDH	Tartaric acid biosynthesis	Ren et al. (2016)
Solanum tuberosum	TALENs	SBE1 and StvacINV22	Sugar metabolism	Ma et al. (2017)
Solanum lycopersicum	CRISPR-Cas9	*DML2*	Fruit ripening	Lang et al. (2017)
Solanum lycopersicum	CRISPR-Cas9	ORRM4	Fruit ripening	Yang et al. (2017)
Dendrobium officinale	CRISPR-Cas9	C3H, C4H, 4CL, CCR, and IRX	Lignocellulose biosynthesis	Kui et al. (2017)
Solanum tuberosum	CRISPR-Cas9	GBSS	Starch development and tuber quality.	Andersson et al. (2017)
Solanum lycopersicum	CRISPR-Cas9	PSY	Yellow fruit color	Hayut et al. (2017)
Ipomoea nil	CRISPR-Cas9	*InDFR-B*	Anthocyanin biosynthesis decreased and white flowers.	Watanabe et al. (2017)
Solanum lycopersicum	CRISPR-Cas9	SGR1, LCY-E, Beta lycopene cyclase, LCY-B1, LCY-B2	5.1-fold increase in β-lycopene content.	Li et al. (2018)
Solanum lycopersicum	CRISPR-Cas9	Self-pruning, ovate, fruit weight 2.2, lycopene beta cyclase	Fruit size, number, lycopene content.	Zsögön et al. (2018)
Fragaria vesca, Fragaria ananassa	CRISPR-Cas9	MYB10, CHS	Anthocyanin biosynthesis decreased.	Xing et al. (2018)
Solanum lycopersicum	CRISPR-Cas9	lncRNA1459	Repressed fruit ripening, lycopene, ethylene, and carotenoid biosynthesis decreased.	Li et al. (2018)
Lactuca sativa	CRISPR-Cas9	uORF of LsGGP2	Increased ascorbate content.	Zhang et al. (2018)

TABLE 6.1 *(Continued)*

Crop Species	Editing Tool	Gene Target	Trait Target	References
Daucus carota	CRISPR-Cas9	DcPDS, DcMYB113-like	Albino plants and purple color depigmented plants.	Xu et al. (2019)
Solanum lycopersicum	CRISPR-Cas9	Pectate lyase (PL), polygalacturonase 2a (PG2a), and beta-galactanase (TBG4)	Cell wall, fruit color, firmness, and weight.	Wang et al. (2019)
Solanum lycopersicum	CRISPR-Cas9	APETALA2a (AP2a), non-ripening (NOR) and fruitful (FUL1/TDR4 and FUL2/MBP7)	Fruit development and ripening.	Wang et al. (2019)
Ipomoea batatas	CRISPR-Cas9	GBSS1, SBEII	Improvement of starch quality	Wang et al. (2019)
Citrullus lanatus	CRISPR-Cas9	*AGA2*	Reduced soluble sugar content; increased raffinose content.	Guo et al. (2019)
Solanum tuberosum	CRISPR-Cas9	Sbe1, Sbe2	Amylose starch with no detectable starch.	Zhao et al. (2021)
Capsicum frutescens	CRISPR-Cas9	*CaERF28*	Enhanced resistance to *Colletotrichum truncatum*.	Mishra et al. (2021)
Solanum tuberosum	CRISPR-Cas9	*GBSS1*	Amylose-free starch in potato tubers.	Toinga-Villafuerte et al. (2022)

In plants, the photosynthetic tissues contain different carotenoids that are important for photoreception and photoprotection against harmful UV rays, which further increase the nutritional content of important metabolites (Gautam et al., 2021). Various fruit and vegetable crops have elevated levels of secondary metabolites with therapeutic effects on human health, such as anthocyanins, which have been shown to prevent certain malignancies, age-related degenerative disorders, and cardiovascular disease. Furthermore, anthocyanins are associated with the color development of fruits and vegetables when combined with carotenoids and chlorophylls. The distribution of anthocyanin pigment in fruits, vegetables, and flowers is a significant predictor of maturity and quality. Model plant and horticulture crop studies have identified the conserved anthocyanin biosynthesis and regulatory mechanism. One such carotenoid in tomatoes is lycopene which is targeted for quality improvement using multiplex CRISPR/Cas9 by targeting five genes involved in lycopene biosynthesis such as stay green 1 (*SGR1*), l beta-lycopene cyclase (*Blc),* lycopene β-cyclase 1 (*LCY-B1*), lycopene β-cyclase 2 (*LCY-B2*) and lycopene ε-cyclase (*LCY-E*). Further, it was evident that by targeting *SGR1* alone, lycopene content was increased to 5.1-fold suggesting more part of single gene *SGR1* in comparison to the other four genes targeted for carotenoid biosynthesis (Li et al., 2018).

Tomato has shown early success in genome targeting for HR compared to other crops. For instance, the yellow-colored flavonoid naringenin chalcone (NarCh) amount regulates peel color in tomatoes, and its absence results in tomatoes having a pink color. Research revealed that the pink-colored peel is produced in recessive mutants, with the gene responsible for this function being *SlMYB12*, which encodes the *R2R3-MYB* transcription factor (Adato et al., 2009; Li et al., 2013). Targeting the gene *SlMYB12* with CRISPR-Cas9 led to the development of pink tomatoes that are more desirable in Asian markets (Deng et al., 2018). In another study, ZNF-mediated editing of the tomato gene *leafy cotyledon 1-like 4 (L1L4)* was found to be associated with a reduced content of oxalic acid, which acts as an anti-nutritional product and is involved in the accumulation of other metabolites such as citric acid, fructose, β-carotene, polyphenols, and antioxidants (Gago et al., 2017). Among other developments of CRISPR-Cas9 adaptations in horticultural crops processing was the enhancement of wine flavor, where the gene L-donate dehydrogenase (IdnDH) in grapes (*Vitis* vinifera) was knocked out to increase the accumulation levels of tartaric acid (TA) by up to 36%, playing a significant role in wine flavor development (Ren et al., 2016).

6.3 IMPROVED SHELF-LIFE

Central to important characteristics that impact the quality of fruits, flowers, and vegetables is the shelf life. Even after the food is harvested, it continues to respire (take up oxygen and create carbon dioxide), and substantial levels of ethylene can be released at higher temperatures, particularly in the case of "climacteric fruits" (Shipman et al., 2021; Irfan et al., 2023). The previous debate largely focused on the fruit development and harvesting stages, and when it comes to the post-harvest process, one can only discuss logistics and other strategies such as cold storage and ecologically friendly conditions. Global climate change and world population expansion are the two key causes driving attempts to reduce horticultural product postharvest loss and waste management for efficient use of resources. With the advancement of current technology, we can extend the shelf life of horticulture produce by employing approaches that can minimize such a problem and give a better solution. Using targeted genome editing technology significant genes can be targeted that are known to associate with post-harvest characteristics such as the production of plant hormone ethylene which plays a critical role in floral withering and fruit ripening.

Carnation and petunia flower senescence are delayed as a result of suppressing ethylene synthesis and delaying ethylene signal transduction. Thus, by targeting the key genes involved in ethylene biosynthesis, new horticulture crops with an enhanced shelf life may be produced through genome-editing technology. Managing all biological processes after harvesting demands optimizing storage and handling conditions, which may vary depending on product type or variety, as well as understanding how the preharvest environment influences fruit biology following harvest (Kader, 2011). However, post-harvest environmental factors such as temperature, humidity, and oxygen-to-carbon dioxide ratio need to be regulated to slow down the maturation and senescence processes in contemplation to increase the shelf life while maintaining product quality (Kader & Saltveit, 2002; Kays, 2010). At low temperatures, respiration is inhibited, extending produce shelf life as well as limiting water loss, shrinkage, and pathogen emergence caused by physiological diseases and injuries during post-harvesting. For example, browning occurs in freshly cut or bruised apples, potatoes, spinach, and lettuce due to high temperatures, cold injuries, and exposure to insufficient oxygen and carbon dioxide levels (Liang et al., 2020). Meanwhile, three major enzymes are involved in enzymatic browning: peroxidase (POD), polyphenol oxidase (PPO), and phenylalanine ammonia-lyase (PAL) (Li et al., 2021; Shipman et al., 2021).

Adaptation of CRISPR-mediated genome editing in the management of post-harvest challenges has been demonstrated in horticultural crops such as apples, tomatoes, potatoes, mushrooms, tobacco, and grapes (Andersson et al., 2017; Osakabe et al., 2018; Kaur et al., 2020). Using a similar approach, browning after wounding was reduced by knocking off the *PPO* genes, which is a reasonably straightforward target and an established successful technique for decreasing postharvest waste (González et al., 2020). Innate potato® and Arctic Apple® are the non-browning, commercially available products (Richael, 2021; Stowe & Dhingra, 2021). A similar strategy was found effective in non-browning mushrooms developed by CRISPR-Cas9 (Waltz, 2016). Besides the above-defined targets, many postharvest phenotypes impart the quality and shelf-life of crops and are poorly understood at the molecular level. Such traits involve the association of multi-allelic genes and how their expression is altered by the pre- and post-harvest environmental factors. Furthermore, post-harvest chilling damage in horticultural crops includes tissue browning followed by rotting fungal infections, which exhibit a wide range of symptoms across species and contribute to postharvest losses (PHLs) (Rai et al., 2022).

Multiple enzymes involved in cell wall softening during the ripening process, such as Polygalacturonate (PG), pectin methyl esterase (PME), pectate lyase (PL), and β-galactosidase (β-Gal), may also serve as target genes for genome editing approaches (Payasi et al., 2009; Friedman, 2020). For instance, the expression of PG is known to negatively impact the firmness and shelf life of strawberries and tomatoes (Pose et al., 2013). However, extending shelf life can be achieved by enhancing anthocyanin biosynthesis in tomatoes and other fruits, thereby reducing postharvest losses and waste while enhancing the fruit's appeal through healthy characteristics and vibrant color. The transcription factor MADS-box RIN is recognized for its central role in tomato fruit ripening (Kumar et al., 2016; Irfan et al., 2016, 2022) and has been a target for CRISPR-Cas9 in various cultivars like Ailsa Craig, Mamirio, and Golden Bell to prolong shelf life by slowing down the ripening process (Ito et al., 2015; Jung et al., 2018). RIN-deficient tomato mutants ripened at a slower pace and exhibited reduced red pigmentation due to inhibited lycopene production. In another successful application, CRISPR-Cas9 was utilized to suppress ethylene production and lycopene accumulation by targeting long non-coding RNA1459 (lncRNA1459), resulting in delayed tomato maturation (Li et al., 2018). Tomato lines with extended shelf life and improved storage performance were achieved by employing the HR-directed repair mechanism via CRISPR-Cas9 to target the *ALC* (Alcobaca) gene, which harbors a non-synonymous allelic substitution of the non-ripening (Nor) gene (Yu et al., 2017).

6.4 IMPROVED RESISTANCE TO POST-HARVEST PATHOGENS

Another major concern that affects the shelf life and quality of post-harvest horticultural produce from field to processing or storage are bacteria, molds, and various fungal agents. The active respiration states of harvested horticulture produce along with various abiotic factors such as relative humidity, temperature, and oxygen availability in storage places along light conditions effectively contribute to the receptivity to post-harvest pathogenic infections (Gouda et al., 2020). Among post-harvest infectious agents, the majority of pathogens cause fruit and vegetables to rot at a very rapid pace contributing to major post-harvest losses (Wenneker et al., 2020). At the post-harvesting management level, several agricultural technologies are generally used to deal with such losses such as the application of insecticides and pesticides, use of better packaging materials in transportation, storage at ultra-low temperatures, and maintaining overall aseptic conditions during the processing step. But it is well evident from the frequent episodes of post-harvest losses that it's difficult to maintain all those management techniques as well as their resultant cost. Therefore, keeping in mind the economics and environmental risks of such applications, it's more convenient to use modern biotechnological techniques such as genetic engineering (GE) and genome editing to deal with post-harvest pathogens. Currently, genome editing has been a method of choice to develop such varieties of fruits and vegetables with increased capacity for post-harvest biotic stress resistance. However, the prerequisite for the implementation of genome editing in any crop requires a well-defined molecular roadmap and target genes. For example, in citrus, *Xanthomonas citri* subsp. *Citri* causes citrus canker that further results in severe economic damage and disease showing its emergence from pre- to post-harvest phases. Meanwhile, using the CRISPR-Cas9 approach, a gene knockout of *LATERAL ORGAN BOUNDARIES (CsLOB1)* TFs involved in disease manifestation was found effective in alleviating the overall levels of citrus canker (Wang et al., 2019). Similar genes *DLOB9* and *DLOB10* targeted by CRISPR-Cas9 in Duncan grapefruit (*Citrus paradisi*) resulted in increased resistance to citrus canker (Jia et al., 2017). Furthermore, in tomatoes and peas, the development of powdery mildew long-lasting resistance using genome editing was achieved by the loss of function of *MLO* genes, which is generally caused by many *Erysiphales* fungi.

Another major post-harvest disease in tomatoes is caused by *Botrytis cinerea*, which accounts for huge losses. However, using CRISPR-Cas9 to target the *SlMAPK3* gene controlling susceptibility to *Botrytis cinerea* was knocked out to produce tomato mutants with improved post-harvest resistance. Development of gain-of-function mutants by targeting *MLO* alleles using genome editing tools demonstrated the potential for horticultural crop improvement while exhibiting heritable disease resistance mechanisms such as powdery mildew resistance. For example, in Grape (*Vitis vinifera*), CRISPR targeted manipulation of *MLO* genes such as *VvMLO3* and *VvMLO4* showed increased resistance to fungal disease caused by *Erysiphe necator* demonstrating successful use of CRISPR-induced targeted mutations (Wan et al., 2020). Further, plant disease resistance genes of post-harvest pathogens such as *Botrytis cinerea* and *Xanthomonas citri* have also been modified in other horticultural crops like apples, strawberries, and grapes (Tyagi et al., 2021; Kumari et al., 2022). end end

6.5 POSSIBLE IMPLICATIONS OF GENOME EDITING IN HORTICULTURE

Genetic variations can occur naturally as a consequence of spontaneous mutations or be introduced by plant breeders through chemical and physical procedures. Genome editing, on the other hand, may be thought of as sequence-specific biological mutagenesis that outperforms previous techniques concerning variety, efficiency, and specificity. CRISPR-mediated genome alterations, for example, can cause small insertions/deletions, large-fragment deletions, gene replacement, and precise nucleotide replacements in target locations. Furthermore, this technology is constantly evolving as the endonuclease Cpf151, as well as newly discovered Cas9 variations, might find alternative PAM sequences, increasing the number of genome-wide locations that can be modified. Another important aspect of genome-edited crops is their treatment as a different system than GMOs in many countries including the USA and Japan. This difference arises because genome-edited crops may be free of exogenous DNA, unlike GMOs. As a result, utilizing genome editing technology, desirable characteristics may be promptly introduced into commercial or heirloom varieties without losing other features, and new lines with specified improvements will be produced. In addition, the wild species also provide very distinctive and plentiful crop improvement features that could be manipulated by genome editing technologies. For example, a wild relative of tomato *Solanum pimpinellifolium* has now been recognized as an important asset for the introgression of many important traits to commercial cultivars such as vital genes regulating plant architecture, fruit development, and resistance to biotic and abiotic stresses.

The targeted mutations in cis-regulatory genes, which might provide quantitative variation for breeding selection, are another essential component of genome editing. Because gene activity may be influenced not only by modifying the DNA sequence of the cis-regulatory region but also by its epigenetic state. Using a similar approach, research can effectively target the genes of interest and manipulate their epigenetic status by editing the regulatory genes with the CRISPR-Cas9 tool. In one such study, a CRISPR-Cas9-based demethylase tool was designed with deactivated Cas9 (dCas9) fused with the catalytic domain of the *ten-eleven translocation 1 (TET1CD)* and found effective in demethylating the target genes with the help of gRNA molecules, leading to transcriptional upregulation (Choudhury et al., 2016). The resultant epigenetic mutants may then be crossed with different wild-type mutants for the development of epigenetic recombinant inbred lines (epiRILs) and produce populations with genetically similar individuals but epigenetically distinct. Therefore, genome editing tools, especially CRISPR-Cas9, could be an important asset to generate novel alleles and epialleles, providing essential sources of phenotypic diversity for crop development.

6.6 CONCLUSIONS

The prerequisite for the effective application of genome editing tools requires high-fidelity genomic information and well-characterized gene functions. The majority of the features mentioned above were discovered through mutant studies in model plants. As a result, the generated information can only indicate possible gene targets. For many decades, a lack of understanding of the molecular characteristics of many horticultural crops hampered breeding effectiveness;

nevertheless, current technical breakthroughs in the field of omics sciences have supplemented well with modern genome editing technologies. For example, genome sequencing of more than 181 horticultural crops had already been completed including many fruits, vegetables, and medicinal and ornamental crops. In addition, the availability of transcriptome profiling of other horticultural crops is also providing important information relevant to crop improvement programs. This enormous amount of genomic data will undoubtedly aid in unraveling the important molecular regulation of critical features in various species of crop plants and, as a consequence, identifying vital genes that could be targeted by genome editing. With the help of available reference genomes along with transcriptome data of these horticultural crops, a wide spectrum of target genes could be used to design the next generation of crops with improved genome editing strategies, especially the CRISPR-Cas system. Meanwhile, challenges for the widespread application of genome editing in horticultural crops include the scarcity of inadequately described gene targets as well as recalcitrant tissue culture regeneration and transformation behaviors. With well-defined gene targets and standardized transformation protocols, the latest genome-editing technologies, notably CRISPR/Cas, promise to be more efficient and accurate in customizing horticultural crops. Through mutant breeding, these technologies have the potential to create new types. Another advantage of using genome editing approaches over transgenic strategies is the development of new traits without introducing any foreign genes. As a result, crop improvements achieved by genome editing are regulated differently than transgenic crops, making them more acceptable in places where public opinion is unfavorable. The multiple benefits provided by genome editing technology, including site-specific targeting, efficiency, and labor savings, will undoubtedly make them suitable for any horticulture crop improvement program. Meanwhile, a crucial factor to address remains the public's comprehension and acceptance of new horticultural crops developed through the utilization of genome editing tools. To do so at the policy level, a clear distinction should be drawn to categorize GMOs and genome-edited crops as separate innovations. Finally, the comprehensive integration of traditional breeding practices with genome editing approaches will be more fruitful in all domains of crop improvement.

KEYWORDS

- **CRISPR-Cas9**
- **crop improvement**
- **genome editing**
- **horticulture crops**
- **molecular characteristics**
- **post-harvest losses**
- **shelf-life**

REFERENCES

Adato, A., Mandel, T., Mintz-Oron, S., Venger, I., Levy, D., Yativ, M., Domínguez, E., et al., (2009). Fruit-surface flavonoid accumulation in tomato is controlled by a SlMYB12-regulated transcriptional network. *PLoS Genetics, 5*, e1000777.

Ahmad, S., Tang, L., Shahzad, R., Mawia, A. M., Rao, G. S., Jamil, S., Wei, C., et al., (2021). CRISPR-based crop improvements: A way forward to achieve zero hunger. *Journal of Agricultural and Food Chemistry, 69*, 8307–8323.

Andersson, M., Turesson, H., Nicolia, A., Fält, A. S., Samuelsson, M., & Hofvander, P., (2017). Efficient targeted multiallelic mutagenesis in tetraploid potato (*Solanum tuberosum*) by transient CRISPR-Cas9 expression in protoplasts. *Plant Cell Reports, 36*, 117–128.

Čermák, T., Baltes, N. J., Čegan, R., Zhang, Y., & Voytas, D. F., (2015). High-frequency, precise modification of the tomato genome. *Genome Biology, 16*, 1–15.

Choudhury, S. R., Cui, Y., Lubecka, K., Stefanska, B., & Irudayaraj, J., (2016). CRISPR-dCas9 mediated TET1 targeting for selective DNA demethylation at BRCA1 promoter. *Oncotarget, 7*, 46545.

Deng, L., Wang, H., Sun, C., Li, Q., Jiang, H., & Du, M., (2018). Efficient generation of pink-fruited tomatoes using CRISPR/Cas9 system. *J. Genet. Genomics, 45*, 51–54. doi: 10.1016/j.jgg.2017.10.002.

Friedman, H., (2019). Candidate genes to extend fleshy fruit shelf life. *Plant Breeding Reviews, 43*, 61–94.

Gago, C., Drosou, V., Paschalidis, K., Guerreiro, A., Miguel, G., & Antunes, D., (2017). Targeted gene disruption coupled with metabolic screen approach to uncover the leafy cotyledon1-like4 (L1L4) function in tomato fruit metabolism. *Plant Cell Rep., 36*, 1065–1082. doi: http://10.1007/s00299-017- 2137-9.

Gautam, V., Rana, G. S., Kumar, P., & Watpade, S., (2021). Development of carotenoids rich grains by genome engineering. In: *Genome Engineering for Crop Improvement.* Wiley Library Publications. doi: 10.1002/9781119672425.ch5.

González, M. N., Massa, G. A., Andersson, M., Turesson, H., Olsson, N., Fält, A. S., Storani, L., et al., (2020). Reduced enzymatic browning in potato tubers by specific editing of a polyphenol oxidase gene via ribonucleoprotein complexes delivery of the CRISPR/Cas9 system. *Frontiers in Plant Science, 10*, 1649.

Gouda, M. B., Zhang, C., Wang, J., Peng, S. J., Chen, Y. R., Luo, H. B., & Yu, L. J., (2020). ROS and MAPK cascades in the post-harvest senescence of horticultural products. *J. Proteomics Bioinform, 13*, 1–7.

Guo, S., Zhao, S., Sun, H., Wang, X., Wu, S., Lin, T., Ren, Y., et al., (2019). Resequencing of 414 cultivated and wild watermelon accessions identifies selection for fruit quality traits. *Nature Genetics, 51*, 1616–1623.

Hu, B., Li, D., Liu, X., Qi, J., Gao, D., Zhao, S., Huang, S., et al., (2017). Engineering non-transgenic gynoecious cucumber using an improved transformation protocol and optimized CRISPR/Cas9 system. *Molecular Plant, 10*, 1575–1578.

Irfan, M., Ghosh, S., Meli, V. S., Kumar, A., Kumar, V., Chakraborty, N., Chakraborty, S., & Datta, A., (2016). Fruit ripening regulation of α-mannosidase expression by the MADS box transcription factor RIPENING INHIBITOR and ethylene. *Frontiers in Plant Science*, *7*.

Irfan, M., Kumar, P., Ahmad, M. F., & Siddiqui, M. W., (2023). Biotechnological interventions in reducing losses of tropical fruits and vegetables. *Current Opinion in Biotechnology, 79*, 102850. doi.org/10.1016/j.copbio.2022.102850.

Irfan, M., Kumar, P., Kumar, V., & Datta, A., (2022). Fruit ripening specific expression of β-D-N-acetylhexosaminidase (β-Hex) gene in tomato is transcriptionally regulated by ethylene response factor SlERF.E4. *Plant Science, 323*, 111380.

Ito, Y., Nishizawa-Yokoi, A., Endo, M., Mikami, M., & Toki, S., (2015). CRISPR/Cas9-mediated mutagenesis of the RIN locus that regulates tomato fruit ripening. *Biochem. Biophys. Res. Commun., 467*, 76–82.

Jeong, S. Y., Ahn, H., Ryu, J., Oh, Y., Sivanandhan, G., Won, K. H., Park, Y. D., et al., (2019). Generation of early-flowering Chinese cabbage (*Brassica Rapa* spp. pekinensis) through CRISPR/Cas9-mediated genome editing. *Plant Biotechnology Reports, 13*, 491–499.

Jia, H., Zhang, Y., Orbović, V., Xu, J., White, F. F., Jones, J. B., & Wang, N., (2017). Genome editing of the disease susceptibility gene CsLOB1 in *Citrus* confers resistance to *Citrus* canker. *Plant Biotechnology Journal, 15*, 817–823.

Jiang, F., & Doudna, J. A., (2015). The structural biology of CRISPR-Cas systems. *Current Opinion in Structural Biology, 30*, 100–111.

Jiang, F., & Doudna, J. A., (2017). CRISPR-Cas9 structures and mechanisms. *Annu. Rev. Biophys., 46*, 505–529.

Jung, Y. J., Lee, G. J., Bae, S., & Kang, K. K., (2018). Reduced ethylene production in tomato fruits upon CRSPR/Cas9-mediated leMADS-RIN mutagenesis. *Horticultural Science and Technology,* 396–405. https://doi.org/10.12972/kjhst.20180039.

Kader, A. A., & Saltveit, M. E., (2002). In: Bartz, J. A., & Brecht, J. K., (eds.), *Postharvest Physiology and Pathology of Vegetables* (Vol. 123, pp. 1–6). CRC Press.

Kader, A. A., (2011). In: Kader, A. A., (eds.), *Postharvest Technology of Horticultural Crops* (Vol. 3529, pp. 39–48). University of California Agriculture and Natural.

Karmakar, S., Das, P., Panda, D., Xie, K., Baig, M. J., & Molla, K. A., (2022). A detailed landscape of CRISPR-Cas-mediated plant disease and pest management. *Plant Science, 323*, 111376.

Kaur, N., Awasthi, P., & Tiwari, S., (2020). Fruit crops improvement using CRISPR/Cas9 system. In: *Genome Engineering via CRISPR-Cas9 System* (pp. 131–145). Academic Press.

Kays, S. J., (2010). Quality maintenance of fresh produce. *Acta Hortic., 875*, 27–31.

Khan, Z., Saboor, T., Ashfaq, M., Saddique, A., & Khanum, P., (2021). Disease resistance in crops through CRISPR/Cas. In: *CRISPR Crops* (pp. 151–175). Springer, Singapore.

Kui, L., Chen, H., Zhang, W., He, S., Xiong, Z., Zhang, Y., Yan, L., et al., (2017). Building a genetic manipulation tool box for orchid biology: Identification of constitutive promoters and application of CRISPR/Cas9 in the orchid, *Dendrobium officinale*. *Frontiers in Plant Science, 7*, 2036.

Kumar, V., Irfan, M., Ghosh, S., Chakraborty, N., Chakraborty, S., & Datta, A., (2016). Fruit ripening mutants reveal cell metabolism and redox state during ripening. *Protoplasma, 253*, 581–594. https://doi.org/10.1007/s00709-015-0836-z.

Kumari, C., Sharma, M., Kumar, V., Sharma, R., Kumar, V., Sharma, P., Kumar, P., & Irfan, M., (2022). Genome editing technology for genetic amelioration of fruits and vegetables for alleviating post-harvest loss. *Bioengineering, 9*, 176.

Lang, Z., Wang, Y., Tang, K., Tang, D., Datsenka, T., Cheng, J., Zhang, Y., Handa, A. K., & Zhu, J. K., (2017). Critical roles of DNA demethylation in the activation of ripening-induced genes and inhibition of ripening-repressed genes in tomato fruit. *Proceedings of the National Academy of Sciences, 114*, 4511–4519.

Li, S., Chen, K., & Grierson, D., (2021). Molecular and hormonal mechanisms regulating fleshy fruit ripening. *Cells, 10*, 1136.

Li, X. W., Li, J. W., Zhai, Y., Zhao, Y., Zhao, X., Zhang, H. J., Su, L. T., Wang, Y., & Wang, Q. Y., (2013). A R2R3-MYB transcription factor, GmMYB12B2, affects the expression levels of flavonoid biosynthesis genes encoding key enzymes in transgenic *Arabidopsis* plants. *Gene, 532*, 72–79.

Li, X., Wang, Y., Chen, S., Tian, H., Fu, D., Zhu, B., Luo, Y., & Zhu, H., (2018). Lycopene is enriched in tomato fruit by CRISPR/Cas9-mediated multiplex genome editing. *Frontiers in Plant Science, 9*, 559.

Liang, S. M., Kuang, J. F., Ji, S. J., Chen, Q. F., Deng, W., Min, T., Shan, W., Chen, J. Y., & Lu, W. J., (2020). The membrane lipid metabolism in horticultural products suffering chilling injury. *Food Quality and Safety, 4*, 9–14.

Ma, J., Xiang, H., Donnelly, D. J., Meng, F. R., Xu, H., Durnford, D., & Li, X. Q., (2017). Genome editing in potato plants by agrobacterium-mediated transient expression of transcription activator-like effector nucleases. *Plant Biotechnology Reports, 11*, 249–258.

Malnoy, M., Viola, R., Jung, M. H., Koo, O. J., Kim, S., Kim, J. S., Velasco, R., & Nagamangala, K. C., (2016). DNA-free genetically edited grapevine and apple protoplast using CRISPR/Cas9 ribonucleoproteins. *Frontiers in Plant Science, 7*, 1904.

Mishra, R., Mohanty, J. N., Mahanty, B., & Joshi, R. K., (2021). A single transcript CRISPR/Cas9 mediated mutagenesis of CaERF28 confers anthracnose resistance in chilli pepper (*Capsicum annuum* L.). *Planta, 254*, 1–17.

Mushtaq, M., Ahmad Dar, A., Skalicky, M., Tyagi, A., Bhagat, N., Basu, U., Bhat, B. A., et al., (2021). CRISPR-based genome editing tools: Insights into technological breakthroughs and future challenges. *Genes, 12*, 797.

Naito, Y., Hino, K., Bono, H., & Ui-Tei, K., (2015). CRISPR direct: Software for designing CRISPR/Cas guide RNA with reduced off-target sites. *Bioinformatics, 31*, 1120–1123.

Nishitani, C., Hirai, N., Komori, S., Wada, M., Okada, K., Osakabe, K., Yamamoto, T., & Osakabe, Y., (2016). Efficient genome editing in apple using a CRISPR/Cas9 system. *Scientific Reports, 6*, 1–8.

O'Brien, A., & Bailey, T. L., (2014). GT-Scan: Identifying unique genomic targets. *Bioinformatics, 30*, 2673–2675.

Osakabe, Y., Liang, Z., Ren, C., Nishitani, C., Osakabe, K., Wada, M., Komori, S., et al., (2018). CRISPR–Cas9-mediated genome editing in apple and grapevine. *Nature Protocols, 13*, 2844–2863.

Payasi, A., Mishra, N. N., Chaves, A. L. S., & Singh, R., (2009). Biochemistry of fruit softening: An overview. *Physiology and Molecular Biology of Plants*, 103–113.

Posé, S., Paniagua, C., Cifuentes, M., Blanco-Portales, R., Quesada, M. A., & Mercado, J. A., (2013). Insights into the effects of polygalacturonase FaPG1 gene silencing on pectin matrix disassembly, enhanced tissue integrity, and firmness in ripe strawberry fruits. *Journal of Experimental Botany, 64*, 3803–3815.

Rai, A., Kumari, K., & Vashistha, P., (2022). Umbrella review on chilling injuries: Post-harvest issue, cause, and treatment in tomato. *Scientia Horticulturae, 293*, 110710.

Ren, C., Liu, X., Zhang, Z., Wang, Y., Duan, W., & Li, S., (2016). CRISPR/Cas9- mediated efficient targeted mutagenesis in chardonnay (*Vitis vinifera* L.). *Sci. Rep., 6*, 32289, doi: 10.1038/srep32289.

Richael, C. M., (2021). Development of the genetically modified innate® potato. *Plant Breeding Reviews, 44*, 57–78.

Shipman, E. N., Yu, J., Zhou, J., Albornoz, K., & Beckles, D. M., (2021). Can gene editing reduce postharvest waste and loss of fruit, vegetables, and ornamentals? *Horticulture Research, 8*, 1. https://doi.org/10.1038/s41438-020-00428-4.

Stajič, E., Kiełkowska, A., Murovec, J., & Bohanec, B., (2019). Deep sequencing analysis of CRISPR/Cas9 induced mutations by two delivery methods in target model genes and the CENH3 region of red cabbage (*Brassica oleracea* var. capitata f. rubra). *Plant Cell, Tissue, and Organ Culture, 139*, 227–235.

Stowe, E., & Dhingra, A., (2021). Development of the Arctic® apple. *Plant Breeding Reviews, 44*, 273–296.

Tashkandi, M., Ali, Z., Aljedaani, F., Shami, A., & Mahfouz, M. M., (2018). Engineering resistance against tomato yellow leaf curl virus via the CRISPR/Cas9 system in tomato. *Plant Signaling & Behavior, 13*, e1525996.

Toinga-Villafuerte, S., Vales, M. I., Awika, J. M., & Rathore, K. S., (2022). CRISPR/Cas9-mediated mutagenesis of the granule-bound starch synthase gene in the potato variety Yukon gold to obtain amylose-free starch in tubers. *International Journal of Molecular Sciences, 23*, 4640.

Tran, M. T., Doan, D. T. H., Kim, J., Song, Y. J., Sung, Y. W., Das, S., Kim, E. J., et al., (2021). CRISPR/Cas9-based precise excision of SlHyPRP1 domain (s) to obtain salt stress-tolerant tomato. *Plant Cell Reports, 40*, 999–1011.

Tuncel, A., & Qi, Y., (2022). CRISPR/Cas mediated genome editing in potato: Past achievements and future directions. *Plant Science*, 111474.

Tyagi, S., Kumar, R., Kumar, V., Won, S. Y., & Shukla, P., (2021). Engineering disease resistant plants through CRISPR-Cas9 technology. *GM Crops & Food, 12*, 125–144.

Waltz, E., (2016). Gene-edited CRISPR mushroom escapes US regulation. *Nature, 532*, 293.

Wan, D. Y., Guo, Y., Cheng, Y., Hu, Y., Xiao, S., Wang, Y., & Wen, Y. Q., (2020). CRISPR/Cas9-mediated mutagenesis of VvMLO3 results in enhanced resistance to powdery mildew in grapevine (*Vitis vinifera*). *Horticulture Research, 7*, 116. https://doi.org/10.1038/s41438-020-0339-8.

Wang, D., Samsulrizal, N. H., Yan, C., Allcock, N. S., Craigon, J., Blanco-Ulate, B., Ortega-Salazar, I., et al., (2019). Characterization of CRISPR mutants targeting genes modulating pectin degradation in ripening tomato. *Plant Physiology, 179*, 544–557.

Wang, H., Wu, Y., Zhang, Y., Yang, J., Fan, W., Zhang, H., Zhao, S., Yuan, L., & Zhang, P., (2019). CRISPR/Cas9-based mutagenesis of starch biosynthetic genes in sweet potato (ipomoea batatas) for the improvement of starch quality. *International Journal of Molecular Sciences, 20*, 4702.

Wang, L., Chen, S., Peng, A., Xie, Z., He, Y., & Zou, X., (2019). CRISPR/Cas9-mediated editing of CsWRKY22 reduces susceptibility to *Xanthomonas citri* subsp. *citri* in Wanjincheng orange (*Citrus sinensis* (L.) Osbeck). *Plant Biotechnology Reports, 13*, 501–510.

Wang, R., Tavano, E. C. D. R., Lammers, M., Martinelli, A. P., Angenent, G. C., & De Maagd, R. A., (2019). Re-evaluation of transcription factor function in tomato fruit development and ripening with CRISPR/Cas9-mutagenesis. *Scientific Reports, 9*, 1–10.

Watanabe, K., Kobayashi, A., Endo, M., Sage-Ono, K., Toki, S., & Ono, M., (2017). CRISPR/Cas9-mediated mutagenesis of the dihydroflavonol-4-reductase-B (DFR-B) locus in the Japanese morning glory *Ipomoea* (Pharbitis) nil. *Scientific Reports, 7*, 1–9.

Wenneker, M., & Thomma, B. P., (2020). Latent postharvest pathogens of pome fruit and their management: From single measures to a systems intervention approach. *European Journal of Plant Pathology, 156*, 663–681.

Xie, S., Shen, B., Zhang, C., Huang, X., & Zhang, Y., (2014). sgRNAcas9: A software package for designing CRISPR sgRNA and evaluating potential off-target cleavage sites. *PloS One, 9*(6), e100448.

Xing, S., Jia, M., Wei, L., Mao, W., Abbasi, U. A., Zhao, Y., Chen, Y., et al., (2018). CRISPR/Cas9-introduced single and multiple mutagenesis in strawberry. *Journal of Genetics and Genomics, 45*(12), 685–687. https://doi.org/10.1016/j.jgg.2018.04.006.

Xu, J., Hua, K., & Lang, Z., (2019). Genome editing for horticultural crop improvement. *Horticulture Research, 6*, 113. https://doi.org/10.1038/s41438-019-0196-5.

Xu, Z. S., Feng, K., & Xiong, A. S., (2019). CRISPR/Cas9-mediated multiply targeted mutagenesis in orange and purple carrot plants. *Molecular Biotechnology, 61*, 191–199.

Yang, Y., Zhu, G., Li, R., Yan, S., Fu, D., Zhu, B., Tian, H., Luo, Y., & Zhu, H., (2017). The RNA editing factor SlORRM4 is required for normal fruit ripening in tomato. *Plant Physiology, 175*, 1690–1702.

Yu, Q., Wang, B., Li, N., Tang, Y., Yang, S., Yang, T., et al., (2017). CRISPR/Cas9-induced targeted mutagenesis and gene replacement to generate long-shelf-life tomato lines. *Sci. Rep., 7*, 11874. doi: 10.1038/s41598-017- 12262-1.

Zhang, B., Yang, X., Yang, C., Li, M., & Guo, Y., (2016). Exploiting the CRISPR/Cas9 system for targeted genome mutagenesis in petunia. *Scientific Reports, 6*, 1–8.

Zhang, H., Si, X., Ji, X., Fan, R., Liu, J., Chen, K., Wang, D., & Gao, C., (2018). Genome editing of upstream open reading frames enables translational control in plants. *Nature Biotechnology, 36*, 894–898.

Zhao, X., Jayarathna, S., Turesson, H., Fält, A. S., Nestor, G., González, M. N., Olsson, N., et al., (2021). Amylose starch with no detectable branching developed through DNA-free CRISPR-Cas9 mediated mutagenesis of two starch branching enzymes in potato. *Scientific Reports, 11,* 1–13.

Zsögön, A., Čermák, T., Naves, E. R., Notini, M. M., Edel, K. H., Weinl, S., Freschi, L., et al., (2018). De novo domestication of wild tomato using genome editing. *Nature Biotechnology, 36*, 1211–1216.

CHAPTER 7

REGULATING POSTHARVEST PATHOGEN INFECTION AND DECAY USING BIOTECHNOLOGICAL TOOLS

MUKESH KUMAR MEENA[1,2]

[1]*National Institute of Plant Genome Research, New Delhi, India*

[2]*Leibniz Institute of Plant Biochemistry, Halle (Saale), Germany*

ABSTRACT

Regulating postharvest pathogen infection and decay using biotechnological tools explores the critical issues surrounding postharvest losses in fruits and vegetables caused by pathogens and their management. Despite the pivotal role of plants in sustaining life on Earth, they face continuous challenges from environmental stresses, including biotic factors like bacteria, fungi, and insect pests, which lead to preharvest and postharvest crop losses. With the global population expected to reach 10 billion by 2050, there is a pressing need to address these losses, estimated at 30%–50% or more, to meet increasing food demands. This chapter delves into the causes of postharvest diseases, including latent infections and infections initiated during and after harvest, highlighting the importance of pathogen detection and identification. Additionally, economic and social challenges posed by postharvest factors, such as environmental pollution from pesticide usage and health risks associated with pesticide residues, are discussed. Conventional methods like chemical treatments are commonly employed for disease control, but their adverse effects necessitate the exploration of alternative strategies. The chapter then explores

Bioengineered Fruit and Vegetables: Trends in Postharvest Quality Improvement.
Mohammad Irfan and Pankaj Kumar (Eds.)

novel biotechnological approaches, including genetic engineering, RNA interference, and genome editing, as promising solutions to mitigate post-harvest losses sustainably. While challenges in the commercialization of biotechnological approaches exist, the precision and potential of CRISPR-Cas-based genome editing offer promising avenues for future research and crop improvement to ensure global food security.

7.1 INTRODUCTION

Plants harbor photosynthetic machinery that converts carbon dioxide into sugar. All living organisms, including humans, are directly or indirectly dependent on them for food, oxygen, and other essential products. Due to their lack of mobility, plants continuously face challenges from environmental stresses (abiotic and biotic) that negatively impact crop production and disrupt the food supply. Biotic stress factors such as bacteria, fungi, and insect pests play a critical role in preharvest and postharvest crop losses. With the global population expected to reach 10 billion by 2050 (FAO report), the demand for food, fodder, and fiber is projected to increase by 70%. This growth can only be managed through increased production and reduced postharvest crop loss. This chapter will focus on postharvest losses of fruits and vegetables caused by pathogens and their management. It is important to note that postharvest losses (PHLs) are not solely due to pathogen infections but also result from poor practices during harvesting, packaging, and transport. Effective postharvest disease management strategies require pathogen detection, identification, and diagnosis. Additionally, factors like poor cultivar selection (susceptible to diseases), postharvest storage conditions (temperature, relative humidity, and air composition), handling methods, pest control, and postharvest hygiene significantly contribute to increased postharvest diseases. According to the infection process, postharvest diseases are classified into two categories: (i) Latent or quiescent infections occur when the pathogen infects the host tissues before harvest but remains dormant until the host tissues' physiological status becomes favorable for their growth and infection. During fruit ripening, significant changes in metabolism trigger the reactivation of quiescent infections (Kumar et al., 2016; Tayal et al., 2022). Examples include anthracnose in tropical fruits (Mango, Banana) caused by *Colletotrichum* spp. and gray mold of strawberries caused by *Botrytis cinerea*. (ii) The other major category of postharvest disease infections is initiated during and after the harvest of fruits and vegetables. Microscopic to large surface wounds created by mechanical injuries during produce handling and insect infestations are significant contributors to the infection. Examples include blue and green mold caused by *Penicillium* spp. in citrus fruits, grapes, and apples due to wound injuries.

7.2 ECONOMICAL AND SOCIAL CHALLENGES OF POSTHARVEST DISEASES

Fruits and vegetables play an important role in food and nutrition due to their high sugar, mineral, fiber, vitamins, and other important metabolite content. Furthermore, they contribute significantly to the enhancement of farmers' income and living standards worldwide. Although fruits and vegetables are critical to sustaining the global food supply chain, they are considered low shelf-life crops compared to cereals and pulses due to high moisture content (70%–95%), high sugar content (fruits), and soft texture. All these morphological and physiological features make them prone to postharvest diseases caused by microbial and insect pathogens between the periods of harvest and consumption. Postharvest losses (PHLs) of fruits and vegetables are a major concern in lower-middle-income countries and are estimated to be about 30%–50% or more (Kumari et al., 2022; Irfan et al., 2023). Apart from direct economic implications, diseased fruits and vegetables pose a potential health risk to consumers. Many fungal pathogen species such as *Alternaria, Aspergillus, Colletotrichum,* and *Penicillium* are known to produce carcinogenic mycotoxins and can cause potential health risks to humans and animals (Klich, 2007). One of the most commonly used strategies for disease control is fungicides and pesticides spraying on fruits and vegetables during preharvest and postharvest stages. These pesticides cause environmental pollution (air, soil, and water) due to foliar spray and rain runoff effects (Malaj et al., 2014; Queyrel et al., 2016; Sharma et al., 2022). Furthermore, human consumption of pesticide residue directly (inadequate washing of pesticide-sprayed fruits/vegetables) or indirectly (consumption of pesticide-contaminated water) leads to several neurological disorders, liver cancer, or pregnancy-related issues (Allsop et al., 2015; McGlynn et al., 2006), indicating the social challenges involved in postharvest diseases and management strategies.

7.3 CAUSES OF POSTHARVEST DISEASES AND PRODUCE LOSS

In general, postharvest diseases caused by bacteria, fungi, and insects are the major contributors to postharvest produce loss. Some examples of the economically important pathogens that cause severe postharvest disease are (i) bacteria: *Erwinia*, *Pseudomonas*, and *Xanthomonas*; (ii) fungi: *Alternaria*, *Aspergillus*, *Botrytis*, *Colletotrichum*, *Fusarium*, *Penicillium*, and *Sclerotinia*; and (iii) insects: *Spodoptera*, *Helicoverpa*, *Manduca*, and *Aphis*, etc. Except for chewing insects, bacteria and fungi are weak pathogens. They largely depend on microscopic wound sites on fruits and vegetables to invade and initiate infections. These wounds could be caused by improper handling, insects, birds, and animals. Chewing insects such as *Spodoptera* and *Helicoverpa* partially or completely damage the fruit/vegetables, causing severe produce loss. The postharvest disease management strategy largely depends on the correct identification of the pathogen causing postharvest disease. Table 7.1 provides the names of the postharvest diseases with their causing pathogens.

7.3.1 POST-HARVEST BACTERIAL DISEASES

Plant pathogenic bacteria cause significant produce loss in several fruit and vegetable plant species due to their broad host range (Ansari et al., 2022). Some of the agriculturally important bacteria belonging to genera *Erwinia, Pseudomonas, Ralstonia, Xanthomonas,* and *Xylella* are devastating plant pathogens (Mansfield et al., 2012). They enter plant tissues through wounds or natural openings (stomata, lenticels). Wounds can be caused by mechanical injuries during handling or feeding (insects/birds/animals) of the produce. The speed of bacterial infection largely depends on favorable environmental conditions. Factors that play a positive role in bacterial infections include high humidity, irregular watering schedule, temperature, and contaminated soil. Under favorable conditions, bacteria multiply quickly, leading to infection and tissue damage. One economically important postharvest disease is bacterial soft rots, caused by various species of *Bacillus, Erwinia, Lactobacillus, Pseudomonas,* and *Xanthomonas* in vegetables (Bhat et al., 2012; Sundin et al., 2016). Symptoms of the disease manifest as water-soaked spots on vegetable pods. Fruits are less susceptible to bacterial soft rot due to their acidic

TABLE 7.1 Economically Important Postharvest Diseases and Pathogens of Fruits and Vegetables

Disease	Pathogen			
	Bacterial	**Fungal**	**Insect**	**References**
		Fruits: Mango		
Mango black spot	*Xanthomonas axonopus*pv. *mangiferaeindicae*	–	–	Sanahuja et al. (2016)
Stem end rot	–	*Lasiodiplodia theobromae.*	–	Zhan et al. (2023)
Anthracnose	–	*Colletotrichum gloeosporioides*	–	Kamle& Kumar (2016)
Black soft rot	–	*Phomopsis mangiferae*	–	Ko et al. (2009)
Fruit infestation	–	–	*Bactrocera dorsalis*, *Sternochetus mangiferae*	Hossain et al. (2020); Silva et al. (2022)
Papaya				
Bacterial canker	*Erwinia* spp.	–	–	Hanagasaki et al. (2021)
Stem end rot	–	*botryodiplodia theobromae*	–	Rahman et al. (2008)
Anthracnose	–	*Gloeosporium papaya*	–	Waghunde et al. (2014)
Phomopsis fruit rot	–	*Phomopsis caricae-papayae*	–	Ventura et al. (2004)
Fruit infestation	–	–	*Paracoccus marginatus*, *Pseudaulacaspis pentagona*	Muniappan et al. (2008); Pantoja et al. (2002)
		Banana		
Banana bacterial wilt	*Xanthomonas campestris* pv. *musacearum*	–	–	–
Anthracnose	–	*Colletotrichum musae*	–	Intan Sakinah et al. (2013)
Sigatoka leaf spot	–	*Mycosphaerella fijiensisis*	–	Soares et al. (2021)
Banana weevil	–	–	*Cosmopolites sordidus*	Gold et al. (2002)

TABLE 7.1 *(Continued)*

Disease	Pathogen			
	Bacterial	**Fungal**	**Insect**	**References**
		Pineapple		
Bacterial heart rot	*Dickeyazeae*	–	–	Young et al. (2022)
Stem end rot	–	*Thielaviopsis paradoxa*	–	Hewajulige& Wijesundera (2014)
Pineapple black rot	–	*Ceratocystis paradoxa*	–	Hubert et al. (2014)
Fruit borer	–	–	*Strymon megarus*	Inclán et al. (2008)
		Citrus Fruits		
Citrus canker	*Xanthomonas* spp.	–	–	Brunings & Gabriel (2003)
Fruit anthracnose	–	*Colletotrichum gloeosporioides*	–	McGovern et al. (2012)
Blue mold	–	*Penicillium italicum*	–	Ismail & Zhang (2004)
Green mold	–	*Penicillium digitatum*	–	Pitt & Hocking (2009)
Stem end rot	–	*Phomopsis citri*	–	Mahadevakumar et al. (2014)
Fruit thrips	–	–	*Scirtothrips citri*	Tanigoshi & Moffitt (1984)
		Grapes		
Pierce's disease	*Xylella fastidiosa*	–	–	Kyrkou et al. (2018)
Anthracnose	–	*Elsinoë ampelina*	–	Li et al. (2021)
Botrytis bunch rot	–	*Botrytis cinerea*	–	Elad et al. (2016)
Blue mold rot	–	*Penicillium digitatum*	–	Pitt & Hocking (2009)
Grape mealybug	–	–	*Pseudococcus maritimus*	Liu (2013)
		Apple		
Blister spot	*Pseudomonas syringae* pv. *Papulans*	–	–	Huang et al. (1997)

TABLE 7.1 *(Continued)*

Disease	Pathogen			
	Bacterial	**Fungal**	**Insect**	**References**
Apple scab	–	*Venturia inaequalis*	–	Bowen et al. (2011)
Blue mold	–	*Penicillium expansum*	–	Khan & Javaid (2021)
Grey mold	–	*Botrytis cinerea*	–	Nybom et al. (2020)
Apple maggot	–	–	*Rhagoletis pomonella*	Bostanian et al. (1999)
Vegetables: Solanaceae (Tomato, Potato)				
Bacterial soft rot	*Erwinia* spp., *Xanthomonas* spp.	–	–	Akbar et al. (2015); Emma et al. (2013)
Scab	*Streptomyces scabies*	–	–	Ismail et al. (2020)
Grey mold	–	*Botrytis cinerea*	–	Williamson et al. (2007)
Fusarium rot	–	*Fusarium* spp.	–	Petrasch et al. (2019)
Sclerotium rot	–	*Sclerotium rolfsii*	–	Daami-Remadi et al. (2010)
Late blight	–	*Phytophthora infestans*	–	Dong & Zhou (2022)
Fruit worm	–	–	*Helicoverpa armigera*	Singh et al. (2017)
Hornworms	–	–	*Manduca* spp.	BK et al. (2021)
Brassicaceae (Cabbage, Broccoli)				
Black rot	*Xanthomonas campestris*	–	–	Vicente & Holub (2013)
Bacterial soft rot	*Erwinia* spp.	–	–	Bhat et al. (2012)
Anthracnose	–	*Colletotrichum* spp.	–	Yu et al. (2022)
Alternaria leaf spot	–	*Alternaria brassicae*	–	Nowicki et al. (2012)
Watery soft rot	–	*Sclerotinia sclerotiorum*	–	Bolton et al. (2006)
Beet armyworm	–	–	*Spodoptera exigua*	Pashley (1988)

TABLE 7.1 *(Continued)*

Disease	Pathogen			
	Bacterial	**Fungal**	**Insect**	**References**
Cabbage looper	–	–	*Trichoplusia ni*	Mpumi et al. (2020)
Cabbageworm	–	–	*Pieres rapae*	Hasan & Ansari (2010)
Legumes (Bean, Chickpea, Lentils)				
Bacterial brown spot	*Pseudomonas* spp.	–	–	Muedi et al. (2015)
Bacterial blight	*Xanthomonas* spp.	–	–	Akhavan et al. (2013)
Ascochyta blight	–	*Ascochyta* spp.	–	Davidson & Kimber (2007)
Gray mold	–	*Botrytis cinerea*	–	Brauna-Morževska et al. (2019)
Watery soft rot	–	*Sclerotinia* spp.	–	Antwi-Boasiako et al. (2022)
Pea aphid	–	–	*Aphis* spp.	Sandhi & Reddy (2020)
Pod borers	–	–	*Helicoverpa armigera*	Kambrekar (2016)
Bulb (Onion, Garlic)				
Bulb rot	*Pseudomonas* spp.	–	–	Abd-Alla et al. (2011)
Blue mold rot	–	*Penicillium* spp.	–	Dugan et al. (2017)
White rot	–	*Sclerotinia* spp.	–	Amin et al. (2014)
Bulb rot	–	*Fusarium* spp.	–	Gálvez & Palmero (2022)
Onion maggot	–	–	*Delia antiqua*	Sable &Chandaragi (2021)

pH. Another significant bacterial disease is bacterial canker, caused by *Xanthomonas* species in fruits such as mangoes, citrus, and stone fruits (Sanahuja et al., 2016; Islam et al., 2019). *Xanthomonas* invade through weak lenticels during fruit ripening. Symptoms on the fruit start as small water-soaked specks around lenticels or small star-shaped lesions, which eventually blacken and develop cracks.

7.3.2 POST-HARVEST FUNGAL DISEASES

Plant fungal pathogens are most devastating in postharvest crop losses. The mode of invasion is similar to bacterial pathogens as they also invade through microscopic wounds and infect fruits and vegetables. Additionally, environmental factors such as temperature, relative humidity, light conditions, and soil health affect their growth and infection progression (Ansari et al., 2022). Most important economically plant fungi belong to phylum Ascomycota, and it was observed that the asexual stage of fungus (the anamorph) is majorly involved in postharvest diseases in comparison to the sexual stage (the teleomorph). Some important genera of anamorphic pathogens are *Alternaria, Aspergillus, Botrytis, Colletotrichum, Fusarium, Geotrichum, Penicillium, Phomopsis, Rhizoctonia, Sclerotium,* and *Sclerotinia*. We will discuss a few most important fungal pathogens involved in the postharvest disease of fruits and vegetables. (i) Anthracnose: *Colletotrichum* spp. causes anthracnose fungal disease in mango, banana, citrus, and vegetables from *Brassicaceae* and cucurbits (Kamle & Kumar, 2016; Intan Sakinah et al., 2013; McGovern et al., 2012). Symptoms include black lesions on fruits, pulp becomes hard, cracks, and decays at ripening. (ii) Gray mold: *Botrytis* spp. causes fungal disease gray mold in berries, stone fruit, legumes, and leafy vegetables (Vandendriessche et al., 2012; Eckert & Ogawa, 1988). Being a necrotrophic fungus, *Botrytis* leads to gray-brown lesions on leaves and fruits. Infected areas soon take on a velvety, gray-brown coat of spores. Rotted fruit tissues become soft and mushy. (iii) Watery white rot: *Sclerotinia sclerotiorum* causes watery white rot fungal disease in vegetables such as cabbage, cauliflower, lettuce, broccoli, pea, and bean (Uloth et al., 2013). Warm and humid weather supports fungal growth on infected vegetable pods, which eventually covers whole pods with white, thick mycelial mat and black-colored sclerotia (Bolton et al., 2006).

7.3.3 POST-HARVEST INSECT PESTS MEDIATED DAMAGE

Insect pests are challenging pathogens to deal with and cause severe post-harvest produce loss. Insect pests are broadly divided into two categories: Chewing insects (*Spodoptera, Helicoverpa, Menduca, Trichoplusia*, and *Pieres*) and sucking insects (*Thrips, Myzus, Aphis*). (i) Chewing insects have 4 developmental stages in their life cycle: egg, larval, pupa, and adult (moth). Among these developmental stages, the larval stage causes the most damage to plant leaves and fruits as they are produced in large numbers with high energy demand to grow further. *Spodoptera* spp. and *Helicoverpa* spp. are generalist insect pests, and they have a large host range. *Spodoptera* spp. causes leaf and fruit infestation in apple, grape, orange, papaya, peach, strawberry, and vegetables from the *Brassicaceae* family (Pashley, 1988). *Manduca sexta* is a specialist insect and the host range is confined to plants from the *Solanaceae* family, which includes potato, tomato, pepper, and eggplant (del Campo & Renwick, 1999). (ii) Sucking insect has 3 developmental stages in their life cycle: egg, nymph, and adult. The nymph stage concentrates in very high densities on young plant tissues and by using specialized mouthparts called stylets, pierce the plant tissues, and suck out the phloem sap which causes water stress and wilting of plant tissues. *Myzus* spp. has a very broad host range including plants from *Solanaceae, Compositae, Brassicaceae,* and *Cucurbitaceae* (Heathcote, 1962).

7.4 CONVENTIONAL AND NOVEL METHODS OF POSTHARVEST DISEASE MANAGEMENT

Postharvest losses (PHLs) in fruits and vegetables are mainly caused by bacterial, fungal infection, and insect infestation. Producers use various strategies to contain PHLs according to geographical region and pathogen types. Due to globalization and trade liberalization, producers need to put serious effort according to the guidelines of the target region and country. Here we will discuss point-wise conventional and novel methods for post-harvest disease management.

7.4.1 CHEMICAL TREATMENT

Chemical fungicides and pesticides are commonly used methods for postharvest disease control in fruits and vegetables worldwide. Two types of fungicides are commonly used depending on the crop's developmental stage. (i) For pathogens that infect fruits and vegetables before harvest, systemic and protectant fungicides are applied to inactivate new and already established infections at the field level. Some examples of commonly used protectant fungicides are mancozeb, Captan, and chlorothalonil. (ii) The second type of fungicides is applied at the postharvest level; they prevent the spread of already established infection at the surface tissue or protect against future infections occurring during storage and transportation of the produce. Mostly second types of fungicides are fungistatic rather than fungicidal and they are applied on fruits and vegetables as sprays, dips, box liners, and wax coating. Some examples are benomyl, thiabendazole, prochloraz, and imazalil. A range of insecticides is commonly used to control insect infection and diseases on fruits and vegetables. The most effective insecticides are neurotoxic and target the nervous system of the insects. Some commonly used insecticides are Endosulfan, Diazinon, Methomyl, and Permethrin.

7.4.2 DISINFECTANT

Fruits and vegetables are prone to physical injuries during harvesting, sorting, packaging, and transportation due to their soft texture. As mentioned earlier, even microscopic injuries are enough for pathogen invasion and infection. Apart from precautions during the handling of the produce, surface disinfection techniques could minimize infection and disease development. Surface disinfection could be done by immersing the fruits and vegetables in disinfection solutions like calcium hypochlorite, formaldehyde, alcohol, and electrolyzed water (Eckert, 1990; Rahman et al., 2016). Fumigation treatments are also found to be effective with steam, carbon dioxide, ozone, and ammonia (Weavers & Wickramanayake, 2001).

7.4.3 ESSENTIAL OILS AND PLANT EXTRACTS

Many medicinal plants produce a large number of metabolically active defense compounds such as phenols, flavonoids, alkaloids, terpenoids, tannins, and coumarins. These compounds exhibit strong antimicrobial activity and are used for postharvest disease control against various pathogens. Furthermore, their biodegradability and environmentally friendly nature make them a more suitable and sustainable method for postharvest disease control. They are commonly referred to as green pesticides. Some common examples include menthol from various species of mint (*Mentha* spp.), eugenol from clove oil (*Syzygium aromaticum*), 1,8-cineole from rosemary (*Rosmarinus officinale*), and eucalyptus (*Eucalyptus globulus*) oils. The aforementioned compounds (menthol, eugenol, 1,8-cineole) also exhibit insect repellent and insecticidal activity against insect pests (Isman & Machial, 2006).

7.4.4 BIOLOGICAL *METHODS*

Many developed countries, specifically in the European Union, have set up very strict regulations for pesticide usage, including the type and amount of pesticides. Exporters from developing countries face numerous challenges when trying to sell their produce to developed countries. Therefore, biological control methods could provide a strong alternative to pesticides. Biological control methods are based on microbes such as bacteria, fungi, and viruses (bacteriophages) (Droby et al., 2000), and these methods are as effective as chemical control methods. Some most important bio-agent microbes are *Trichoderma viride*, *Pseudomonas fluorescens,* and *Bacillus subtilis* which produce biologically active compounds like elicitors, antibiotics, and activators of plant immune response. Biocontrol methods rely on the antagonistic nature of bio-agents against plant pathogens and are more effective when applied before infection. For example, *Trichoderma* exhibits strong antifungal properties against *Botrytis cinerea, Sclerotinia sclerotiorum,* and other important plant pathogens.

7.5 EMERGING BIOTECHNOLOGICAL APPROACHES IN POSTHARVEST DISEASE MANAGEMENT

International market regulation, increasing pesticide resistance in pathogens, and the harmful effect of toxic chemicals on human health and the environment are some of the important socioeconomic factors that are pushing scientists to develop novel and sustainable methods for postharvest disease management. Nowadays, biotechnology approaches represent our best resource to reduce preharvest and PHLs through science-based sustainable methods. Over the last couple of decades, agriculture biotechnologists have made significant progress in understanding plant defense mechanisms and exploited the potential of genetic engineering (GE) to efficiently transfer resistance traits in crop plants. Conventional breeding methods are alternative approaches and have been used for several decades to reduce preharvest and PHLs in selected fruit and vegetable plant species. Time-consuming and labor-intensive procedures are the major limiting factors in their implementation (Mathiazhagan et al., 2021). To overcome the limitation of conventional breeding of fruits and vegetables, a variety of biotechnological approaches have been developed and successfully employed such as genetic transformation and transgenics, RNA interference (RNAi), and genome editing. Over the last decade, many reports have been published that emphasize the potential of biotechnological approaches in preharvest and postharvest disease management. We will discuss above-mentioned biotechnological approaches in subsections.

7.5.1 TRANSGENIC APPROACH

The transgenic approach allows the manipulation and expression of target genes from various sources (self or foreign) and their stable insertion into the plant genome to gain desirable traits for crop improvement. The organisms generated by genetic manipulation are categorized as genetically modified organisms (GMOs). Bt cotton was among the first commercialized GM crop plants approved for commercial cultivation in the United States in 1995. Bt cotton expressed a *Bacillus thuringiensis cry* gene which codes for an endotoxin to protect transgenic cotton plants from bollworm invasion (Perlak et al., 1990). This approach has been used to develop disease-resistant plants in tropical fruits and vegetable crops to prevent yield losses due to various pathogens. Citrus crops are considered one of the major cash crops worldwide, with major producers including Brazil, Mexico, China, the United States, and some European countries. Economic losses caused by postharvest diseases represent a significant challenge for citrus fruit producers. Citrus canker, caused by *Xanthomonas citri* sub spp. *citri* (Xcc), is one of the most economically important citrus diseases. Symptoms include circular water-soaked lesions, leading to defoliation and fruit drop (Brunings & Gabriel, 2003). In the absence of effective phytosanitary methods, biotechnology approaches based on the use of plant receptors that recognize pathogen-associated molecular patterns (PAMPs) molecules and trigger plant immune responses are strategies adopted to confer broad-spectrum resistance. *Nicotiana benthamiana FLS2* (*NbFLS2*) was overexpressed in Hamlin sweet orange and Carrizo citrange. Transgenic plants displayed ROS overproduction and activation of defense marker (*PR1* and *WRKY22*) genes in response to flg22Xcc along with strong resistance against disease (Hao et al., 2016). In another approach, plant immune signaling protein Mitogen activate protein kinase *CsMAPK1* was expressed under the control of a *Xanthomonas*-inducible promoter (PR5) in Troyer citrange plants that lead to increased ROS production, decrease *Xanthomonas citri* growth and reduced citrus canker disease symptoms (de Oliveira et al., 2013). Banana (*Musa* spp.) is one of the most significant economically important fruit crops in tropical regions, having commercial production in many developing countries. The banana fruit industry is severely affected by Sigatoka leaf spot and gray mold fungal diseases caused by *Mycosphaerella fijiensis* and *Botrytis cinerea*, respectively. To address this, Grand Naine banana was transformed using a triple gene construct, containing endo chitinase gene *ThEn-42* from *Trichoderma harzianum*, tomato superoxide dismutase Cu, *Zn-SOD,* and grape stilbene synthase (Vishnevetsky et al., 2011). Transgenic plants displayed significant resistance against Sigatoka and gray mold (*Botrytis cinerea*) (Dash et al., 2016). Potato is the most common vegetable grown and consumed worldwide. Due to its heterozygous nature, genetic improvement of potatoes is difficult by conventional breeding to generate potato disease-resistant varieties. Potato disease-resistant plants against late blight and black spot bruising were developed using Innate® technology and approved for commercialization (Richael, 2021). Tomato, another economically important *Solanaceae* crop, is widely cultivated and consumed. Tomato production is significantly impacted by various bacterial and fungal diseases. To combat this, a genetic engineering approach was utilized in tomato plants. The Arabidopsis *NPR1* (nonexpresser of PR genes) gene was introduced into a tomato cultivar, resulting in a broader spectrum of enhanced resistance against bacterial wilt (*Ralstonia solanacearum*), bacterial spot (*Xanthomonas campestris*), Fusarium wilt (*Fusarium oxysporum*), and gray leaf spot (*Stemphylium solani*) without any negative impact on growth (Lin et al., 2004). Similar outcomes were observed in carrots (*Daucus carota* L.). Expression of *AtNPR1* in carrots provided broad-spectrum resistance against *Alternaria radicina, Botrytis cinerea, Sclerotinia sclerotiorum,* and *Xanthomonas hortorum* (Wally et al., 2009).

7.5.2 RNA INTERFERENCE (RNAI) TECHNOLOGY

RNA interference (RNAi) or antisense technology functions through targeted gene expression suppression that leads to low protein abundance and influences function. MicroRNA (miRNA) and small interfering RNA (siRNA) are tools for RNAi, and this approach is successfully applied in various fruits and vegetable plants to improve disease resistance (Ghosh et al., 2017). In citrus fruits, major postharvest diseases caused by fungi are green mold (*Penicillium digitatum*) and blue mold (*Penicillium italicum*) (Pitt & Hocking, 2009; Ismail & Zhang, 2004), leading to significant postharvest crop losses. Another important disease is sour rot caused by the fungus *Geotrichum citri-aurantii* (Eckert & Brown, 1986). In this context, an antisense construct was introduced in sweet orange (*Citrus sinensis* L. Osb. cv. Navelina) plants, leading to the downregulation of D-limonene synthase activity. The transgenic plants displayed constitutive upregulation of plant immune signaling genes and increased accumulation of jasmonic acid, leading to resistance against *Penicillium digitatum* and other specialized fungi (Rodríguez et al., 2014, 2015). Extended shelf life could also help to prevent PHLs due to microbial disease. Some efforts are made in bananas to delay fruit ripening and extend shelf life through antisense targeting of two MADS box genes (*MaMADS1* and *MaMADS2*) which are important for banana ripening (Elitzur et al., 2016). An innovative antisense approach was employed in tomatoes to develop resistance against gray mold (*Botrytis cinerea*). During infection, *Botrytis cinerea* delivers small RNAs (*Bc*-sRNAs) into plant cells to suppress host immunity-related genes. Such sRNA effectors are mostly generated by *Botrytis cinerea* Dicer-like proteins. Wang et al. (2016) designed a strategy to express sRNA that targets *Bc-DCL1* and *Bc-DCL2* in tomatoes to silence *Bc-DCL* genes and attenuate fungal pathogenicity and growth. This strategy can be adapted to control multiple fungal diseases occurring simultaneously to reduce postharvest losses. In another approach, orthologs of Arabidopsis susceptibility genes in potato and tomato, DND1, were silenced by RNAi, leading to a significant reduction in *Botrytis cinerea* lesions (Sun et al., 2017). Sweet potato (*Ipomoea batatas* (L.) Lam.) is an important staple food crop in many African countries. The African sweet potato weevil (*Cylas brunneus*) insect pest severely compromises crop production. To address this plant disease, Christiaens et al. (2016) designed a crop protection strategy where gene-specific dsRNA against

24 target genes of *Cylas brunneus* were expressed in transgenic plants, resulting in the effective silencing of target genes and ultimately leading to high mortality of the insects.

7.5.3 GENOME EDITING

Over the last decade, genome editing has emerged as an efficient tool to introduce precise mutations (insertion or deletion) in plants and also provides a better alternative to time-consuming and labor-intensive conventional breeding methods. Scientists are more optimistic about the acceptance of genome-edited crops from regulatory authorities and consumers compared to GMOs because of the lack of foreign DNA in edited crops. The CRISPR-Cas-based genome editing systems are applicable to study gene function and the underlying mechanism by the generation of mutant plants in various crops (dicots and monocots) and serve as excellent tools for crop improvement for desired traits. In recent years, several fruits and vegetable crops have been developed with enhanced disease resistance and other economically important traits using CRISPR-Cas9 technology. To promote the shelf life of bananas, the *MaACO1* (amino-cyclopropane-1-carboxylate oxidase 1) gene was mutated by the CRISPR-Cas9 approach. The mutant banana fruits exhibited reduced ethylene biosynthesis and extended shelf life (Hu et al., 2021). Banana Xanthomonas wilt (BXW) caused by *Xanthomonas campestris* pv. *musacearum (Xcm)* is considered among the most destructive banana diseases in many African countries and causes huge economic losses to producers. To address this challenge, the banana *downy mildew resistance 6 (DMR6)* gene was mutated by the CRISPR-Cas-based approach. DMR6 encodes 2-oxoglutarate Fe(II)-dependent oxygenase (2OGO) enzyme, and it was reported as a susceptibility gene (Low et al., 2020). The *Musa dmr6* mutants of banana showed enhanced resistance to BXW and did not show any adverse effect on plant growth (Tripathi et al., 2021). As mentioned in previous sections, citrus canker is a severe disease in commercial citrus cultivars, and a genome editing approach was employed to generate disease-resistant Duncan grapefruit (Jia et al., 2016). The *CsLOB1* (*Citrus sinensis* Lateral Organ Boundaries) gene is a target gene for citrus canker and is induced by the pathogenicity factor PthA4 released by pathogen *Xanthomonas citri* subspecies *citri* (Xcc). The CRISPR-Cas-based approach was used to disrupt the PthA4 effector binding elements (EBEs) in the *CsLOB1* Promoter. The mutant fruits showed resistance against citrus canker disease (Jia et al., 2016). This finding was successfully translated into Wanjincheng sweet orange to generate citrus canker disease-resistant fruit (Peng et al., 2017). Recently, the CsWRKY22 transcription factor, a negative regulator of citrus canker, was mutated in Wanjincheng orange background to generate disease-resistant plants (Wang et al., 2019). In grapes, Wang et al. (2018) designed a CRISPR-Cas9-mediated strategy to generate the *VvWRKY52* knockouts. The homozygous mutants of *VvWRKY52* exhibited increased resistance to gray mold disease in grapes. Earlier examples were mostly focused on the mutation of target genes to generate knockouts by CRISPR-Cas9; now we will discuss one example of the generation of the dominant version of the target protein. Plant defense hormones jasmonate (JA) and salicylate (SA) display antagonistic behavior in normal conditions, and it is exploited by *Pseudomonas syringae pv. tomato* (*Pto*) DC3000, the causal bacterial pathogen of tomato speck disease. It secretes coronatine (a JA-Ile mimic) that stimulates stomata opening and facilitates bacterial entry in leaf tissues. Ortigosa et al. (2019) uncoupled this antagonism by the generation of dominant JAZ2 (SlJAZ2Δjas) repressors by CRISPR-Cas9. SlJAZ2Δjas expressing plants inhibited COR-mediated stomatal reopening and contained *Pto*DC3000 infection without compromising JA responses against the necrotrophic fungal pathogen (Ortigosa et al., 2019).

7.6 FUTURE CHALLENGES IN THE COMMERCIALIZATION OF BIOTECHNOLOGICAL APPROACHES IN POST-HARVEST DISEASE MANAGEMENT

Genetically modified (GM) crops have shown huge potential for the successful improvement of disease resistance and other economically important traits. According to the USA Food and Drug Administration (USFDA) data, corn, soybean, cotton, canola, potato, papaya, apple, sugar beet, and pineapple are a few examples of GMOs that are successfully permitted for commercial cultivation. Although the application of biotechnological approaches is limited to fewer tropical fruits, including banana, papaya, and citrus fruits, their implementation remains challenging in many other economically important tropical and temperate fruits. The limiting factors are the lack of good genetic transformation methods, the recalcitrant nature, and the unavailability of genomic resources (Arias et al., 2012). Therefore, the development of efficient genetic transformation and genome resources (genome sequencing, gene expression maps) are key factors in the implementation of biotechnological approaches in the improvement of traits in fruits and vegetables. Another concern for fruit GE is the pleiotropic effect, sometimes caused by constitutive expression of foreign genes in plants. Therefore, the use of tissue-specific expression systems, for example, fruit-specific promoters, can be useful for targeted gene expression in fruits or other edible parts (Dutt et al., 2014; Irfan et al., 2016; Irfan, 2022). Despite the potential application, many concerns are raised in GM crops generated by transgenic and RNAi technology due to high off-target effects resulting in modified and undesired phenotypes. Furthermore, these unpredictable pleiotropic effects of GM crops on human health and the environment demand careful field trial studies (Casacuberta et al., 2015; Raman, 2017). In this context, CRISPR-Cas-based genome editing is the most promising approach for future research and GM crop development due to precision and no foreign DNA introduction. These features make genome-edited crops more suitable for regulatory permission and consumer acceptance. In the last couple of years, many countries have prepared and ratified the regulatory guidelines, particularly for CRISPR-Cas-mediated genome-engineered crops, and some key economically important crops have been approved for commercial use such as soybean, canola, rice, maize, mushroom, tomato, and camelina (Pixley et al., 2022; Menz et al., 2020). According to a recent report from Japan, genome-edited tomatoes enriched with γ-aminobutyric acid became the first CRISPR-Cas-edited crop to enter the market in 2021 (Waltz, 2022). Further, the constant emphasis on fundamental research is necessary to understand plant defense mechanisms against known and novel pathogens to identify targets for crop improvement to prevent crop losses at preharvest and postharvest levels.

7.7 CONCLUSION

According to the FAO report, the world population is continuously rising and is projected to reach 10 billion by 2050. With increasing demand and the adverse effects of ongoing climate change on crop production, ensuring global food security will be one of the major challenges in the 21st century for producers, scientists, and policymakers. A significant increase in crop production and postharvest disease management is critical to meet food demand. Postharvest diseases cause significant crop losses of 30%–50% or more in fruits and vegetables. Biotechnological approaches provide an excellent opportunity to develop disease-resistant crop plants to reduce preharvest and postharvest crop losses (Irfan et al., 2023). Among the discussed biotechnological tools, CRISPR-Cas-based genome editing has more advantages due to precision and no foreign DNA insertion. It is evident that scientists and policymakers are more optimistic about developing and releasing CRISPR-edited crops with disease resistance and other economical traits to reduce crop losses and achieve global food security.

KEYWORDS

- **bacterial diseases**
- **biotechnological approaches**
- **biotechnological tools**
- **disinfectant**
- **genome editing**
- **pathogen**
- **post-harvest disease management**

REFERENCES

Abd-Alla, M. H., Bashandy, S. R., Ratering, S., & Schnell, S., (2011). First report of soft rot of onion bulbs in storage caused by *Pseudomonas aeruginosa* in Egypt. *Journal of Plant Interactions*, *6*(4), 229–238.

Akbar, A., Ahmad, M., Khan, S. Z., & Ahmad, Z., (2015). Characterization of the causal organism of soft rot of tomatoes and other vegetables and evaluation of its most aggressive isolates. *American Journal of Plant Sciences*, *6*(4), 511.

Akhavan, A., Bahar, M., Askarian, H., Lak, M. R., Nazemi, A., & Zamani, Z., (2013). Bean common bacterial blight: Pathogen epiphytic life and effect of irrigation practices. *SpringerPlus*, *2*, 1–9.

Allsop, M., Huxdorff, C., Johnston, P., Santillo, D., & Thompson, K., (2015). *Pesticides and Our Health a Growing Concern.* Greenpeace Research Laboratories, School of Biosciences, Innovation Centre, University of Exeter, Exeter, UK.

Amin, M., Tadele, S., & Selvaraj, T., (2014). White rot (*Sclerotium cepivorum*-Berk) an aggressive pest of onion and garlic in Ethiopia: An overview. *Journal of Agricultural Biotechnology and Sustainable Development*, *6*(1), 6–15.

Ansari, S., Kumar, V., Bhatt, D. N., Irfan, M., & Datta, A., (2022). N-Acetylglucosamine sensing and metabolic engineering for attenuating human and plant pathogens. *Bioengineering, 9*(2), 64. https://doi.org/10.3390/bioengineering9020064.

Antwi-Boasiako, A., Wang, Y., Dapaah, H. K., & Zhao, T., (2022). Mitigating against sclerotinia diseases in legume crops: A comprehensive review. *Agronomy*, *12*(12), 3140.

Arias, R. S., Borrone, J. W., Tondo, C. L., Kuhn, D. N., Irish, B. M., & Schnell, R. J., (2012). Genomics of tropical fruit tree crops. *Genomics of Tree Crops*, 209–239.

Bhat, K. A., Bhat, N. A., Mohiddin, F. A., Sheikh, P. A., & Wani, A. H., (2012). Studies on pectinase activities of isolates of *Erwinia carotovora* and *Rhizopus* sp. causing soft rot in cabbage (*Brassica oleracea* var capitata L.). *African Journal of Agricultural Research*, *7*(45), 6062–6067.

BK, S. K., Quarcoo, F., Kpomblekou-A, K., & Mortley, D., (2021). Organic tomato production in alabama: Host preference of the tomato hornworm (*Manduca quinquemaculata)* and performance of selected biopesticides. *American Journal of Entomology*, *5*(1), 10–17.

Bolton, M. D., Thomma, B. P., & Nelson, B. D., (2006). *Sclerotinia sclerotiorum* (Lib.) de Bary: Biology and molecular traits of a cosmopolitan pathogen. *Molecular Plant Pathology*, *7*(1), 1–16.

Bostanian, N. J., Vincent, C., Chouinard, G., & Racette, G., (1999). Managing apple maggot, *Rhagoletis pomonella* [*Diptera: Tephritidae*], by perimeter trapping. *Phytoprotection*, *80*(1), 21–33.

Bowen, J. K., Mesarich, C. H., Bus, V. G., Beresford, R. M., Plummer, K. M., & Templeton, M. D., (2011). *Venturia inaequalis:* The causal agent of apple scab. *Molecular Plant Pathology*, *12*(2), 105–122.

Brauna-Morževska, E., Bankina, B., & Kaņeps, J., (2019). Botrytis genus fungi as causal agents of legume diseases: A review. *Research for Rural Development*, *2*.

Brunings, A. M., & Gabriel, D. W., (2003). *Xanthomonas citri:* Breaking the surface. *Molecular Plant Pathology*, *4*(3), 141–157.

Casacuberta, J. M., Devos, Y., Du Jardin, P., Ramon, M., Vaucheret, H., & Nogué, F., (2015). Biotechnological uses of RNAi in plants: Risk assessment considerations. *Trends in Biotechnology*, *33*(3), 145–147.

Christiaens, O., Prentice, K., Pertry, I., Ghislain, M., Bailey, A., Niblett, C., & Smagghe, G., (2016). RNA interference: A promising biopesticide strategy against the African Sweetpotato weevil *Cylas brunneus*. *Scientific Reports*, *6*(1), 1–11.

Daami-Remadi, M., Jabnoun-Khiareddine, H., Sdiri, A., & El Mahjoub, M., (2010). Effect of temperature on *Sclerotium rolfsii* mycelial growth and rot severity on potato tubers. *The African Journal of Plant Science and Biotechnology*, *4*, 54–58.

Dash, P. K., & Rai, R., (2016). Translating the "banana genome" to delineate stress resistance, dwarfing, parthenocarpy and mechanisms of fruit ripening. *Frontiers in Plant Science*, *7*, 1543.

Davidson, J. A., & Kimber, R. B., (2007). Integrated disease management of *Ascochyta* blight in pulse crops. *Ascochyta Blights of Grain Legumes*, 99–110.

De Oliveira, M. L. P., De Lima Silva, C. C., Abe, V. Y., Costa, M. G. C., Cernadas, R. A., & Benedetti, C. E., (2013). Increased resistance against *Citrus* canker mediated by a *Citrus* mitogen-activated protein kinase. *Molecular Plant-Microbe Interactions*, *26*(10), 1190–1199.

Del Campo, C. M. L., & Renwick, J. A. A., (1999). Dependence on host constituents controlling food acceptance by *Manduca sexta* larvae. *Entomologia Experimentalis et Applicata*, *93*(2), 209–215.

DONG, S. M., & ZHOU, S. Q., (2022). Potato late blight caused by *Phytophthora infestans:* From molecular interactions to integrated management strategies. *Journal of Integrative Agriculture*, *21*(12), 3456–3466.

Droby, S., Wilson, C., Wisniewski, M., & El Ghaouth, A., (2000). Biologically based technology for the control of postharvest diseases of fruits and vegetables. *Microbial Food Contamination* (pp. 187–206). CRC Press, Boca Raton, FL.

Dugan, F. M., Lupien, S. L., Vahling-Armstrong, C. M., Chastagner, G. A., & Schroeder, B. K., (2017). Host ranges of *Penicillium* species causing blue mold of bulb crops in Washington State and Idaho. *Crop Protection*, *96*, 265–272.

Dutt, M., Dhekney, S. A., Soriano, L., Kandel, R., & Grosser, J. W., (2014). Temporal and spatial control of gene expression in horticultural crops. *Horticulture Research*, *1*, 14047.

Eckert, J. W., & Brown, G. E., (1986). Postharvest *Citrus* diseases and their control. *Fresh Citrus Fruits*, 315–360.

Eckert, J. W., & Ogawa, J. M., (1988). The chemical control of postharvest diseases: Deciduous fruits, berries, vegetables, and root/tuber crops. *Annual Review of Phytopathology*, *26*(1), 433–469.

Eckert, J. W., (1990). Recent developments in the chemical control of postharvest diseases. In: *Symposium on Tropical Fruit in International Trade* (Vol. 269, pp. 477–494).

Elad, Y., Vivier, M., & Fillinger, S., (2016). Botrytis, the good, the bad and the ugly. *Botrytis–The Fungus, the Pathogen, and its Management in Agricultural Systems*, 1–15.

Elitzur, T., Yakir, E., Quansah, L., Zhangjun, F., Vrebalov, J., Khayat, E., & Friedman, H., (2016). Banana MaMADS transcription factors are necessary for fruit ripening and molecular tools to promote shelf-life and food security. *Plant Physiology*, *171*(1), 380–391.

Emma, O., Theresa, N., & Ucheoma, O., (2013). Control of postharvest bacteria l diseases of tomato in Abia State, South Eastern Nigeria. *Journal of Biology, Agriculture and Healthcare*, *3*(19), 24–30.

FAO Report: "How to Feed the World in 2050." https://www.fao.org/ (accessed on 1 March 2024).

Gálvez, L., & Palmero, D., (2022). *Fusarium* dry rot of garlic bulbs caused by *Fusarium proliferatum:* A review. *Horticulturae*, *8*(7), 628.

Ghosh, S., Irfan, M., & Datta, A., (2017). Application of RNA silencing in improving plant traits for industrial use. In: Dalmay, T., (ed.), *Plant Gene Silencing: Mechanisms and Applications* (pp. 128–146). CABI: Wallingford, UK.

Gold, C. S., Pinese, B., & Pena, J. E., (2002). Pests of banana. In: Pena, J. E., Sharp, J., & Wysoki, M., (eds.), *Pests and Pollinators of Tropical Fruit* (pp. 13–56). Wallingford: CABI Int.

Hanagasaki, T., Yamashiro, M., Gima, K., Takushi, T., & Kawano, S., (2021). Characterization of *Erwinia* sp. causing black rot of papaya (*Carica papaya*) first recorded in Okinawa Main Island, Japan. *Plant Pathology*, *70*(4), 932–942.

Hao, G., Pitino, M., Duan, Y., & Stover, E., (2016). Reduced susceptibility to *Xanthomonas citri* in transgenic *Citrus* expressing the FLS2 receptor from *Nicotiana benthamiana. Molecular Plant-Microbe Interactions*, *29*(2), 132–142.

Hasan, F., & Ansari, M. S., (2010). *Cabbage White Butterfly Pieris brassicae (Linn.): Bionomics and Management* (p. 144). LAP Lambert Acad. Publ.

Heathcote, G. D., (1962). The suitability of some plant hosts for the development of the peach-potato aphid, *Myzus persicae* (Sulzer). *Entomologia Experimentalis et Applicata*, *5*(2), 114–118.

Hewajulige, I. G. N., & Wijesundera, R. L. C., (2014). *Thielaviopsis paradoxa*, *Thielaviopsis basicola* (black rot, black root rot). In: *Postharvest Decay* (pp. 287–308). Academic Press.

Hossain, M. S., Sarkar, B. C., Hossain, M. M., Mian, M. Y., Rajotte, E. G., Muniappan, R., & O'Rourke, M. E., (2020). Comparison of biorational management approaches against mango fruit fly (*Bactrocera dorsalis* Hendel) in Bangladesh. *Crop Protection*, *135*, 104807.

Hu, C., Sheng, O., Deng, G., He, W., Dong, T., Yang, Q., & Bi, F., (2021). CRISPR/Cas9-mediated genome editing of MaACO1 (aminocyclopropane-1-carboxylate oxidase 1) promotes the shelf life of banana fruit. *Plant Biotechnology Journal*, *19*(4), 654.

Huang, T. C., Burr, T. J., Smith, C. A., & Matteson, M. C., (1997). Management of apple blister spot caused by streptomycin-resistant populations of *Pseudomonas syringae* pv. papulans. In: *Pseudomonas syringae Pathovars and Related Pathogens* (pp. 590–595). Dordrecht: Springer Netherlands.

Hubert, J., Fourrier, C., Laplace, D., & Ioos, R., (2014). First report of pineapple black rot caused by *Ceratocystis paradoxa* on *Ananas comosus* in French Guiana. *Plant Disease*, *98*(11), 1584–1584.

Inclán, D. J., Bermúdez, F. J., Alvarado, E., Ellis, M., Williams, R. N., & Acosta, N., (2008). Comparison of biological and conventional insecticide treatments for the management of the pineapple fruit borer, *Strymon megarus* (*Lepidoptera: Lycaenidae*) in costa Rica. *Ecological Engineering*, *34*(4), 328–331.

Intan, S. M. A., Suzianti, I. V., & Latiffah, Z., (2013). First report of *Colletotrichum gloeosporioides* causing anthracnose of banana (*Musa* spp.) in Malaysia. *Plant Disease*, *97*(7), 991–991.

Irfan, M., Ghosh, S., Meli, V. S., Kumar, A., Kumar, V., Chakraborty, N., Chakraborty, S., & Datta, A., (2016). Fruit ripening regulation of α-mannosidase expression by the MADS box transcription factor RIPENING INHIBITOR and ethylene. *Frontier Plant Science, 7*, 10.

Irfan, M., Kumar, P., Ahmad, M. F., & Siddiqui, M. W., (2023). Biotechnological interventions in reducing losses of tropical fruits and vegetables. *Current Opinion in Biotechnology, 79*, 102850. doi.org/10.1016/j.copbio.2022.102850.

Irfan, M., Kumar, P., Kumar, V., & Datta, A., (2022). Fruit ripening specific expression of β-D-N-acetylhexosaminidase (β-Hex) gene in tomato is transcriptionally regulated by ethylene response factor SlERF.E4. *Plant Science, 323*, 111380.

Islam, M. N., Ali, M. S., Choi, S. J., Hyun, J. W., & Baek, K. H., (2019). Biocontrol of *Citrus* canker disease caused by *Xanthomonas citri* subsp. citri using an endophytic *Bacillus thuringiensis*. *The Plant Pathology Journal, 35*(5), 486.

Ismail, M., & Zhang, J., (2004). Post-harvest *Citrus* diseases and their control. *Outlooks on Pest Management, 15*(1), 29.

Ismail, S., Jiang, B., Nasimi, Z., Inam-ul-Haq, M., Yamamoto, N., Danso, O. A., & Zheng, A., (2020). Investigation of *Streptomyces scabies* causing potato scab by various detection techniques, its pathogenicity and determination of host-disease resistance in potato germplasm. *Pathogens, 9*(9), 760.

Isman, M. B., & Machial, C. M., (2006). Pesticides based on plant essential oils: From traditional practice to commercialization. *Advances in Phytomedicine, 3*, 29–44.

Jia, H., Orbovic, V., Jones, J. B., & Wang, N., (2016). Modification of the PthA4 effector binding elements in type I Cs LOB 1 promoter using Cas9/sg RNA to produce transgenic Duncan grapefruit alleviating XccΔpthA4: DCs LOB 1.3 infection. *Plant Biotechnology Journal, 14*(5), 1291–1301.

Kambrekar, D. N., (2016). Management of legume podborer, *Helicoverpa armigera* with host plant resistance. *Legume Genomics and Genetics, 7*(5), 1–19.

Kamle, M., & Kumar, P., (2016). *Colletotrichum gloeosporioides*: Pathogen of anthracnose disease in mango (*Mangifera indica* L.). *Current Trends in Plant Disease Diagnostics and Management Practices*, 207–219.

Khan, I. H., & Javaid, A., (2021). Molecular characterization of *Penicillium expansum* associated with blue mold disease of apple in Pakistan. *Pak. J. Bot., 53*(6), 2299–2303.

Klich, M. A., (2007). Aspergillus flavus: The major producer of aflatoxin. *Molecular Plant Pathology, 8*(6), 713–722.

Ko, Y., Liu, C. W., Chen, C. Y., Maruthasalam, S., & Lin, C. H., (2009). First report of stem-end rot of mango caused by *Phomopsis mangiferae* in Taiwan. *Plant Disease, 93*(7), 764.

Kumar, V., Irfan, M., Ghosh, S., Chakraborty, N., Chakraborty, S., & Datta, A., (2016). Fruit ripening mutants reveal cell metabolism and redox state during ripening. *Protoplasma, 253*, 581–594. https://doi.org/10.1007/s00709-015-0836-z.

Kumari, C., Sharma, M., Kumar, V., Sharma, R., Kumar, V., Sharma, P., Kumar, P., & Irfan, M., (2022). Genome editing technology for genetic amelioration of fruits and vegetables for alleviating post-harvest loss. *Bioengineering, 9*(4), 176.

Kyrkou, I., Pusa, T., Ellegaard-Jensen, L., Sagot, M. F., & Hansen, L. H., (2018). Pierce's disease of grapevines: A review of control strategies and an outline of an epidemiological model. *Frontiers in Microbiology, 9*, 2141.

Li, Z., Dos, S. R. F., Gao, L., Chang, P., & Wang, X., (2021). Current status and future prospects of grapevine anthracnose caused by *Elsinoe ampelina*: An important disease in humid grape-growing regions. *Molecular Plant Pathology, 22*(8), 899–910.

Lin, W. C., Lu, C. F., Wu, J. W., Cheng, M. L., Lin, Y. M., Yang, N. S., & Cheng, C. P., (2004). Transgenic tomato plants expressing the *Arabidopsis* NPR1 gene display enhanced resistance to a spectrum of fungal and bacterial diseases. *Transgenic Research, 13*, 567–581.

Liu, Y. B., (2013). Controlled atmosphere treatment for control of grape mealybug, *Pseudococcus maritimus* (Ehrhorn) (*Hemiptera: Pseudococcidae*), on harvested table grapes. *Postharvest Biology and Technology*, *86*, 113–117.

Low, Y. C., Lawton, M. A., & Di, R., (2020). Validation of barley 2OGO gene as a functional orthologue of *Arabidopsis* DMR6 gene in *Fusarium* head blight susceptibility. *Scientific Reports*, *10*(1), 1–13.

Mahadevakumar, S., Yadav, V., Tejaswini, G. S., Sandeep, S. N., & Janardhana, G. R., (2014). First report of *Phomopsis citri* associated with dieback of citrus lemon in India. *Plant Disease, 98*(9), 1281.

Malaj, E., Von Der, O. P. C., Grote, M., Kühne, R., Mondy, C. P., Usseglio-Polatera, P., & Schäfer, R. B., (2014). Organic chemicals jeopardize the health of freshwater ecosystems on the continental scale. *Proceedings of the National Academy of Sciences*, *111*(26), 9549–9554.

Mansfield, J., Genin, S., Magori, S., Citovsky, V., Sriariyanum, M., Ronald, P., & Foster, G. D., (2012). Top 10 plant pathogenic bacteria in molecular plant pathology. *Molecular Plant Pathology*, *13*(6), 614–629.

Mathiazhagan, M., Chidambara, B., Hunashikatti, L. R., & Ravishankar, K. V., (2021). Genomic approaches for improvement of tropical fruits: Fruit quality, shelf life and nutrient content. *Genes*, *12*(12), 1881.

McGlynn, K. A., Abnet, C. C., Zhang, M., Sun, X. D., Fan, J. H., O'Brien, T. R., & Qiao, Y. L., (2006). Serum concentrations of 1, 1, 1-trichloro-2, 2-bis (p-chlorophenyl) ethane (DDT) and 1, 1-dichloro-2, 2-bis (p-chlorophenyl) ethylene (DDE) and risk of primary liver cancer. *Journal of the National Cancer Institute*, *98*(14), 1005–1010.

McGovern, R. J., Seijo, T. E., Hendricks, K., & Roberts, P. D., (2012). New report of *Colletotrichum gloeosporioides* causing postbloom fruit drop on *Citrus* in Bermuda. *Canadian Journal of Plant Pathology*, *34*(2), 187–194.

Menz, J., Modrzejewski, D., Hartung, F., Wilhelm, R., & Sprink, T., (2020). Genome edited crops touch the market: A view on the global development and regulatory environment. *Frontiers in Plant Science*, *11*, 586027.

Mpumi, N., Machunda, R. S., Mtei, K. M., & Ndakidemi, P. A., (2020). Selected insect pests of economic importance to *Brassica oleracea,* their control strategies and the potential threat to environmental pollution in Africa. *Sustainability*, *12*(9), 3824.

Muedi, H. T., Fourie, D., & McLaren, N. W., (2015). Distribution and severity of bacterial brown spot on dry beans in South Africa: An update. *South African Journal of Science*, *111*(3, 4), 1–6.

Muniappan, R., Shepard, B. M., Watson, G. W., Carner, G. R., Sartiami, D., Rauf, A., & Hammig, M. D., (2008). First report of the papaya mealybug, *Paracoccus marginatus* (*Hemiptera*: *Pseudococcidae*), in Indonesia and India. *Journal of Agricultural and Urban Entomology*, *25*(1), 37–40.

Nowicki, M., Nowakowska, M., Niezgoda, A., & Kozik, E., (2012). Alternaria black spot of crucifers: Symptoms, importance of disease, and perspectives of resistance breeding. *Vegetable Crops Research Bulletin*, *76*.

Nybom, H., Ahmadi-Afzadi, M., Rumpunen, K., & Tahir, I., (2020). Review of the impact of apple fruit ripening, texture, and chemical contents on genetically determined susceptibility to storage rots. *Plants*, *9*(7), 831.

Ortigosa, A., Gimenez-Ibanez, S., Leonhardt, N., & Solano, R., (2019). Design of a bacterial speck resistant tomato by CRISPR/Cas9-mediated editing of SlJAZ2. *Plant Biotechnology Journal*, *17*(3), 665–673.

Pantoja, A., Follett, P. A., & Villanueva-Jiménez, J. A., (2002). Pests of pawpaw. In: *Tropical Fruit Pests and Pollinators: Biology, Economic Importance, Natural Enemies, and Control* (pp. 131–156). Wallingford UK: CABI Publishing.

Pashley, D. P., (1988). Current status of fall armyworm host strains. *Florida Entomologist*, 227–234.

Peng, A., Chen, S., Lei, T., Xu, L., He, Y., Wu, L., & Zou, X., (2017). Engineering canker-resistant plants through CRISPR/Cas9-targeted editing of the susceptibility gene CsLOB1 promoter in *Citrus*. *Plant Biotechnology Journal*, *15*(12), 1509–1519.

Perlak, F. J., Deaton, R. W., Armstrong, T. A., Fuchs, R. L., Sims, S. R., Greenplate, J. T., & Fischhoff, D. A., (1990). Insect resistant cotton plants. *Bio/technology*, *8*(10), 939–943. Nature Publishing Company.

Petrasch, S., Silva, C. J., Mesquida-Pesci, S. D., Gallegos, K., Van, D. A. C., Papin, V., & Blanco-Ulate, B., (2019). Infection strategies deployed by *Botrytis cinerea, Fusarium acuminatum,* and *Rhizopus stolonifer* as a function of tomato fruit ripening stage. *Frontiers in Plant Science*, *10*, 223.

Pitt, J. I., & Hocking, A. D., (2009). *Fungi and Food Spoilage* (Vol. 519, p. 388). New York: Springer.

Pixley, K. V., Falck-Zepeda, J. B., Paarlberg, R. L., Phillips, P. W., Slamet-Loedin, I. H., Dhugga, K. S., & Gutterson, N., (2022). Genome-edited crops for improved food security of smallholder farmers. *Nature Genetics*, *54*(4), 364–367.

Queyrel, W., Habets, F., Blanchoud, H., Ripoche, D., & Launay, M., (2016). Pesticide fate modeling in soils with the crop model STICS: Feasibility for assessment of agricultural practices. *Science of the Total Environment*, *542*, 787–802.

Rahman, M. A., Mahmud, T. M. M., Kadir, J., Abdul, R. R., & Begum, M. M., (2008). Major postharvest fungal diseases of papaya cv. 'Sekaki' in Selangor, Malaysia. *Pertanika J. Trop. Agric. Sci.*, *31*(1), 27–34.

Rahman, S. M. E., Khan, I., & Oh, D. H., (2016). Electrolyzed water as a novel sanitizer in the food industry: Current trends and future perspectives. *Comprehensive Reviews in Food Science and Food Safety*, *15*(3), 471–490.

Raman, R., (2017). The impact of genetically modified (GM) crops in modern agriculture: A review. *GM Crops & Food*, *8*(4), 195–208.

Richael, C. M., (2021). Development of the genetically modified Innate® potato. *Plant Breeding Reviews*, *44*, 57–78.

Rodríguez, A., Shimada, T., Cervera, M., Alquézar, B., Gadea, J., Gómez-Cadenas, A., & Peña, L., (2014). Terpene down-regulation triggers defense responses in transgenic orange leading to resistance against fungal pathogens. *Plant Physiology*, *164*(1), 321–339.

Rodríguez, A., Shimada, T., Cervera, M., Redondo, A., Alquézar, B., Rodrigo, M. J., & Pena, L., (2015). Resistance to pathogens in terpene down-regulated orange fruits inversely correlates with the accumulation of D-limonene in peel oil glands. *Plant Signaling & Behavior*, *10*(6), e1028704.

Sable, P. A., & Chandaragi, M. K., (2021). Onion maggot (*Delia antiqua)*: A destructive pest of onion. *Indian Horticulture*, *66*(1).

Salunkhe, D. K., & Desai, B. B., (1988). Postharvest biotechnology of fruits. In: *Postharvest Biotechnology of Fruits* (p. 147). CRC Press, Inc.

Sanahuja, G., Ploetz, R. C., Lopez, P., Konkol, J. L., Palmateer, A. J., & Pruvost, O., (2016). Bacterial canker of mango, *Mangifera indica,* Caused by *Xanthomonas citri* pv. mangiferaeindicae, confirmed for the first time in the Americas. *Plant Disease*, *100*(12), 2520.

Sandhi, R. K., & Reddy, G. V., (2020). Biology, ecology, and management strategies for pea aphid (*Hemiptera: Aphididae*) in pulse crops. *Journal of Integrated Pest Management*, *11*(1), 18.

Sharma, S., Shree, B., Aditika, Sharma, A., Irfan, M., & Kumar, P., (2022). Nanoparticle-based toxicity in perishable vegetable crops: Molecular insights, impact on human health and mitigation strategies for sustainable cultivation. *Environmental Research, 212*(A), 113168. doi.org/10.1016/j.envres.2022.113168.

Silva, A. C., Ricalde, M. P., Scalzer, R. R. C., Zilli, J. E., & Lopes, R. B., (2022). Natural occurrence of *Beauveria bassiana* on adults of the invasive mango seed weevil *Sternochetus mangiferae* (*Coleoptera: Curculionidae*) in Brazil. *Journal of Plant Diseases and Protection, 129*(1), 79–84.

Singh, N., Dotasara, S. K., Jat, S. M., & Naqvi, A. R., (2017). Assessment of crop losses due to tomato fruit borer, *Helicoverpa armigera* in tomato. *J. Entomol. Zool. Stud.*, *5*, 595–597.

Soares, J. M., Rocha, A. J., Nascimento, F. S., Santos, A. S., Miller, R. N., Ferreira, C. F., & Amorim, E. P., (2021). Genetic improvement for resistance to black Sigatoka in bananas: A systematic review. *Frontiers in Plant Science*, *12*, 657916.

Sun, K., Van, T. A., Van, K. J. A., Wolters, A. M. A., Jacobsen, E., Visser, R. G., & Bai, Y., (2017). Silencing of DND1 in potato and tomato impedes conidial germination, attachment, and hyphal growth of *Botrytis cinerea. BMC Plant Biology*, *17*, 1–12.

Sundin, G. W., Castiblanco, L. F., Yuan, X., Zeng, Q., & Yang, C. H., (2016). Bacterial disease management: Challenges, experience, innovation, and future prospects: Challenges in bacterial molecular plant pathology. *Molecular Plant Pathology*, *17*(9), 1506–1518.

Tanigoshi, L. K., & Moffitt, L. J., (1984). *Citrus* thrips, *Scirtothrips citri* (*Thysanoptera: Thripid*ae) and 'Washington' navel orange injury. *Environmental Entomology*, *13*(5), 1312–1318.

Tayal, R., Kumar, V., & Irfan, M., (2022). Harnessing the power of hydrogen sulphide (H_2S) for improving fruit quality traits. *Plant Biology*, *24*(4), 594–601.

Tripathi, J. N., Ntui, V. O., Shah, T., & Tripathi, L., (2021). CRISPR/Cas9-mediated editing of DMR6 orthologue in banana (*Musa* spp.) confers enhanced resistance to bacterial disease. *Plant Biotechnology Journal*, *19*(7), 1291.

Uloth, M. B., You, M. P., Finnegan, P. M., Banga, S. S., Banga, S. K., Sandhu, P. S., & Barbetti, M. J., (2013). New sources of resistance to *Sclerotinia sclerotiorum* for crucifer crops. *Field Crops Research*, *154*, 40–52.

Vandendriessche, T., Keulemans, J., Geeraerd, A., Nicolai, B. M., & Hertog, M. L., (2012). Evaluation of fast volatile analysis for detection of *Botrytis cinerea* infections in strawberry. *Food Microbiology*, *32*(2), 406–414.

Ventura, J. A., Costa, H., & Tatagiba, J. D. S., (2004). Papaya diseases and integrated control. *Diseases of Fruits and Vegetables: Volume II: Diagnosis and Management*, 201–268.

Vicente, J. G., & Holub, E. B., (2013). *Xanthomonas campestris* pv. campestris (cause of black rot of crucifers) in the genomic era is still a worldwide threat to brassica crops. *Molecular plant Pathology*, *14*(1), 2–18.

Vishnevetsky, J., White, T. L., Palmateer, A. J., Flaishman, M., Cohen, Y., Elad, Y., & Perl, A., (2011). Improved tolerance toward fungal diseases in transgenic *Cavendish banana* (*Musa* spp. AAA group) cv. grand Nain. *Transgenic Research*, *20*, 61–72.

Waghunde, R. R., Jadhav, P. B., Padhiar, B. V., Mehta, B. P., Patel, A. N., & Sabalpara, A. N., (2014). *Machrophomina* and stem end rot-new post-harvest diseases of papaya in Gujarat. *Journal of Plant Disease Sciences*, *9*(1), 98, 99.

Wally, O., Jayaraj, J., & Punja, Z. K., (2009). Broad-spectrum disease resistance to necrotrophic and biotrophic pathogens in transgenic carrots (*Daucus carota* L.) expressing an *Arabidopsis* NPR1 gene. *Planta*, *231*, 131–141.

Waltz, E., (2022). GABA-enriched tomato is first CRISPR-edited food to enter market. *Nat. Biotechnol.*, *40*(1), 9–11.

Wang, L., Chen, S., Peng, A., Xie, Z., He, Y., & Zou, X., (2019). CRISPR/Cas9-mediated editing of CsWRKY22 reduces susceptibility to *Xanthomonas Citri* subsp. *Citri* in Wanjincheng orange (*Citrus sinensis* (L.) Osbeck). *Plant Biotechnology Reports*, *13*, 501–510.

Wang, M., Weiberg, A., Lin, F. M., Thomma, B. P., Huang, H. D., & Jin, H., (2016). Bidirectional cross-kingdom RNAi and fungal uptake of external RNAs confer plant protection. *Nature Plants*, *2*(10), 1–10.

Wang, X., Tu, M., Wang, D., Liu, J., Li, Y., Li, Z., & Wang, X., (2018). CRISPR/Cas9-mediated efficient targeted mutagenesis in grape in the first generation. *Plant Biotechnology Journal*, *16*(4), 844–855.

Weavers, L. K., & Wickramanayake, G. B., (2001). Disinfection and sterilization using ozone. In: Block, S. S., (ed.), *Disinfection, Sterilization, and Preservation* (5th edn., pp. 205–214). Philadelphia: Lippincott Williams and Wilkins.

Williamson, B., Tudzynski, B., Tudzynski, P., & Van Kan, J. A., (2007). *Botrytis cinerea:* The cause of grey mould disease. *Molecular Plant Pathology*, *8*(5), 561–580.

Young, A. J., Pathania, N., Manners, A., & Pegg, K. G., (2022). Heart rot of Australian pineapples caused by *Dickeya zeae*. *Australasian Plant Pathology*, *51*(5), 525–533.

Yu, L., Lan, G., Yang, Y., Tang, Y., Li, Z., She, X., & He, Z., (2022). First report of anthracnose caused by *Colletotrichum fructicola* on *Brassica parachinensis* in China. *Crop Protection*, *154*, 105842.

Zhan, X., Khan, R. A. A., Zhang, J., Chen, J., Yin, Y., Tang, Z., & Liu, T., (2023). Control of postharvest stem-end rot on mango by antifungal metabolites of *Trichoderma pinnatum* LS029-3. *Scientia Horticulturae*, *310*, 111696.

CHAPTER 8

PROMOTER ENGINEERING FOR IMPROVING FRUIT AND VEGETABLE TRAITS

EHASANULLAH KHAN,[1] KAMLESH KUMAR SONI,[2] and MOHAMMAD IRFAN[3]

[1]*Department of Biotechnology, Jamia Millia Islamia, New Delhi, India*

[2]*Institute of Plant and Microbial Biology, Academia Sinica, Taipei, Taiwan*

[3]*Plant Biology Section, School of Integrative Plant Science, Cornell University, Ithaca, New York, USA*

ABSTRACT

Genetic engineering aims to enhance desirable traits in target plants through the precise introduction of specific genetic regions. Central to this endeavor is the modulation of gene expression, which is influenced by the strength of the promoter fused to the gene of interest. Promoters, located upstream of the transcription start site, contain essential cis-acting elements and binding sites for transcription factors crucial for gene regulation. The dynamic nature of promoter activity, regulated by internal and external environmental variations, underscores its significance in genetic engineering strategies. Specifically, in the context of fruit and vegetable traits, promoters play a crucial role in regulating processes such as fruit ripening, senescence, and postharvest characteristics. Through promoter engineering, researchers can target and modulate gene expression to

Bioengineered Fruit and Vegetables: Trends in Postharvest Quality Improvement.
Mohammad Irfan and Pankaj Kumar (Eds.)

enhance desired traits, such as prolonged shelf-life, improved nutritional quality, and enhanced stress tolerance. Techniques such as in-silico prediction, PCR-based isolation, and bioinformatics analysis facilitate the identification and characterization of promoters for tailored gene expression. This chapter discusses various types of promoters, including constitutive, tissue-specific, fruit-specific, flower-specific, seed-specific, and synthetic promoters, highlighting their structural features and functional significance in genetic engineering approaches. Moreover, the chapter underscores the importance of cautious promoter selection to achieve desired expression patterns while minimizing adverse environmental effects, emphasizing the critical role of promoter engineering in the development of genetically modified crops with enhanced agricultural productivity and sustainability.

8.1 INTRODUCTION

The application of genetic engineering (GE) has been playing a significant role in improving agricultural products to enhance the food and feed values (Irfan & Datta, 2017). The main aim of GE is to introduce the specific genetic region to improve the desirable traits in the target plant for both temporal and spatial expression (Kumar et al., 2016a). Considering the success of GE in developing stable transgenics depends on the expression of the introduced transgene on the host or target plant (Irfan & Datta, 2017). The expression of any gene varies based on the promoter strength fused to that gene. Therefore, the promoter plays a vital role in GE approaches by regulating gene expression. It contains RNA polymerase binding sites and binding sites for the transcription factor important for gene expression. These transcription factors' (TFs) expression, in turn, is regulated under different conditions such as fluctuations in the internal or external cell environment. Thus, binding elements on the promoter and TFs are cryptic components involved in the activation or repression of any gene expression leading to cell functions. The promoter is the stretch of DNA upstream to the transcription start site (TTS) having many cis-acting elements and providing binding sites for proteins (generally TFs) involved in the initiation and regulation of transcription (Kummari et al., 2020). Hence, the identification of cis-element(s) and their corresponding binding transcription factor subsequently helps the modulation of targeted gene expression to achieve a substantial amount of final products with desirable traits. Based on the target tissue and traits, promoters can be selected to develop transgenic plants (Biłas et al., 2016).

During fruit development, cell division and expansion regulate and control the fruit size. After their development, fruits are destined to ripen and senescence (Ghosh et al., 2013; Kumar et al., 2016b). During postharvest, fruit respiration and transpiration continue, regulating the ripening and senescence (Kumari et al., 2022). This eventually affects the shelf-life and postharvest stability of fruits and vegetables. Respiration controls and utilizes the stored carbohydrates, while transpiration regulates the moisture content of the fruits (Kumari et al., 2022; Irfan et al., 2023). Therefore, regulation of these factors is the key process to controlling excessive fruit ripening during postharvest management and storage (Kumari et al., 2022; Irfan et al., 2023). Moreover, other factors involved in ripening are phytohormones such as ethylene, abscisic acid (ABA), carotenoids, and other gaseous signaling molecules such as H_2S, NO, and H_2O_2 (Tayal et al., 2022). Among them, ethylene has long been accepted as the chief natural ripening phytohormone (Grierson, 2013). Cell wall metabolism is also a key factor that regulates the ripening of fruits (Canton et al., 2020). Therefore, control of ripening or ripening-associated fruit softening at biochemical and physiological levels has been a standard practice of biotechnology. However, there are few studies on the regulation of ripening at the molecular or genetic level that are established. For instance, an ethylene response factors (ERFs) family protein SlERF.F12a, which contains EAR motifs, has been demonstrated to negatively regulate the onset of tomato fruit ripening (Deng et al., 2022). Considerable efforts have been made in the last few decades to identify the regulatory mechanism underlying fruit ripening and cell wall metabolism to enhance fruit and vegetable life (Aghdam et al., 2013). Most of the study involves the identification of promoter regions of genes that are activated post-harvesting. By modulating the gene expression via different constitutive and tissue-specific promoters, scientists have targeted various fruit quality traits that control ripening and senescence, particularly at the postharvest level. The knowledge about the mechanism of gene expression is now well understood, and scientists use this knowledge to control gene activity by regulating promoters. Many scientists have studied the region of the promoter which is very active and responsible for putative and constitutive expression of the targeted gene (Hernandez-Garcia & Finer, 2014b).

8.2 PROMOTER: ITS STRUCTURE AND FUNCTION

Since the emergence of biotechnology, promoters have become an important tool to control the expression of target genes (Dasgupta et al., 2020).

Promoters are the region upstream of the gene sequence that is broadly divided into the core promoter region, proximal promoter region, and distal promoter region (Figure 8.1). The core region generally covers from -50 bp to +50 bp and mainly comprises the TATA box, initiator sequence, and a binding site for TFIID, TFIIH, and other subsidiary TFIIs, which help in recruiting the RNA polymerase. Not every but in most promoters, the proximal region contains the CAAT box and GC box. It also contains activator and/or repressor binding elements that are essential in controlling the promoter activity. The distal region of the promoter has some enhancer, insulator, or silencer sequences. These enhancers or silencers can work in a position and orientation-independent manner (Yasmeen et al., 2023). These enhancers can be located upstream or downstream of the gene and sometimes in the intronic region of the gene (Pennacchio et al., 2015). The core and proximal regions of the promoter provide the fundamentals for promoter engineering. The upstream regions have been intensively studied in the last decade (Dutt et al., 2014).

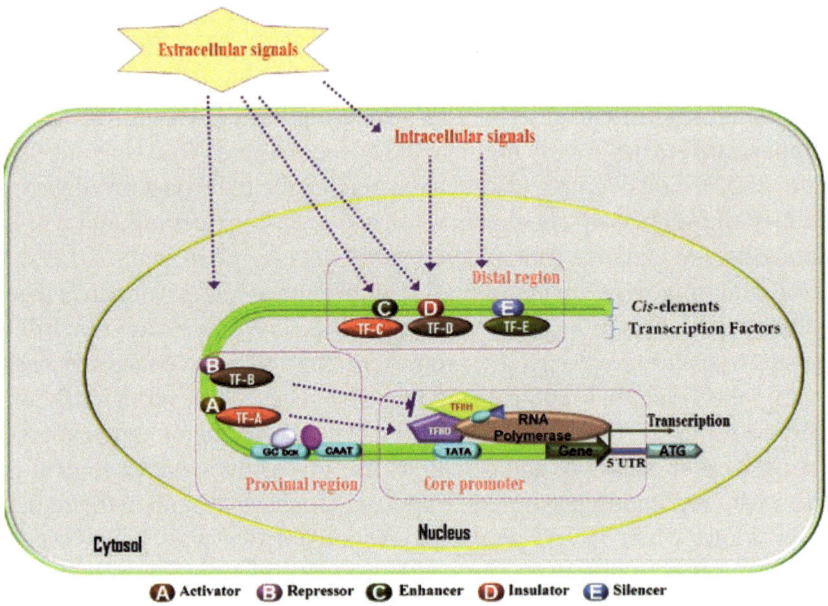

FIGURE 8.1 Pictorial representation of promoter architecture: Core promoters have a TATA box and TFIIs. The proximal region contains many cis-elements that recruit activators and/or repressors. The distal region mainly contains enhancers and silencers.

8.3 PROMOTER ENGINEERING

The fundamentals of promoter engineering involve identification, isolation, and characterization. The promoter region is predicted *in-silico* using bioinformatics tools such as Plant Promoter DB and bidirectional encoder representations from transformers (BERT). After successful bioinformatics analysis of promoters, isolation can be achieved using traditional Polymerase Chain Reaction, also known as PCR, with the promoter's specific oligo primer pair, followed by cloning in a suitable vector and transformation into desired plants (Brown, 2020). Thus, to improve the quality traits of fruits and vegetables, promoter engineering is crucial through targeted expression (Timerbaev & Dolgov, 2019). To date, many fruit-specific promoters have been identified and employed for gene expression in vegetables and fruits such as tomatoes, strawberries, lemons, grapes, bananas, and cucumbers (Erpen et al., 2018; Ghosh et al., 2012; Jiang et al., 2018; Shahan et al., 2019; Unni et al., 2012; Wai et al., 2020). In terms of expression, promoters can be classified into strong and weak promoters. The strong promoter yields maximum mRNA expression than the weak promoters (Liu et al., 2018). In response to external and internal signals and for crop improvement, tissue-specific promoter engineering in plants holds great promise for understanding the complexity of genetics. As mentioned above, a promoter carries many regulatory elements for its tissue specificity, inducibility, and/or constitutive expression. Identification of these *cis*-elements is important, and one can predict the function of a gene based on the known regulatory elements present on its promoter. For instance, apart from the core region, the AEG1 promoter contains tapetum-specific *cis*-elements and cis-elements for root-specific expression. By deleting and/or mutating the root-specific *cis*-elements, the promoter was made tapetum-specific (Soni et al., 2021). In recent years, scientists have provided detailed mechanistic insights into tissue-specific promoters that control the expression of genes in various organs of plants (Bhat et al., 2010a). In the next sections, based on their expression pattern, we will discuss different types of promoters and their usage in GE for improving the fruit quality traits.

8.4 CONSTITUTIVE PROMOTER

In plants, the 35S promoter of cauliflower mosaic virus (CaMV35S) is the most widely used promoter for constitutive expression in specific tissues

(Odell et al., 1985a). In dicotyledonous plants, the 35S promoter allows a high level of expression in tissues. Odell et al. (1985a) demonstrated that despite its viral origin, CaMV35S 5′-fragments could control the expression of genes in tobacco plants. A 343 bp full-length promoter equipped with various regulatory elements such as a TATA box around –28 bp from the transcription initiation site is sufficient to drive high gene expression. It also consists of a CCAAT box at positions –85, –64, –57 (Ow et al., 1987). Another classical example of a strong constitutive promoter was isolated from the cestrum yellow leaf curling virus (CmYLCV). The promoter of CmYLCV is highly active in various tissues of monocotyledonous and dicotyledonous plants. This highly expressed promoter is comparable to or more active than frequently used promoters such as CaMV35S and ubiquitin 1 promoter. The limited host range of the CmYLCV promoter is an attractive tool for the regulation of gene expression in a wide variety of plant species. Higher activity of the CmYLCV promoter in callus and meristematic cells indicates that the promoter is useful for most cells and tissue types (Stavolone et al., 2003). In addition to the above constitutive promoters, polyubiquitin-1 (Ubi-1) from *Zea mays* (maize) is also one of the highest expressed constitutive genes characterized in plants and is induced by a heat shock treatment. In several monocot species, together with a 1 kb intron immediately to the start codon, a 0.9 kb 5′ flanking sequence of *Ubi-1* is sufficient to induce expression of the fused gene in the monocot species (Streatfield et al., 2004). The Ubi-1 promoter consists of *cis*-elements (5′ flanking sequence; 0.9 kb), a TATA box in a consensus position, and two overlapping heat shock domains approximately 0.2 kb upstream of the transcription start site. Additionally, Khan et al. (2015) suggested the potential use of the Rep gene promoter derived from *cotton leaf curl Burewala virus* (CLCuBuV) as a constitutive for the development of transgenic. The study revealed the bidirectional activity of the *Rep* gene promoter and its constitutive expression in the leaf of *Nicotiana tabacum cv*. It was also noted that the activity of the *Rep* promoter was higher than the 35S promoter. Serine carboxypeptidase (SCP) and serine carboxypeptidase-like (SCPL) proteins are hydrolyzing enzymes that catalyze the hydrolysis of the C-terminal peptide bond in proteins or peptides (Breddam, 1986; Feng & Xue, 2006). Studies have shown that the SCP/SCPL gene family is usually expressed constitutively in all tissues in higher plants and has diverse biological processes, with some of them being expressed in specific tissues. Jiang et al. (2018)

characterized the *Arabidopsis AtSCPL30* gene promoter in tobacco (*Nicotiana benthamiana*). PD1 (promoter deletion 1), a wild-type SCLP30 promoter located 2,039 bp upstream to ATG, was found to be constitutive. Through 5′ promoter deletion analysis, it was determined that deletion up to –189 bp was adequate for constitutive expression in various tissues at different developmental stages. Among the deletion promoters (DP), PD2–PD7 exhibited constitutive activities that were two-fold higher than the CaMV35S promoter in transgenic tobacco plants. PD7, with a length of 456 bp, was identified as sufficient for achieving a high level of transgene activity. However, constitutive expression of certain proteins may have detrimental effects on plants. To mitigate these adverse effects, researchers have suggested using tissue-specific promoters to target the expression of the most essential gene products. They have developed tissue-specific promoters to enhance the quality of economically significant plant parts (Dey et al., 2015; Hernandez-Garcia & Finer, 2014a).

8.5 FRUIT-SPECIFIC PROMOTER

Fruit-specific promoters serve as ideal tools to study the regulation of genes involved in the ripening of climacteric and non-climacteric fruits, the nutritional quality of fruits, and other crop improvement programs. To improve the fruit's shelf-life during postharvest storage, the identification of the active promoter region is a critical step in promoter engineering. In recent years, various fruit-specific promoters and their regulatory sites have been identified in various plants (Dutt et al., 2014). Ethylene is a gaseous hormone mainly responsible for fruit ripening. Extensive studies on the ethylene biosynthetic gene and its promoters have been carried out in the last few decades. For instance, E4 and E8 are ethylene-responsive genes, and inhibiting the E4 expression avoids fruit ripening by inhibiting ethylene biosynthesis. On the other hand, E8 influences ethylene synthesis during the flowering and fruiting stage (Yang et al., 2017). Both E4 and E8 become transcriptionally active during the onset of ripening (Deikman et al., 1992; Montgomery et al., 1993). Deletion analysis reveals that minimal 5′ and 3′ regions of the promoter are required for expression during the ripening of tomato fruits (Bhat et al., 2010b; Deikman et al., 1992). In the past, E4 and E8 promoters have been studied extensively to engineer the promoter for fruit quality. Around 1.4 kb of E4 promoter fused with the GUS-reporter gene induces during ethylene response in ripened tomato fruits. Deletion of E4 around 193 bp reduces the GUS activity without affecting ethylene response. The relative to transcription start site positive ethylene-response element found in between nucleotides –161 and –85 (Montgomery et al., 1993). In further analysis of the region between –160 to –90 bp of E4 promoter using the linker scanning mutagenesis (LSM) approach region between –150 to –121 bp is important for ethylene response (Xu et al., 1996). Like E4, E8 first clone from the tomato was also reported as an ethylene-relate biosynthesis gene (Lincoln et al., 1987) E8 is also induced by the ethylene during the onset of ripening. It is spatially and temporally regulated in mature fruits. *1-aminocyclopropane-1-carboxylic acid* (*ACC*) *oxidase* (ACO) and *ACO synthase* (*ACS*) isoforms that involve in the catalysis of enzymes involved in the biosynthesis of ethylene have been studied to identify regulatory cis-element in fruits (Wu et al., 2020). *ACC* and *ACS* are expressed in all vegetative and reproductive tissue variably and produce ethylene (Houben &Van de Poel, 2019). Like E4 and E8 promoters, deletion analysis of *ACC oxidase* gene promoter fused with GUS reporter revealed specific regulatory region during fruits' ripening. The study revealed that *the ACC oxidase gene* promoter requires a 5′ 450 bp sequence for GUS expression. However, that is not enough for fruits to ripen. The sequence required for fruit ripening specificity lies within the –1,966 bp and –1,159 bp sequence (Atkinson et al., 1998b). The *ACS2* and *ACS4* belong to the members of the ACC family and are induced during ripening. The promoter of *ACS2* and *ACS4* contains *cis*-regulatory elements that induce the expression of genes in response to ethylene, wounding, and anaerobiosis. Promoter deletion studies showed that the 373 bp region endorses higher expression of *ACS4* in seeds at different ripening stages (Bhat et al., 2010b). In tomatoes, *the ACS4* promoter has been successfully used to induce the expansins (*EXP1*), which are involved in regulating the texture of the fruit. Overall, the reports suggest that *the ACS* promoter can be a critical target to regulate the genes engaged in the fruit's traits and quality.

N-glycans and N-glycan processing enzymes have been identified as key regulators of fruit ripening in both climacteric and non-climacteric fruits (Ghosh et al., 2011; Meli et al., 2010). For example, *α-Mannosidase (α-Man)* and *β-D-N-acetylhexosaminidase (β-Hex)* are N-glycan processing enzymes specific to ripening, associated with fruit softening. Their suppression using RNAi technology results in enhanced fruit shelf life of tomato and capsicum (Ghosh et al., 2011, 2017; Meli et al., 2010). However, the molecular mechanism underlying their regulation was not understood. Detailed analyses of the promoters of *α-Man* and *β-Hex* genes revealed their positive regulation by the master regulator of fruit ripening inhibitor (RIN), SIERF.E4, and the ripening hormone ethylene (Irfan et al., 2014, 2016, 2022). These examples suggest that *α-Man* and *β-Hex* promoters can be utilized in various crop improvement programs to increase fruit shelf life and postharvest stability by expressing the genes in a fruit ripening-specific manner. *Histidine decarboxylase* (*HDC*), involved in fruit development, encodes an enzyme that catalyzes histidine to histamine (Kumar et al., 2016). Characterization of the *HDC* gene (*SlHDC-A*) promoter in tomato by GUS assay confirms its specificity during fruit development and records 10-fold higher activity than the 35S promoter. Further 5′ deletion analysis followed by quantification of GUS identified the region between –880 bp and –577 bp as essential for its fruit-specific expression, as the deletion of this region abolished the promoter activity (Kim et al., 2019).

8.6 FLOWER-SPECIFIC PROMOTER

Other than the fruit-specific promoter, the flower-specific promoter could be another target of interest, which is relevant to improving the quality of fruits and vegetables. Anthocyanin, a water-soluble pigment, contributes to the color of leaves, flowers, fruits, and stems. It protects plants from UV and high light, and oxidative stress by scavenging reactive oxygen species (ROS) and antimicrobial agents (Tian et al., 2017). Dihydroflavonol reductase (DFR) is a key enzyme involved in anthocyanin biosynthesis and proanthocyanidin synthesis. It is putatively expressed in all parts of the plant organs such as seeds, leaves, roots, flowers, berry skin, and flesh. Higher expression of DFR promotes anthocyanin and enhances the red pigmentation in tobacco (Luo et al., 2016). The DFR promoter contains several DNA binding (DBD) motifs that are influenced by light, calcium, and sucrose. The sequence located between 725 and 233 is involved in the DFR expression of fruits (Gollop et al., 2002). Another fragment analysis of the DFR promoter suggests that a specific fragment (F4) is critical to control the anthocyanin expression in *Malus*.

8.7 SEED SPECIFIC PROMOTER

Plants store reserve food materials such as carbohydrates, protein, and lipids in the seeds. Seed reserves are important sources of plants, especially during their germination and development. These phases of plant life determine seed quality. Seeds are also important for human and animal feed. Foreign protein expression in seeds provides numerous advantages over other expression systems in terms of high storage of expressed recombinant protein (Horvath et al., 2000). The seed storage protein (SSP) regulates protein storage in seeds and has been extracted from crop plants and examined for biological roles and promoter regulatory areas (Bhunia et al., 2014b). The use of the *Arabidopsis thaliana* β-glucosidase 19 gene (*AtBGLU*) promoter has resulted in seed-specific expression of GUS in transgenic *Arabidopsis* and tobacco (Liu et al., 2021). The expression of a substantial amount of certain proteins in specific parts of the seeds is important for improved seed-specific traits without altering their characteristics. Controlled regulation of genes for such proteins at targeted tissues is achieved by promoters. Hence, the identification and modulation of seed-specific promoters to synthesize the required quantity of proteins will be important. The promoter of *S2* albumin has been used for seed-specific expression of many genes in several horticulture species. For instance, in a transient expression assay in tobacco, the *S2* albumin promoter showed higher GUS expression in the seed (Bhunia et al., 2014b). The expression analysis of the Seed Maturation Protein (*CaSMP*) gene of *Coffea arabica* showed a preference for the endosperm. Furthermore, *in-silico* analysis of the 5′ upstream region revealed that the CaSMP promoter contains many *cis*-elements responsible for driving its activity in the seed and endosperm. However, the heterologous expression of this promoter fused with *the GUS* reporter showed its ovule-specific expression in immature tomato fruit (Quintero et al., 2018). During the specific time of maturity, legume seeds accumulate large quantities of storage proteins. *Pisumsativum* code for multigene family namely *legA-legC*. The promoters of *legA*, *legB*, and *legC* have also been well-studied in various plants. The deletion analysis of *the leg* gene promoter reveals that the minimal 549 bp upstream flanking region, consisting of *leg* box element and a 28 bp conserved sequence, is crucial for seed-specific expression (Shirsat et al., 1989). Late embryogenesis abundant (LEA) protein accumulates in almost all higher seed plants. *DC3*, a *LEA* class gene, is highly expressed in developing carrot (*Daucus carota*) embryos. A GUS gene driven by the DC3 promoter in transgenic tobacco is induced by various factors such as ABA, salt shock, hypoxia, ethylene, and MeJA. Through deletion mutation analysis, it has been determined that the region between 599 and 218 is essential for ABA and MeJA-dependent induction of *GUS* expression (Siddiqui et al., 1998). The *Pisum sativum lectin* (*psl*) gene of pea shows high expression during mid-mature seeds. The *psl* promoter contains three overlapping TGCA motifs that facilitate the binding of the basic leucine zipper TGA1a protein. A trimer of a 22 bp sequence, containing a TGAC-like motif, when fused with the 35S minimal promoter, drives high expression of a reporter gene in seeds with weak activity in the leaf and stem of transgenic tobacco. Increased activity is observed in the endosperm and embryo during the mid-mortification stage. Further, histochemical analysis substantiated a similar expression to that noticed by the 2,000 bp fragment *psl* promoter suggesting the 22 bp fragment contains all the essential motifs and sequence necessary to drive the *psl* gene expression in seeds (De Pater et al., 1993). The seed-specific promoter from foxtail millet *F128* was characterized in foxtail millet and *Zea mays*. It is an ortholog of *TaPR60* from hexaploid wheat and *OsPR602* from rice (Pan et al., 2015). A promoter: reporter histochemical assay using *the GUS* gene detected F128 promoter activity higher in the early stage of seed development (Pan et al., 2015). Recently, SSP2 (seed-specific promoter 2) from peanuts was characterized in transgenic *Arabidopsis*. GUS reporter gene assay fused with SSP2 dominantly expressed in mature seeds, suggesting the potential of SSP2 to regulate the expression of key genes in mature seeds (Yuan et al., 2019). Moreover, Tang et al. (2021) functionally characterized LEAFY COTYLEDON1 (LEC1) as a seed-specific promoter from peanuts. The GUS histochemical staining revealed seed-specific expression at different developmental stages. The study revealed that regulatory regions covering 82 bp of 5′ UTR and 2,228 bp promoter can facilitate the expression, particularly in embryos.

Soybean *β*-conglycinin is a major component of SSP in soybean. It is a heterotetrameric protein composed of α, α', and *β* subunits. The differential spatio-temporal accumulation of *β*-conglycinin subunits detected in embryos. The α' present in cotyledon and embryo axis *β* accumulates in cotyledon but not in the axis. However, α subunit protein is present in both cotyledon and axis but with low accumulation (Imoto et al., 2008). Fragment analysis of *β*-conglycinin promoter reveals an 84 bp short sequence between 245 bp and 161 bp contains *the cis*-regulatory elements necessary for promoter activity in the cotyledons and the embryonic axis (Imoto et al., 2008). In beans, *the* β-phaseolingene (phas), encodes a major SSP, and is exclusively expressed in the cotyledons of developing embryos. A 295 bp proximal transcription site contains *cis*-elements sufficient for seed-specific expression (van der Geest & Hall, 1996).

8.8 SYNTHETIC PROMOTER

Promoter engineering has been a promising approach to benefit agriculture for desirable fruit and vegetable production. The success of this technology depends on the expression of the promoter to achieve the desired amount of gene product in transgenics. The CaMV 35S promoter has been extensively used for high levels of gene expression in plant development of transgenics with high levels of desirable products in all tissue normal phenotypes of plants is not a challenge (Kahali et al., 2017), and the application of native promoters has not proven to be a potential tool to obtain desirable traits as most of them are neither tightly regulated nor specific (Rushton et al., 2002). Synthetic promoters can be an effective tool to achieve desirable traits in targeted tissues with less energy cost. The synthetic promoter is a stretch of DNA sequences that are artificially synthesized to drive targeted gene expression in specific tissues or under specific conditions. The *cis*-acting elements present on native promoters are synthesized and fused with the minimal promoter (Figure 8.2).

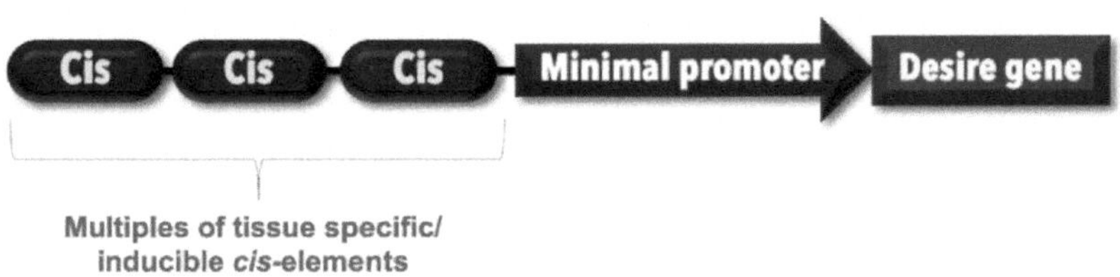

FIGURE 8.2 Generalized representation of a synthetic promoter where copies of desirable cis-elements are fused upstream to the minimal promoter to achieve the specific expression of the target gene.

The activity of synthetic promoters can be tuned by changing the number of *cis*-elements. For example, if we want to create a promoter under the control of ethylene, one can fuse multiple copies of Ethylene Responsive

Elements or ERFs to the minimal promoter. Synthetic promoters carry only elements of desire, and unwanted elements are eliminated to avoid untargeted gene expression (Ruston et al., 2016). Synthetic promoters have successfully expressed the required levels of products when induced by abiotic and biotic stresses. Designing the synthetic promoters and their application could be potential tools for the production of fruits and vegetables with desirable traits (reviewed by Khan et al., 2023) (Table 8.1).

TABLE 8.1 List of Promoters with Potential Applications to Improve Crop Plants and Economically Important Tissues

Promoter	Origin	Plant Used	References
		Constitutive Promoter	
CaMV35S	Cauliflower mosaic virus	Apple, broccoli, citrus, chrysanthemum, cocoa, collard, grape, Indian mustard, Lilium, *Nicotiana glutinosa*, papaya, peach, petunia, plum, poplar, rose, strawberry, tomato, torenia.	Chauhan et al. (2021); Dasgupta et al. (2020); Dutt et al. (2014); Jiang et al. (2016); Odell et al. (1985b)
AtSCPL30	*Arabidopsis thaliana*	Nicotiana	Jiang et al. (2018)
StPat	*Potato*	Potato	Miroshnichenko et al. (2020)
CsGAPC2, CsEF1	*Citrus sinensis*	Tobacco	Corte et al. (2020)
KST1	Potato	Potato, tobacco, cucumber, grape, barley	Kelly et al. (2017)
UEP	*Jatropha curcas*	*Jatropha curcas*, *Arabidopsis*	Tao et al. (2015)
UEP1	Oil palm	Oil palm	Masura et al. (2010)
Rep	Cotton leaf curl Burewala virus	*Nicotiana*	Khan et al. (2015)
Ubi-1	*Zea mays*	*Zea mays*	Streatfield et al. (2004)
		Fruit-Specific Promoter	
E4	Tomato	Tomato	Deikman et al. (1992)
E8	Tomato	Tomato	Montgomery et al. (1993)

TABLE 8.1 *(Continued)*

Promoter	Origin	Plant Used	References
		Constitutive Promoter	
ACC-oxidase	Peach, apple, tomato, banana	Tomato, banana	Atkinson et al. (1998a); Houben & Van de Poel (2019); Moon & Callahan (2004); Wang & Peng (2001)
α-Man	Tomato	Tomato	Irfan et al. (2014)
β-Hex	Tomato	Tomato	Irfan et al. (2016)
CitSEP	Citrus	Citrus plum	Dasgupta et al. (2020)
PamMybAp	Plum	Plum citrus	Dasgupta et al. (2020)
SlHDC-A	Tomato	Tomato	Kim et al. (2019)
		Flower Specific Promoter	
DFR	Malus	Malus	Tian et al. (2017)
BAN215-6	Brassica	Tobacco	Kim et al. (1997)
PchsA	*Petuniahybrida*	*Petuniahybrida*	Hong et al. (2003)
EcSWEET5	*Eucalyptus*	Tobacco	Unnikrishnan &Shankaranarayana (2018)
SlPS	Tomato	Tomato	Avilés-Arnaut &Délano-Frier (2012)
		Seed Specific Promoter	
AtBGLU	*Arabidopsis*	*Arabidopsis*	Liu et al. (2021)
S2 albumin	Soybean	Tobacco	Bhunia et al. (2014a)
GmPM9	Soybean	Soybean	Kumar et al. (2016a)
CaSMP	Coffea	Coffea	Quintero et al. (2018)
Leg	Pea	Pea	Shirsat et al. (1989)
DC3	Carrot	Tobacco	Siddiqui et al. (1998)
F128	Foxtail millet	*Arabidopsis*	Pan et al. (2015)
LEC1	Peanut	*Arabidopsis*	Tang et al. (2021)

8.9 CONCLUSIONS

To control the expression of a gene in a particular plant, tissue, or organ, the selection of a suitable promoter is one of the most important criteria for developing GM crops. Additionally, it is important to understand the future aspects and consequences of the selection of a promoter and its engineering without compromising environmental and biosafety concerns. The specific promoter would be a great choice for genetic engineering for higher expression of genes in a particular tissue or organ. The constitutive promoter is not always a good option for regulating the targeted expression, as it may compete for resources such as energy and building blocks for protein and RNA synthesis. To minimize adverse environmental effects on specific tissues of GM crops, the selection of tissue-specific promoters is critical.

KEYWORDS

- **fruit-specific promoter**
- **genetic engineering**
- **genetically modified**
- **promoter**
- **promoter engineering**
- **synthetic promoter**

REFERENCES

Aghdam, M. S., Sevillano, L., Flores, F. B., & Bodbodak, S., (2013). The contribution of biotechnology to improving post-harvest chilling tolerance in fruits and vegetables using heat-shock proteins. *The Journal of Agricultural Science, 153*, 7–24.

Atkinson, R. G., Bolitho, K. M., Wright, M. A., Iturriagagoitia-Bueno, T., Reid, S. J., & Ross, G. S., (1998a). Apple ACC-oxidase and polygalacturonase: Ripening-specific gene expression and promoter analysis in transgenic tomato. *Plant Mol. Biol., 38*, 449–460.

Atkinson, R. G., Bolitho, K. M., Wright, M. A., Iturriagagoitia-Bueno, T., Reid, S. J., & Ross, G. S., (1998b). Apple ACC-oxidase and polygalacturonase: Ripening-specific gene expression and promoter analysis in transgenic tomato. *Plant Molecular Biology, 38*, 449–460.

Avilés-Arnaut, H., & Délano-Frier, J. P., (2012). Characterization of the *Tomato prosystemin* promoter: Organ-specific expression, hormone specificity and methyl jasmonate responsiveness by deletion analysis in transgenic tobacco plants F. *Journal of Integrative Plant Biology, 54*(1), 15–32.

Bhat, D. S., Anjanasree, K. N., Lenka, S. K., & Bansal, K. C., (2010b). Isolation and characterization of fruit-specific promoters ACS4 and EXP1 from tomato (*Solanum lycopersicum* L). *Journal of Plant Biochemistry and Biotechnology, 19*, 51–57.

Bhat, D., Anjanasree, K., Lenka, S., & Bansal, K., (2010a). Isolation and characterization of fruit-specific promoters ACS4 and EXP1 from tomato (*Solanum lycopersicum* L). *Journal of Plant Biochemistry and Biotechnology, 19*, 51–57.

Bhunia, R. K., Chakraborty, A., Kaur, R., Gayatri, T., Bhattacharyya, J., Basu, A., Maiti, M. K., & Sen, S. K., (2014a). Seed-specific increased expression of 2S albumin promoter of sesame qualifies it as a useful genetic tool for fatty acid metabolic engineering and related transgenic intervention in sesame and other oil seed crops. *Plant Mol. Biol., 86*, 351–365.

Bhunia, R. K., Chakraborty, A., Kaur, R., Gayatri, T., Bhattacharyya, J., Basu, A., Maiti, M. K., & Sen, S. K., (2014b). Seed-specific increased expression of 2S albumin promoter of sesame qualifies it as a useful genetic tool for fatty acid metabolic engineering and related transgenic intervention in sesame and other oil seed crops. *Plant Molecular Biology, 86*, 351–365.

Biłas, R., Szafran, K., Hnatuszko-Konka, K., & Kononowicz, A. K., (2016). Cis-regulatory elements used to control gene expression in plants. *Plant Cell, Tissue, and Organ Culture (PCTOC), 127*, 269–287.

Breddam, K., (1986). Serine carboxypeptidases: A review. *Carlsberg Research Communications, 51*, 83–128.

Brown, T. A., (2020). *Gene Cloning and DNA Analysis: An Introduction*. John Wiley & Sons.

Canton, M., Drincovich, M. F., Lara, M. V., Vizzotto, G., Walker, R. P., Famiani, F., & Bonghi, C., (2020). Metabolism of stone fruits: Reciprocal contribution between primary metabolism and cell wall. *Frontiers in Plant Science, 11*.

Chauhan, A., Modgil, M., & Rajam, M. V., (2021). Establishment of *Agrobacterium tumefaciens* – mediated genetic transformation of apple pathogen *Marssonina coronaria* using marker genes under the control of CaMV 35S promoter. *Microbiological Research, 253*, 126878.

Corte, L. E. D., Mendes, B. M. J., Filho, F. A. A. M., Grosser, J. W., & Dutt, M., (2020). Functional characterization of full-length and 5′ deletion fragments of *Citrus* sinensis-derived constitutive promoters in *Nicotiana benthamiana. In Vitro Cellular & Developmental Biology – Plant, 56*, 280–289.

Dasgupta, K., Hotton, S., Belknap, W., Syed, Y., Dardick, C., Thilmony, R., & Thomson, J. G., (2020). Isolation of novel *Citrus* and plum fruit promoters and their functional characterization for fruit biotechnology. *BMC Biotechnology, 20*, 43.

De Pater, S., Pham, K., Chua, N. H., Memelink, J., & Kijne, J., (1993). A 22-bp fragment of the pea lectin promoter containing essential TGAC-like motifs confers seed-specific gene expression. *The Plant Cell, 5*, 877–886.

Deikman, J., Kline, R., & Fischer, R. L., (1992). Organization of ripening and ethylene regulatory regions in a fruit-specific promoter from tomato (*Lycopersicon esculentum*) 1. *Plant Physiology, 100*, 2013–2017.

Deng, H., Chen, Y., Liu, Z., Liu, Z., Shu, P., Wang, R., & Liu, M., (2022). SlERF. F12 modulates the transition to ripening in tomato fruit by recruiting the co-repressor TOPLESS and histone deacetylases to repress key ripening genes. *The Plant Cell, 34*(4), 1250–1272.

Dey, N., Sarkar, S., Acharya, S., & Maiti, I. B., (2015). Synthetic promoters in planta. *Planta, 242*, 1077–1094.

Dutt, M., Dhekney, S. A., Soriano, L., Kandel, R., & Grosser, J. W., (2014). Temporal and spatial control of gene expression in horticultural crops. *Horticulture Research, 1*, 14047.

Erpen, L., Tavano, E. C. R., Harakava, R., Dutt, M., Grosser, J. W., Piedade, S. M. S., Mendes, B. M. J., & Mourão, F. F. A. A., (2018). Isolation, characterization, and evaluation of three *Citrus* sinensis-derived constitutive gene promoters. *Plant Cell Rep., 37*, 1113–1125.

Feng, Y., & Xue, Q., (2006). The serine carboxypeptidase like gene family of rice (*Oryza sativa* L. ssp. japonica). *Functional & Integrative Genomics, 6*, 14–24.

Ghosh, A., Shekhawat, U. K. S., Ganapathi, T. R., & Bapat, V. A., (2012). Analysis of banana fruit-specific promoters using transient expression in embryogenic cells of banana cultivar *Robusta* (AAA Group). *Journal of Plant Biochemistry and Biotechnology, 21*, 189–197.

Ghosh, S., Irfan, M., & Datta, A., (2017). Applications of RNA silencing in improving plant traits for industrial use. In: Dalmay, T., (ed.), *Plant Gene Silencing: Mechanisms and Applications* (pp. 128–146). CAB International 2017, Biotechnology Series 5. ISBN: 1780647670, 9781780647678, https://doi.org/10.1079/9781780647678.0128.

Ghosh, S., Meli, V. S., Kumar, A., Thakur, A., Chakraborty, N., Chakraborty, S., & Datta, A., (2011). The N-glycan processing enzymes alpha-mannosidase and beta-D-N-acetylhexosaminidase are involved in ripening-associated softening in the non-climacteric fruits of capsicum. *J. Exp. Bot., 62*, 571–582.

Ghosh, S., Singh, U. K., Meli, V. S., Kumar, V., Kumar, A., Irfan, M., Chakraborty, N., et al., (2013). Induction of senescence and identification of differentially expressed genes in tomato in response to monoterpene. *PLoS One, 8*(9), e76029. https://doi.org/10.1371/journal.pone.0076029.

Gollop, R., Even, S., Colova-Tsolova, V., & Perl, A., (2002). Expression of the grape dihydroflavonol reductase gene and analysis of its promoter region1. *Journal of Experimental Botany, 53*, 1397–1409.

Grierson, D., (2013). Ethylene and the control of fruit ripening. In: Seymour, G. B., Poole, M., Giovannoni, J. J., & Tucker, G. A., (eds.), *The Molecular Biology and Biochemistry of Fruit Ripening* (pp. 43–73). Blackwell Publishing Ltd., Oxford.

Hernandez-Garcia, C. M., & Finer, J. J., (2014a). Identification and validation of promoters and cis-acting regulatory elements. *Plant Science, 217*, 109–119.

Hernandez-Garcia, C. M., & Finer, J. J., (2014b). Identification and validation of promoters and cis-acting regulatory elements. *Plant Science: An International Journal of Experimental Plant Biology, 217, 218*, 109–119.

Hong, Y., Jiang, H., Xiao, L., Huang, H., & Zhang, W., (2003). Cloning and sequence analysis of flower tissue-specific expression gene promoter PchsA from petunia hybrida. *Hunan Nong Ye Da Xue Xue Bao, 29*, 474–477.

Horvath, H., Huang, J., Wong, O., Kohl, E., Okita, T., Kannangara, C. G., & Von, W. D., (2000). The production of recombinant proteins in transgenic barley grains. *Proceedings of the National Academy of Sciences, 97*, 1914–1919.

Houben, M., & Van De, P. B., (2019). 1-aminocyclopropane-1-carboxylic acid oxidase (ACO): The enzyme that makes the plant hormone ethylene. *Frontiers in Plant Science,* 10.

Imoto, Y., Yamada, T., Kitamura, K., & Kanazawa, A., (2008). Spatial and temporal control of transcription of the soybean β-conglycinin α subunit gene is conferred by its proximal promoter region and accounts for the unequal distribution of the protein during embryogenesis. *Genes & Genetic Systems, 83*, 469–476.

Irfan, M., & Datta, A., (2017). Improving food nutritional quality and productivity through genetic engineering. *International Journal of Cell Science & Molecular Biology, 2*(1), 555576. https://doi.org/10.19080/IJCSMB.2017.02.555576.

Irfan, M., Ghosh, S., Kumar, V., Chakraborty, N., Chakraborty, S., & Datta, A., (2014). Insights into transcriptional regulation of β-D-N-acetylhexosaminidase, an N-glycan-processing enzyme involved in ripening-associated fruit softening. *Journal of Experimental Botany, 65*, 5835–5848.

Irfan, M., Ghosh, S., Meli, V. S., Kumar, A., Kumar, V., Chakraborty, N., Chakraborty, S., & Datta, A., (2016). Fruit ripening regulation of α-mannosidase expression by the MADS box transcription factor RIPENING INHIBITOR and ethylene. *Frontiers in Plant Science, 7*.

Irfan, M., Kumar, P., Ahmad, M. F., & Siddiqui, M. W., (2023). Biotechnological interventions in reducing losses of tropical fruits and vegetables. *Current Opinion in Biotechnology, 79*, 102850. doi.org/10.1016/j.copbio.2022.102850.

Irfan, M., Kumar, P., Kumar, V., & Datta, A., (2022). Fruit ripening specific expression of β-D-N-acetylhexosaminidase (β-Hex) gene in tomato is transcriptionally regulated by ethylene response factor SlERF.E4. *Plant Science, 323*, 111380.

Jiang, M., Jiang, J. J., He, C. M., & Guan, M., (2016). Broccoli plants over-expressing a cytosolic ascorbate peroxidase gene increase resistance to downy mildew and heat stress. *Journal of Plant Pathology, 98*, 413–420.

Jiang, P., Zhang, K., Ding, Z., He, Q., Li, W., Zhu, S., Cheng, W., et al., (2018). Characterization of a strong and constitutive promoter from the *Arabidopsis* serine carboxypeptidase-like gene AtSCPL30 as a potential tool for crop transgenic breeding. *BMC Biotechnology, 18*, 59.

Kelly, G., Lugassi, N., Belausov, E., Wolf, D., Khamaisi, B., Brandsma, D., Kottapalli, J., et al., (2017). The *Solanum tuberosum* KST1 partial promoter as a tool for guard cell expression in multiple plant species. *J. Exp. Bot., 68*, 2885–2897.

Khan, A., Nasim, N., Pudhuvai, B., Koul, B., Upadhyay, S. K., Sethi, L., & Dey, N., (2023). Plant synthetic promoters: Advancement and prospective. *Agriculture, 13*, 298. https://doi.org/10.3390/agriculture13020298.

Khan, Z. A., Abdin, M. Z., & Khan, J. A., (2015). Functional characterization of a strong Bi-directional constitutive plant promoter isolated from cotton leaf curl Burewala virus. *PloS One, 10*, e0121656.

Kim, A. Y., Kim, H. M., Ma, S. H., Park, S. Y., Dat, M. T., Jang, G., & Joung, Y. H., (2019). The promoter of tomato histidine decarboxylase A is fruit-specific, and its expression is stably maintained in fruits during ripening. *Plant Biotechnology Reports, 13*, 43–50.

Kim, H. U., Park, B. S., Jin, Y. M., & Chung, T. Y., (1997). Promoter sequences of two homologous pectin esterase genes from Chinese cabbage (*Brassica campestris* L. ssp. pekinensis) and pollen-specific expression of the GUS gene driven by a promoter in tobacco plants. *Molecules and Cells, 7*, 21–27.

Kumar, R., Jiwani, G., Pareek, A., SravanKumar, T., Khurana, A., & Sharma, A. K., (2016). Evolutionary profiling of group II pyridoxal-phosphate-dependent decarboxylases suggests expansion and functional diversification of histidine decarboxylases in tomato. *The Plant Genome, 9*, plantgenome2015.2007.0057.

Kumar, V., Chattopadhyay, A., Ghosh, S., Irfan, M., Chakraborty, N., Chakraborty, S., & Datta, A., (2016a). Improving nutritional quality and fungal tolerance in soybean and grass pea by expressing an oxalate decarboxylase. *Plant Biotechnology Journal, 14*, 1394–1405. https://doi.org/10.1111/pbi.12503.

Kumar, V., Irfan, M., Ghosh, S., Chakraborty, N., Chakraborty, S., & Datta, A., (2016b). Fruit ripening mutants reveal cell metabolism and redox state during ripening. *Protoplasma, 253*, 581–594. https://doi.org/10.1007/s00709-015-0836-z.

Kumari, C., Sharma, M., Kumar, V., Sharma, R., Kumar, V., Sharma, P., Kumar, P., & Irfan, M., (2022). Genome editing technology for genetic amelioration of fruits and vegetables for alleviating post-harvest loss. *Bioengineering*, *9*(4), 176. https://doi.org/10.3390/bioengineering9040176.

Kummari, D., Palakolanu, S. R., Kishor, P. B. K., Bhatnagar-Mathur, P., Singam, P., Vadez, V., & Sharma, K. K., (2020). An update and perspectives on the use of promoters in plant genetic engineering. *Journal of Biosciences, 45*.

Lincoln, J. E., Cordes, S., Read, E., & Fischer, R. L., (1987). Regulation of gene expression by ethylene during *Lycopersicon esculentum* (tomato) fruit development. *Proceedings of the National Academy of Sciences of the United States of America, 84*, 2793–2797.

Liu, Q., Liu, M., & Wu, W., (2018). Strong/weak feature recognition of promoters based on position weight matrix and ensemble set-valued models. *Journal of Computational Biology, 25*, 1152–1160.

Liu, S., Liu, C., Wang, X., & Chen, H., (2021). Seed-specific activity of the *Arabidopsis* β-glucosidase 19 promoter in transgenic *Arabidopsis* and tobacco. *Plant Cell Reports, 40*, 213–221.

Luo, P., Ning, G., Wang, Z., Shen, Y., Jin, H., Li, P., Huang, S., et al., (2016). Disequilibrium of flavonol synthase and dihydroflavonol-4-reductase expression associated tightly to white vs. red color flower formation in plants. *Frontiers in Plant Science, 6*, 1257.

Masura, S. S., Parveez, G. K. A., & Ismail, I., (2010). Isolation and characterization of oil palm constitutive promoter derived from ubiquitin extension protein (uep1) gene. *New Biotechnology, 27*, 289–299.

Meli, V. S., Ghosh, S., Prabha, T. N., Chakraborty, N., Chakraborty, S., & Datta, A., (2010). Enhancement of fruit shelf life by suppressing *N* -glycan processing enzymes. *Proceedings of the National Academy of Sciences, 107*, 2413–2418.

Miroshnichenko, D., Firsov, A., Timerbaev, V., Kozlov, O., Klementyeva, A., Shaloiko, L., & Dolgov, S., (2020). Evaluation of plant-derived promoters for constitutive and tissue-specific gene expression in potato. *Plants (Basel, Switzerland), 9*.

Montgomery, J., Goldman, S., Deikman, J., Margossian, L., & Fischer, R. L., (1993). Identification of an ethylene-responsive region in the promoter of a fruit ripening gene. *Proceedings of the National Academy of Sciences, 90*, 5939–5943.

Moon, H., & Callahan, A. M., (2004). Developmental regulation of peach ACC oxidase promoter–GUS fusions in transgenic tomato fruits. *Journal of Experimental Botany, 55*, 1519–1528.

Odell, J. T., Nagy, F., & Chua, N. H., (1985a). Identification of DNA sequences required for activity of the cauliflower mosaic virus 35S promoter. *Nature, 313*, 810–812.

Odell, J. T., Nagy, F., & Chua, N. H., (1985b). Identification of DNA sequences required for activity of the cauliflower mosaic virus 35S promoter. *Nature, 313*, 810–812.

Ow, D. W., Jacobs, J. D., & Howell, S. H., (1987). Functional regions of the cauliflower mosaic virus 35S RNA promoter determined by use of the firefly luciferase gene as a reporter of promoter activity. *Proceedings of the National Academy of Sciences, 84*, 4870–4874.

Pan, Y., Ma, X., Liang, H., Zhao, Q., Zhu, D., & Yu, J., (2015). Spatial and temporal activity of the foxtail millet (*Setaria italica*) seed-specific promoter pF128. *Planta, 241*, 57–67.

Pennacchio, L. A., et al., (2013). "Enhancers: Five essential questions." Nature reviews. *Genetics 14*(4), 288–295. doi: 10.1038/nrg3458.

Peremarti, A., Twyman, R. M., Gómez-Galera, S., Naqvi, S., Farré, G., Sabalza, M., Miralpeix, B., et al., (2010). Promoter diversity in multigene transformation. *Plant Molecular Biology, 73*, 363–378.

Quintero, F. O., Pinto, L. G., Barsalobres-Cavallari, C. F., Arcuri, Md. L. C., Pino, L. E., Peres, L. E. P., Maluf, M. P., & Maia, I. G., (2018). Identification of a seed maturation protein gene from *Coffea arabica* (CaSMP) and analysis of its promoter activity in tomato. *Plant Cell Reports 37*, 1257–1268.

Rushton, P. J., (2016). What have we learned about synthetic promoter construction. In: Hehl, R., (ed.), *Plant Synthetic Promoters: Methods and Protocols* (pp. 1–13). Springer: New York, NY, USA. ISBN 978-1-4939-6396-6.

Rushton, P. J., Reinstädler, A., Lipka, V., Lippok, B., & Somssich, I. E., (2002). Synthetic plant promoters containing defined regulatory elements provide novel insights into pathogen- and wound-induced signaling. *Plant Cell, 14*, 749–762. [Google Scholar] [CrossRef] [Green Version].

Shahan, R., Li, D., & Liu, Z., (2019). Identification of genes preferentially expressed in wild strawberry receptacle fruit and demonstration of their promoter activities. *Horticulture Research, 6*.

Shirsat, A., Wilford, N., Croy, R., & Boulter, D., (1989). Sequences responsible for the tissue specific promoter activity of a pea legumin gene in tobacco. *Molecular and General Genetics MGG, 215*, 326–331.

Siddiqui, N. U., Chung, H. J., Thomas, T. L., & Drew, M. C., (1998). Abscisic acid-dependent and -independent expression of the carrot late-embryogenesis-abundant-class gene Dc3 in transgenic tobacco seedlings. *Plant Physiol., 118*, 1181–1190.

Soni, K. K., Mehrotra, A. K., & Burma, P. K., (2021). Engineering the AEG1 promoter from cotton to develop male sterile lines. *Plant Cell Rep.,* 40, 1789–1792. https://doi.org/10.1007/s00299-021-02704-6.

Stavolone, L., Kononova, M., Pauli, S., Ragozzino, A., De Haan, P., Milligan, S., Lawton, K., & Hohn, T., (2003). Cestrum yellow leaf curling virus (CmYLCV) promoter: A new strong constitutive promoter for heterologous gene expression in a wide variety of crops. *Plant Molecular Biology, 53*, 703–713.

Streatfield, S. J., Magallanes-Lundback, M. E., Beifuss, K. K., Brooks, C. A., Harkey, R. L., Love, R. T., Bray, J., et al., (2004). Analysis of the maize polyubiquitin-1 promoter heat shock elements and generation of promoter variants with modified expression characteristics. *Transgenic Research, 13*, 299–312.

Tang, G., Xu, P., Li, P., Zhu, J., Chen, G., Shan, L., & Wan, S., (2021). Cloning and functional characterization of seed-specific LEC1A promoter from peanut (*Arachis hypogaea* L.). *PloS One, 16*, e0242949.

Tao, Y. B., He, L. L., Niu, L. J., & Xu, Z. F., (2015). Isolation and characterization of an ubiquitin extension protein gene (JcUEP) promoter from *Jatropha curcas. Planta, 241*, 823–836.

Tayal, R., Kumar, V., & Irfan, M., (2022). Harnessing the power of hydrogen sulfide (H2S) for improving fruit quality traits. *Plant Biology, 24*, 594–601. https://doi.org/10.1111/plb.13372.

Tian, J., Chen, M. C., Zhang, J., Li, K. T., Song, T. T., Zhang, X., & Yao, Y. C., (2017). Characteristics of dihydroflavonol 4-reductase gene promoters from different leaf colored *Malus* crabapple cultivars. *Horticulture Research, 4*, 17070.

Timerbaev, V., & Dolgov, S., (2019). Functional characterization of a strong promoter of the early light-inducible protein gene from tomato. *Planta, 250*, 1307–1323.

Unni, S. C., Vivek, P. J., Maju, T. T., Varghese, R. T., & Soniya, E. V., (2012). Molecular cloning and characterization of fruit specific promoter from *Cucumis sativus* L. *American Journal of Molecular Biology, 2*, 132–139.

Unnikrishnan, B. V., & Shankaranarayana, G. D., (2018). Functional characterization of a reproductive tissue specific promoter from *Eucalyptus camaldulensis. Genome, 61*, 777–786.

Van, D. G. A. H., & Hall, T. C., (1996). A 68 bp element of the beta-phaseolin promoter functions as a seed-specific enhancer. *Plant Mol. Biol., 32*, 579–588.

Wai, A. H., Naing, A. H., Lee, D. J., Kim, C. K., & Chung, M. Y., (2020). Molecular genetic approaches for enhancing stress tolerance and fruit quality of tomato. *Plant Biotechnology Reports, 14*, 515–537.

Wang, X. L., & Peng, X. X., (2001). [Cloning of promoter of banana fruit ripening-related ACO1 and primary study on its function]. *Sheng Wu Gong Cheng Xue Bao = Chinese Journal of Biotechnology, 17*, 428–431.

Wu, Y. Y., Liu, X. F., Fu, B. L., Zhang, Q. Y., Tong, Y., Wang, J., Wang, W. Q., et al., (2020). Methyl jasmonate enhances ethylene synthesis in kiwifruit by inducing NAC genes that activate ACS1. *Journal of Agricultural and Food Chemistry, 68*, 3267–3276.

Xu, R., Goldman, S., Coupe, S., & Deikman, J., (1996). Ethylene control of E4 transcription during tomato fruit ripening involves two cooperative cis elements. *Plant Mol Biol., 31*(6), 1117–1127. doi: 10.1007/BF00040829. PMID: 8914528.

Yang, L., Hu, G., Li, N., Habib, S., Huang, W., & Li, Z., (2017). Functional characterization of SlSAHH2 in tomato fruit ripening. *Frontiers in Plant Science, 8*.

Yasmeen, et al., (2023). Designing artificial synthetic promoters for accurate, smart, and versatile gene expression in plants. *Plant Communications*. https://doi.org/10.1016/j.xplc.2023.100558.

Yuan, C., Li, C., Yan, C., Zhao, X., Wang, J., Sun, Q., & Shan, S., (2019). Isolation and characterization of a novel seed-specific promoter from peanut (*Arachis hypogaea* L.). *Molecular Biology Reports, 46*, 3183–3191.

CHAPTER 9

EPIGENETIC REGULATION OF PRE- AND POSTHARVEST QUALITY TRAITS IN FRUITS AND VEGETABLES

ANSHUL KUMAR,[1,3] SUNAKSHI GAUTAM,[2,3]
SHUBHAM SMAKARIA,[2] NITIKA NEGI,[4] and DIKSHA ABROL[5]

[1]*Department of Fruit Science, Dr. Y. S. Parmar University of Horticulture and Forestry, Nauni, Solan, Himachal Pradesh, India*

[2]*Department of Food Technology, Dr. Y. S. Parmar University of Horticulture and Forestry, Nauni, Solan, Himachal Pradesh, India*

[3]*M.S. Swaminathan School of Agriculture, Shoolini University, Bhajol, Solan, Himachal Pradesh, India*

[4]*Department of Entomology, College of Agriculture, CSK HPKV, Palampur, Himachal Pradesh, India*

[5]*Department of Entomology, Dr. Y. S. Parmar University of Horticulture and Forestry, Nauni, Solan, Himachal Pradesh, India*

ABSTRACT

Fruits and vegetables undergo metabolic processes post-harvest, rendering them vulnerable to deterioration, influenced by genetic, environmental, and epigenetic factors. Epigenetic modifications, including DNA methylation, histone modifications, chromatin remodeling, and small RNA regulation, play pivotal roles in regulating gene expression without altering the DNA sequence, impacting fruit ripening, senescence, and stress responses.

Bioengineered Fruit and Vegetables: Trends in Postharvest Quality Improvement.
Mohammad Irfan and Pankaj Kumar (Eds.)

These epigenetic changes affect various physiological and biochemical pathways, influencing traits such as flavor, texture, and shelf life. Strategies for controlling post-harvest quality encompass a range of pre- and post-harvest interventions, including genetic and epigenetic modulation, to mitigate losses and enhance food availability. Understanding the epigenetic regulation of fruit and vegetable quality presents opportunities for breeding novel cultivars with improved traits and minimizing post-harvest deterioration without genetic alteration. Thus, this chapter provides a comprehensive overview of epigenetic mechanisms in fruits and vegetables, offering insights for future research aimed at advancing our understanding and application of epigenetic regulation for the benefit of agriculture and food security.

9.1 INTRODUCTION

Fruits and vegetables respire after harvest and consume all the reserves present in them during the post-harvest phase, thus becoming very susceptible to deterioration (Pedreschi & Lurie, 2015). The quality of any fresh commodity is predetermined by various factors like biochemical reactions that generate by-products, physiological events that define the organoleptic traits of the commodity, and most importantly, genetic events that can influence the overall quality of the fresh commodity (Zhou et al., 2020). Fruit development involves both genetic and epigenetic regulations and is affected by environmental conditions as well (Tang et al., 2020). These quality traits after harvest show a composite effect produced due to environmental conditions such as time, temperature, humidity, nutrients, soil condition, etc., and genetic factors as well (Shipman et al., 2021).

The metabolically active plants undergo ripening and senescence, which should be controlled for a longer shelf life (Brasil & Siddiqui, 2018). To achieve this, there are various pre- and post-harvest treatments. Additionally, genetic information and regulation can help prolong quality traits and maintain quality for a longer period of time. Within this genetic makeup, epigenetic mechanisms influence heritable traits such as stress response and, most importantly, crop ripening (Farinati et al., 2017). Post-harvest losses are crucial for researchers as they lead to crop wastage, impacting the economy and the well-being of farmers. The pre-harvest and post-harvest stages collectively affect quality, determining the extent of post-harvest losses of a crop (Chen et al., 2021). These factors encompass all quality traits like flavor, texture, acids, sugars, etc. Most flavor-related compounds develop during the ripening stage of fruits and vegetables and degrade during senescence (Cao et al., 2021), which can be influenced by epigenetic regulations.

Epigenetics may be defined as phenotypic variation that occurs without altering the DNA sequence, is potentially heritable, and happens due to changes in gene expression (Kumar et al., 2022; Tang et al., 2020; Bollati & Baccarelli, 2010; Eccleston et al., 2007). In 1942, Conrad Waddington described the term "epigenetics" for the first time (Basdeki & Hagidimitriou, 2019). Epigenetic modifications include DNA methylation, histone modification through acetylation, nucleosome remodeling, non-coding small RNAs (Kumar et al., 2022; Gu et al., 2021; Murr, 2010), chromatin modifications, chromatin remodeling (Basdeki & Hagidimitriou, 2019; Reinders et al., 2015), etc. These epigenetic processes are associated with the regulation of biosynthesis, accumulation, and degradation of biochemical and bioactive compounds (Anwar et al., 2021). Due to epigenetic modifications, phenotypic characteristics differ, even with similar genomes in an organism (Tiffon, 2018).

Basically, this area has been considered remarkable in the genetic improvement of plants (Samantara et al., 2021). In the case of agricultural crops, there are hundreds of genes that are directly involved during the ripening phase of fruits and vegetables. These genes can modify any kind of physiological and biochemical signals, such as metabolism of carbohydrates, organic acids, restructuring of cell walls, rate of response towards stress, production of ethylene, and influence organoleptic properties and palatability of commodities (Garcia-Gomez et al., 2021). The regulation of fruit ripening is influenced by transcription factors (TFs) (Fortes et al., 2017). To modulate the growth of plants and development to adapt under environmental stresses, understanding these mechanisms is very important (Kumar et al., 2016).

So, these modifications can help scientists alter gene expression based on the requirements of environmental conditions (Bucher et al., 2018), specifically playing a key role in the ripening process of fruits and vegetables (Tang et al., 2020). Plant hormones and TFs are responsible for regulating and controlling ripening in both climacteric and non-climacteric fruits. There are many studies supporting the contribution of epigenetic regulations playing a crucial role in reprogramming global transcriptomics

(Liu et al., 2020). In terms of nutrition, dietary polyphenolic compounds influence the epigenetic process. These compounds include catechin, caffeic acid, chlorogenic acid, and many more (Francini & Sebastiani, 2013).

Therefore, keeping in view the importance and significant role of epigenetic regulation in fruits and vegetable quality, the content of the chapter has been designed. This chapter discusses all aspects of epigenetic regulation and summarizes the most recent studies on pre and postharvest epigenetic regulation in fruits and vegetables. Thus, provide extensive information on the topic and help in development of novel solutions to minimize pre and postharvest losses (PHLs) in fruits and vegetables.

9.2 IMPORTANT PRE- AND POST-HARVEST QUALITY TRAITS OF FRUITS AND VEGETABLE

The term quality in fruits and vegetables traditionally refers to the extrinsic characteristics, which include parameters like firmness, size, color, and shape. In combination with the sensorial characteristics, these visual attributes play a direct role in influencing the acceptance and rejection by consumers, thus having a direct impact on economic values (Ziv & Fallik, 2021). Every fruit and vegetable has specific quality characteristics that determine their postharvest quality. Textural attributes such as crispness, juiciness, firmness, mealiness, along with flavor and appearance, are primary attributes in fruits like apples that are responsible for their marketability. Certain factors are essential to maintain the quality of fresh produce, specifically vegetables and fruit. This category broadly includes pre-harvest factors (which are generally environmental and cultural practices applied before harvesting the produce), harvest factors (including harvesting time, methods applied for harvesting, and most importantly, maturity indices), and post-harvest factors (including unit operations like washing, pre-cooling, sorting, grading, packing, and transporting) (Ramjan & Ansari, 2018).

The cultural practices include adequate availability of nutrients from soil and fertilizers, supply of water, proper spacing, and disease-free conditions directly influence the quality of vegetables and fruits. Whereas the environmental factors impose more impact on plant growth compared to nutrients. Environmental conditions have a direct impact on the synthesis of photosynthetic products like sugars and pigments (Desta et al., 2020). Pre-harvest factors have a direct influence on the post-harvest performance of produce as they alter phenotype and influence genotypic traits (Pedreschi & Lurie, 2015). Before harvesting, the ripening phase plays an important role in regulating further factors, i.e., harvesting and post-harvesting. During the ripening phase, metabolic changes within the fruits occur, such as the accumulation of pigments like anthocyanin and carotenoids with the degradation of chlorophyll, breakdown of the cell wall, catabolic reactions due to the breakdown of polymeric components into their monomeric forms, thus producing metabolites with low molecular weight (Pott et al., 2020).

These pre- and post-harvest attributes are very important in reducing losses and play a prominent role in defining and maintaining sustainable post-harvest management of fruits and vegetables (Ziv & Fallik, 2021). In addition to these, post-harvest factors include stress from environmental/abiotic and biotic factors that directly affect the sensory characteristics of fruits and vegetables (Pott et al., 2020). The stages of fruit ripening and senescence are two complex stages in the crop's lifespan that are of utmost importance as they determine the quality of produce after harvesting. These stages influence color, flavor, quality traits, sweetness, and the level of bioactive components regulated by multiple genes during spatiotemporal programming (Chen et al., 2021). Signaling mechanisms trigger plant development and involve intra-organ molecules such as peptides, phytohormones, RNAs, and TFs (Sparks et al., 2013). Once a horticultural commodity is harvested, the metabolic changes cannot be controlled, but the rate of degradation due to catabolic reactions can be slowed down. Factors like relative humidity, temperature, storage, and transportation conditions play a crucial role in reducing the rate of physiological reactions (Elik et al., 2019). The optimal combination of pre- and post-harvest factors directly influences the quality and shelf life of fresh produce (Desta et al., 2022).

9.3 EPIGENETIC REGULATIONS IN FRUITS AND VEGETABLES

There is evidence that reports the direct impact of patterns of gene expressions on the setting of fruits as well as on their development. These genetic expressions are produced due to a combined effect of exogenous factors like climate, environment, and endogenous factors such as hormones (Farinati et al., 2017). Epigenetic mechanisms affect different developmental phases in fruits and vegetables. Any type of heritable modifications in the regulation as well as the expression of genes that can be transmitted through the replication of DNA or propagation of cells without causing any change in the primary DNA sequence are termed "Epigenetic" (Anwar et al., 2021). Avramidou et al. (2015), who examined the epigenetic diversity of sweet cherries, found that it was lower than the genetic diversity. Substantial DNA methylation variation in the near-isogenic East African Highland banana (EAHB) is reflected in the phenotypic diversity of this species and is passed down from generation to generation via vegetative and seed reproduction (Kitavi et al., 2020). Epigenetic modifications in plants include DNA methylation, histone modifications, and small RNAs (Kumar et al., 2022; Murr, 2010), DNA cytosine methylation (Kumar et al., 2016, 2022), chromatin modifications, chromatin remodeling (Basdeki & Hagidimitriou, 2019; Reinders et al., 2015), etc.

This means that genetic expression plays a key role in the quality and quantity of fruits and vegetables. Environmental factors can induce changes in heritable traits through epigenetic mechanisms (Li et al., 2022). Understanding these complex structures can help reduce agricultural produce wastage to a certain extent. By developing new crops with longer shelf lives and the ability to maintain quality traits throughout the supply chain, wastage can be managed effectively (Shipman et al., 2021). The ripening process of fruits involves optimizing levels of secondary metabolites like phenolic acids, flavonoids, and tannins. The phenylpropanoids pathway, responsible for flavonoid biosynthesis, is linked to enzymes and plays a crucial role in fruit quality and economic value. DNA methylation levels influence phenotypic changes during fruit ripening and senescence. Adding methyl groups (CH3) to cytosine at the fifth carbon or adenine at the sixth nitrogen can result in different phenotypes without altering the nucleotide sequence (Pu et al., 2020; Fragou et al., 2019; Rottach et al., 2009). Specific enzymes called DNA methyltransferase catalyze these additions (Basdeki & Hagidimitriou, 2019; Molfetas et al., 1994). During the transition period from white to red, the highest levels of DNA methylase gene expression were observed in strawberries (Pu et al., 2020; Gu et al., 2016).

Histone modifications play an essential role in epigenetic inheritance and gene regulation. DNA in plants is wrapped around eight histone molecules and can be covalently modified at various amino acids with a varying number of added methyl groups (mono-, di-, or trimethylation) and acetylation (Song et al., 2014; Margueron & Reinberg, 2010; Turner, 2007), mono-ubiquitination of lysine residues (Berr et al., 2011), and phosphorylation (Zhao et al., 2019). Histone modifications function as a "histone code" for gene activity, promoting either active transcription or gene silencing (Zhao et al., 2019). The Flowering locus C (FLC) gene regulates the transition from vegetative growth to flowering, and FLC must be silenced for flowering to occur (Deal & Henikoff, 2011). Precocious trifoliate orange, known for its short juvenile phase and a citrus homolog gene called LEAFY (LFY), considered to be the central regulator of the flowering network that controls flowering and gene expression in precocious trifoliate oranges, is severely altered, which causes abnormalities in the fruit (Basdeki & Hagidimitriou, 2019; Zhang et al., 2014). The acetylation of histones is generally related to the activation of gene-controlled activity by histone deacetylases (HDACs) and histone acetyltransferases (HATs) (Tang et al., 2020). Whereas, for alterations in gene expressions, methylation of nucleotides and modification in the histone configuration are used (Samantara et al., 2021).

Methylation of transposable elements (TEs) is another way of epigenetic regulation in plants. TEs methylation is the most stable and consistent (Basdeki & Hagidimitriou, 2019; Niederhuth & Schmitz, 2013). DNA methylation in TEs appears to affect expression by influencing nearby genes (Farinati et al., 2017; Hollister & Gaut, 2009). High levels of methylation of nrDNA sequences were noticed in TEs and promoters. The DNA cytosine level and pattern of ITS1, 5.8S, and ITS2 were unique for most tissues/developmental phases (Ince & Karaca, 2021). Short noncoding RNA molecules (sRNAs) also play a vital role in gene regulation and development. They serve as molecular postcodes, directing AGO proteins to specific nucleic acids. Plants degrade sRNA-targeted mRNAs, which reduces gene expression. In contrast, cytosine methylation occurs in sRNA-targeted DNA sequences, a pathway known as RNA-directed DNA methylation (RdDM). Cytosine methylation can inhibit transcription, and thus sRNAs are effective gene expression regulators (Fei et al., 2021). Numerous miRNAs with essential roles in regulating quality development in a wide range of horticultural crops have been revealed (He et al., 2022). The cis-acting regulatory elements of Dicer-Like (DCL), Argonaute (AGO), and RNA-dependent RNA polymerase (RDR) are directly linked to the post-transcriptional gene silencing (PTGS) mechanism that mediates RNA silencing. The RNAi pathway in bananas was studied using multiple sequence alignment and a phylogenetic tree, which led to identifying 3 MaDCL, 13 MaAGO, and 5 MaRDR candidate genes. Arabidopsis thaliana has RNAi-silencing genes that are an identical copy of these genes. Key TFs such as the ERF, Dof, C2H2, TCP, GATA, and MIKC MADS families were found through network and sub-network analyses of TFs and candidate RNAi gene families (Ahmed et al., 2021). Cao et al. (2021) identified the alcohol acyltransferase PpAAT1 and transcription factor(TF) PpNAC1 as an activator of *PpAAT1* expression and ester production.

Methylation of DNA is involved in gene regulation at various stages of development (Zhang et al., 2014; Bitonti et al., 2002). Cell type – and potentially gene-specific DNA methylation levels can vary over time (Basdeki & Hagidimitriou, 2019; Tagliasacchi et al., 2007). MdMYB1/A/10, an R2R3-type MYB gene, regulates anthocyanin biosynthesis in apple (Kumar et al., 2022b; Jiang et al., 2020; Bai et al., 2016). Ma et al. (2021) compared DNA methylation and the transcriptome between mature fruits of 'Honeycrisp' and 'Qinguan' and found that the CG context had the highest level of DNA methylation, followed by the CHG and CHH contexts, whereas distribution of DMRs varied among genic regions and TEs. MYB1 allele transcript levels correlate with anthocyanin accumulation and are significantly higher in red fruit peel sectors of cultivars than in green peel sectors of cultivars. This is because red fruit peel sectors contain more anthocyanin (Basdeki & Hagidimitriou, 2019; Telias et al., 2011). High levels of methylation, greater than 60% of the MYB10 promoter, have been found in the apple varieties Honeycrisp and Royal Gala (Basdeki & Hagidimitriou, 2019; Telias et al., 2011). However, Cheng et al. (2014) observed six different types of anthocyanins in peach and wild relative flowers and noted that the structural diversity of these pigments can be attributed to glycosylation or methylation. DNA methylation differed between diploids and polyploids, showing that epigenetic patterns in potatoes are altered by polyploidization (Marfil et al., 2018). More than 1,000 common genes were identified by Kondhare et al. (2021), suggesting a shared gene regulatory network may be involved in potato tuber development.

There has been a rise in the use of epigenetic modifications in grafting, bud dormancy, and fruit development and ripening (Chachar et al., 2022).

When DNA is methylated in its promoter regions, transcriptional gene silencing generally occurs (Basdeki & Hagidimitriou, 2019; Jones, 2012). Transcription in plants is regulated by epigenetic modifications such as sRNA-directed DNA methylation and histone modifications (Simon & Meyers, 2011). Overexpression of ethylene response factor D7 (SlERF. D7) was found to promote ripening, while silencing of SlERF.D7 resulted in delayed ripening in tomato fruits (Gambhir et al., 2022). Fruit development and ripening in Capsicum were found to be governed by ethylene and ethylene-related genes, as well as by several epigenetic mechanisms, including non-coding RNAs (ncRNAs) and cytosine methylation (Kumar et al., 2018). DNA methylation levels were found to be both high in dormant buds and low during differentiation and cellular reprogramming (Basdeki & Hagidimitriou, 2019; Santamaria et al., 2006; Hasbun et al., 2005), illustrating epigenetic control among distinct developmental stages that probably regulate the expression of related genes (Basdeki & Hagidimitriou, 2019; Santamaria et al., 2011). Hypermethylation of certain chromatin domains was ascertained in adult meristems, in comparison to juvenile meristems, demonstrating that methylation level could have clear effects on growth and development (Basdeki & Hagidimitriou, 2019; Bitonti et al., 2002). It was reported by Zhang et al. (2014) that demethylation treatment (5-AzaC) affected the expression of the genes CiFt, CiTFL1, CiLFY, CiAP1, and CiFLC during the early flowering process in precocious trifoliate orange (Basdeki & Hagidimitriou, 2019). According to a study published in 2022 by Gafni et al., over-expression of MiFT2 and the two cloned MiTFL1 genes suppressed flowering, while overexpression of MiFT1 in Arabidopsis resulted in early flowering. In mango (*Mangifera indica L.*), cold temperature induces flowering through opposing effects on the expression of MiFT1 and the MiTFL1s, whereas high fruit load represses flowering through suppression of MiFT1 upregulation and delay of MiTFL1s downregulation. Methylation responses in male flowers were more responsive to stress than those in female flowers, and differences in the transcription of CpSVP and CpAP1 may be responsible for early flowering in male papaya (Zhou et al., 2020).

Modern 'omics techniques, including epigenomics, make it easier to track down the genes responsible for key quality traits in fruits and vegetables (Table 9.1) (Du et al., 2021). Three short vegetative phase genes, LcSVP1, LcSVP2, and LcSVP3, showed distinct expression patterns, inferring that they play different roles in bud development regulation

(Zhang et al., 2016). High chill conditions correlated with a decrease in the percentage of fully methylated fragments in apple, which occurred during dormancy break and fruit set (Basdeki & Hagidimitriou, 2019; Kumar et al., 2016). Agust et al. (2020) found that the floral repressor CcMADS19 inhibits CiFT2 (floral promoter) activation. In Citrus's tender new leaves, CcMADS19, a floral repressor, was found to be enhanced in repressive H3K27me3 (Mesejo et al., 2022). As part of the induction process for citrus flowers, PtHB13 binds to the promoter of the citrus flower-specific gene PtFLC (Ma et al., 2020). During floral bud dormancy in sweet cherry, it has been reported that the relative expression of genes PavDRM2.1 and PavDRM2.3 increases with chilling accumulation, suggesting a role for RdDM in dormancy regulation (Basdeki & Hagidimitriou, 2019; Rothkegel et al., 2017). Chestnut bud dormancy (Basdeki & Hagidimitriou, 2019) and bud burst, and bud set in chestnut (Santamaria et al., 2011) are regulated by epigenetic mechanisms, specifically DNA methylation and histone H4 acetylation (Basdeki & Hagidimitriou, 2019). Bud dormancy, anthocyanin expression, fertilization, and perpetuation are a few of the developmental processes that rely on methylation (Basdeki & Hagidimitriou, 2019). Methylation of DNA plays an important role in fruit plant development, including bud dormancy, flowering, fertilization, and anthocyanin synthesis (Basdeki & Hagidimitriou, 2019). Flowering behavior and timing in apple, peach, and other Rosaceae family members may be affected by apple winter dormancy, which is governed by the MdFLC-like gene (Lempe et al., 2022a, b; Zhu et al., 2020; Falavigna et al., 2019).

The transcriptional control of these enzymes is correlated with factors including MYB transcription and WD40 repeat proteins (Chen et al., 2021). Here, MYB transcription determines the fate of genes that are activated, while WD40 regulates the TFs (Liu et al., 2016). Other quality traits are also influenced by various genes. For instance, the production of volatile flavors in sweet citrus fruits is due to the presence of terpenes, which is controlled by AP2/ERF transcription factor genes (Cao et al., 2021). In the investigation of pre and postharvest maturation in peach, genomic methods and integrated metabolomics approaches have been combined, confirming the influence of genes in controlling the aromatic profile (Sirangelo et al., 2022). According to El-Sharkawy et al. (2015), various environmental stimulators such as low temperature conditions, light, and humidity directly influence the functioning of proteins and genes. Regarding DNA methylation and gene expression, two candidate genes, MdTSTa and MdMa11, were found to be linked to fruit soluble sugar and acidity levels (Lempe et al., 2022b). Increased transcript levels of carotenoid biosynthetic genes MaLCYB1.1 and MaLCYB1.2 consequently increase total carotenoid accumulation in banana fruits following transient transformation of MaSPL16 (Zhu et al., 2019). MdTSTa upregulation in tomato enhanced the fruit soluble sugar content. Furthermore, transient expression of MdMa11 in tobacco leaves considerably lowered pH (Ma et al., 2021). Lysine crotonylation (Kcr) is a recently discovered post-translational modification that, when co-localized with lysine acetylation, enhances gene expression on histones and non-histone proteins (Contreras-de la Rosa Perla et al., 2022). Liu et al. (2021) identified 1,734 differentially expressed genes (DEGs) and suggested that the hormone-related DEGs had an enhancement of transcription factor binding motifs (MYB2, GAMYB, and AP2/EREBP). Nonetheless, sex differentiation may be coordinated and regulated at multiple levels, including the transcriptional, epigenetic, and phytohormonal levels.

The final phase of fruit development (ripening) begins once the seed has reached maturity (Gillaspy et al., 1993). In climacteric fruit, ACS controls ethylene production during ripening and senescence (Pu et al., 2020; Tatsuki et al., 2006). Tomato fruit ripening was found to be regulated by DNA methylation (Zhou et al., 2019). Expression patterns of the MaACO1, MaEXP2, MaEXP7, and MaEXP8 genes are correlated with the changes of histone H3 and H4 acetylation level during fruit ripening, and MaERF11 binds to promoters of these genes via the GCC-box motif (Han et al., 2016). Yang et al. (2022) found that the transcription factor MaMYB4 acts as a negative regulator of fruit ripening in banana, which is important for the maintenance of moderate quality traits and the reduction of postharvest deterioration. MaMYB4 protein levels decreased gradually as bananas ripened, leading to the rise in ethylene and the softening of the fruit. Fu et al. (2018) identified 17 HDACs in the banana genome, which they classified into the RPD3/HDA1, SIR2, and HD2 sub-families. However, expression profiling showed that MaHDACs were differentially expressed during banana fruit ripening, and of these HDACs, MaHDA6 showed the highest inducement in fruit ripening. In litchi (Litchi chinensis Sonn. cv. Feizixiao), Peng et al. (2017) identified six HATs, 11 HDACs, 48 histone methyltransferase (HMTs), and 22 histone demethylases (HDMs), and found that the expression of 17 HMs was significantly altered during fruit abscission. In dry, ethylene-independent, fleshy fruits, H3K27me3

TABLE 9.1 Influence of Epigenetic Regulatory Mechanism on Some Important Fruit and Vegetable Crops

Crop	Epigenetic Regulatory Mechanism	Role	Improvement in Trait	References
Apple	*R2R3-MYB* and *bHLH-MYC*	• Anthocyanin biosynthesis.	Fruit development.	El-Sharkawy et al. (2015)
	MdMYB10	• DNA methylation.	Fruit pigmentation during ripening.	Kumar et al. (2022); Cho et al. (2020); El-Sharkawy et al. (2015)
	MdTSTa and *MdMa11*	• Based on expression and DNA methylation levels, encode tonoplast-localized proteins.	Associated with fruit soluble sugar and acidity.	Ma et al. (2021)
	MdFLC-like gene	• Winter dormancy.	Flowering	Lempe et al. (2022a, b); Zhu et al. (2020); Falavigna et al. (2019)
Avocado	*PamACS1*, *PamACS2*, and *PamACO*	• Decreased expression of gene.	Inhibition of ethylene metabolism and delayed ripening.	Olivares et al. (2022)
Banana	*MaLCYB1.1* and *MaLCYB1.2*	• Transient transformation of MaSPL16.	Increase total carotenoid.	Zhu et al. (2019)
	MaMYB4	• Protein levels decreased leads to the rise in ethylene and the softening of the fruit.	Ripening	Fu et al. (2018)
	MaHDACs	• Differentially expressed during ripening.	Ripening	Fu et al. (2018)
	mir393	• Involved in the regulation of redox in response to stress.	Response to stress.	Iglesias et al. (2014)
Berry	*R2R3 MYB*	• Influence flavonoids profile.	Accumulation of carotenoids and flavonoids.	Fortes et al. (2017)
Carrot	*CRTISO*	• α- and β-carotene biosynthesis.	Carotenoid regulation	Anwar et al. (2021)

TABLE 9.1 *(Continued)*

Crop	Epigenetic Regulatory Mechanism	Role	Improvement in Trait	References
Cheery	*PavDRM2.1*, *PavDRM2.3*, and *RdDM*	• Relative expression of genes increases with chilling accumulation, suggesting a role for RdDM in dormancy regulation.	Dormancy regulation	Basdeki & Hagidimitriou (2019); Rothkegel et al. (2017)
Citrus	*LEAFY* (*LFY*)	• Central regulator of the flowering network.	Controls flowering	Basdeki & Hagidimitriou (2019); Zhang et al. (2014)
	CcMADS19	• A floral repressor, was found to be enhanced in repressive H3K27me3.	Influence flowering	Mesejo et al. (2022)
Cucumber	*CsACO3* and *MADS-box* gene.	• Downregulation of gene due to CHG-type hypermethylation in the promoter regions.	Mediate the sex expression change induced by high temperature and long-day.	Lai et al. (2018)
Mango	*MiFT1* and *MiTFL1s*	• High fruit load represses flowering through suppression of MiFT1 upregulation and delay of MiTFL1s downregulation.	Influence flowering	Gafni et al. (2022)
Melon	*ACS1*, *ETR1*, and *ACO1*	• DNA methylation.	Triggered fruit ripening	Giordano et al. (2022)
Orange	Methylation sensitive McrBC-qPCR.	• DNA methylation	De-greening	Huang et al. (2018)
Papaya	*CpSVP* and *CpAP1*	• Transcription differences in *CpSVP* and *CpAP1*.	Early flowering in male papaya.	Zhou et al. (2020)
Peach	MYB, bHLH, and WD40	• Determines the spatial and temporal expression of anthocyanin biosynthesis genes.	Accumulation of anthocyanins and pigments.	Garcia-Gomez et al. (2021)
	PpTPS3 expression	• DNA methylation. • Decreased 5 mC level in its promoter.	Higher linalool accumulation and fruit ripening.	Wei et al. (2021)

TABLE 9.1 *(Continued)*

Crop	Epigenetic Regulatory Mechanism	Role	Improvement in Trait	References
Pineapple	AtHD2	• Histone deacetylases. • Elevated HDAC enzyme levels.	Abiotic stress and enhanced dwarfism.	Halim et al. (2022)
Potato	DEGs	• Resistance-related metabolic pathways.	Induce disease resistance.	Shi et al. (2022)
Strawberry	*FaERF9* + *FaMYB98* (FaQR ERF-MYB complex)	• Regulate furaneol synthesis. • Transcription of quinones oxidoreductase gene.	Flavor	Zhang et al. (2018)
Tomato	*RIN MADS* box	• Influence production of alcohol and aldehyde.	Flavor	Cao et al. (2021)
	HDA1 and *HDA3*	• Regulate ethylene bioethylene. • Influence level of carotenoids.	Ripening	Anwar et al. (2021)
	SlALKBH2	• Ripening-associated dynamics of mRNA m6A methylation.	Ripening	Zhuo et al. (2019)
	MdTSTa gene	• Over expression of *MdTSTa*.	Increase the fruit soluble sugar.	Ma et al. (2021)
	S1ERF.D7	• Over expression promote ripening. • Silencing delayed ripening.	Ripening	Gambhir et al. (2022)

has been shown to play a conserved role in repressing ripening genes and their associated genes (Lu et al., 2018). The tomato's ripening process was slowed due to a DNA demethylase defect, and the application of a DNA methylation inhibitor altered the ripening process in many fruit species (Tang et al., 2020).

9.4 STRATEGIES FOR CONTROLLING THE QUALITY OF POST-HARVEST AND PROCESSED FOOD

Fruits and vegetables are appreciated for their health benefits. Fresh-cut fruits and vegetables are a rising concern for consumers worldwide. Product quality involves visual attributes such as size, color, shape, physical and chemical properties such as texture, and flavor and other organoleptic characteristics. Postharvest handling processes don't improve produce quality; they only slow deterioration. Appearance, freshness, and sensory qualities can impact a consumer's buying decision and other sensory perception. A deeper understanding of fruit ripening from a physiological and molecular perspective is required to improve fruit quality and reduce PHLs.

With a constantly growing population and ever-increasing demand for food, the reduction of PHLs is an important strategy to ensure food availability and quality. PHLs represent about 40% of the harvested crop, which is quite significant (FAO, 2011). Quality deterioration and PHLs in fresh fruits and vegetables occur due to a variety of reasons during both pre-harvest and postharvest stages. These losses are much more pronounced in developing nations, where about one third of the fresh produce fails to reach the consumers' plate. Physiological senescence and infection by fungal pathogens are the major reasons for the PHLs (Tian et al., 2016). Different post-harvest strategies are commercially used to prevent spoilage of foods, dating back to prehistoric times. Among the oldest methods, drying, refrigeration, and fermentation are widely used for food preservation. Modern methods include canning, pasteurization, freezing, irradiation, ultrasound, microwave processing, and the addition of chemicals. These methods are used to reduce the respiratory rate, retard ripening and senescence, and inhibit the growth of microbial pathogens, thereby extending the shelf life of the product. Different postharvest strategies (e.g., low temperature, air atmosphere modification, chemical treatments) represent abiotic stresses that activate several metabolic pathways to cope with the stress and attain homeostasis, increasing the shelf life of the commodity. The detrimental effects of various stresses on post-harvest quality have been studied by many researchers with the "omics" perspective. Most of the epigenetic studies in the field of post-harvest research are mainly oriented towards the modulation of pathways involved in senescence (Osorio et al., 2013; Xu et al., 2013; Sam et al., 2014) and response towards various biotic/abiotic stresses (Jones-Rhoades & Bartel, 2004; Dowen et al., 2012).

Several physiological and biochemical changes are linked to fruit ripening and senescence, which play a crucial role in fruit life cycles. Losses from postharvest fruit storage due to this problem are always extremely high. Heat treatment is a popular method for preserving the quality of fruits and vegetables after harvest and preventing spoilage and physiological diseases (Naser et al., 2018). Successfully using heat treatment to mitigate chilling injury in kiwi fruit (Ma et al., 2014), preserve sensory quality in cucumbers (Nasef et al., 2018), and postpone ripening in peach fruit (Ma et al., 2018) has been observed (Huan et al., 2016).

9.5 MODULATION OF PATHWAYS INVOLVED IN SENESCENCE

Cell senescence is an irreversible state of cell cycle arrest; it is characterized by the permanent cell growth arrest in normal and altered physiological processes. Cell senescence can be triggered by various senescence stress factors, such as replicative stress, oxidative stress, and DNA instability, affecting the gene expressions and signal pathways that result in various senescence phenotypes. Epigenetic mechanisms are involved in altering structures of chromatin and gene expression without altering the genome and hence have an important role in cell senescence too (Lowe et al., 2016). Cell senescence is associated with DNA methylation, histone-associated processes (epigenetic), chromatin remodeling, and ncRNA expression.

DNA methylation is regulated by DNA methyltransferase and DNA demethylation enzymes. DNA methylation is used to target two primary functions: (i) regulation of gene expression through methylation and demethylation at specific sites; and (ii) providing genome stability by silencing repeat sequences, particularly TEs (Chan et al., 2005). In some angiosperms, changes corresponding to the extent of DNA methylation were observed during maturation processes by Fraga et al. (2002).

Furthermore, during senescence, TEs with mutagenic potential have been reported in crop plants (Feschotte et al., 2002). Under stress conditions such as pathogen attack or heat stress, TEs become activated and DNA methylation alterations occur (Bilichak et al., 2012). Similarly, Dowen et al. (2012) also reported dynamic changes in DNA methylation, which subsequently lead to the activation of defense-related genes. DNA methylation alterations formed during cell senescence interact with telomere dysfunction and DNA damage, leading to cell cycle arrest (Cheng et al., 2017).

Similarly, histone-associated epigenetic processes are affected by various induction factors such as maturation, reactive oxygen species (ROS), gene mutagenesis, and environmental damage, which lead to changes in chromatin structure and gene expression resulting in cell senescence. Histone alterations due to senescence stress are the most direct response to senescence (Prado et al., 2016). On the other hand, chromatin remodeling complexes use the energy generated through ATP hydrolysis to alter the DNA-histone contacts and thus impact the chromatin status and structure (Clapier et al., 2012). DNA methylation correlated with chromatin structure modifications has an influence on post-harvest physiology events, thus impacting the shelf life and quality of the food commodity.

The ncRNAs are non-protein transcripts; these transcripts were previously considered as transcription noise and junk RNA but are now recognized as essential regulators of gene expression and cellular activities (Ding et al., 2013). The expression of numerous ncRNAs is induced by senescence. These ncRNAs, in turn, influence cell senescence and form a complex network that regulates various senescence phenotypes (Ay et al., 2014). The various epigenetic mechanisms and modifications mentioned above interact with each other, resulting in different types of gene expressions. Histone alterations induced by senescence stress may lead to senescence-induced DNA methylation, chromatin remodeling complexes, and ncRNA expression (Spiegelman et al., 2013). The epigenetic alterations and cell senescence interact, forming a complex network that regulates various senescence phenotypes.

Epigenetic mechanisms targeting cell senescence can be targeted to inhibit the accumulation of senescent cells and increase the shelf life of fresh produce. Various inhibitors of DNA methylation and histone deacetylation, as well as mechanisms targeting histone methylation and ncRNA, are being used and investigated by researchers worldwide.

Epigenetic mechanisms targeting cell senescence can inhibit senescence phenotypes, which can lead to the reactivation of normal cellular activity. New cultivars with convenient shapes and sizes, improved nutrition and flavor, increased shelf life, reduced chilling susceptibility, water loss, discoloration, and greater disease resistance can be developed using these mechanisms, such as reduced chilling-associated internal browning in pineapple and viral resistance in papaya. Fruit ripening and senescence are interrelated physiological processes that are highly complex and significantly influence fruit quality (Klee et al., 2011). These physiological processes involve the reprogramming of multiple genes associated with color, flavor, aroma, texture, and other quality attributes (Lu et al., 2018). Ripe fruits (fleshy) are more susceptible to pathogens and decomposition (Prusky, 1996), which may lead to greater post-harvest losses.

9.6 RESPONSE TOWARDS BIOTIC/ABIOTIC STRESSES

The physiological and molecular aspects of postharvest, particularly ethylene-related stresses, greatly influence fruit quality. Understanding these aspects can be helpful in regulating fruit quality (Ansari & Tuteja, 2013). Several pre-harvest factors, stage of development, and genetic background have a significant influence on the response of the commodity to various postharvest stresses. DNA methylation may change, both at the whole-genome level and at individual genes, in response to environmental stresses (Basdeki & Hagidimitriou, 2019; Michalak et al., 2013). Abiotic stresses during the pre-harvest period, such as drought, light, and field temperatures, influence post-harvest traits (Toivonen & Hodges, 2011). Exposure to moderate levels of pre-harvest stresses has been reported to enhance the upregulation of genes and pathways that improve tolerance and resistance to various postharvest stresses (Bowler & Fluhr, 2000; Toivonen & Hodges, 2011). For example, apples and pears cultivated under deficit irrigation during a certain period had lower water losses postharvest compared to fully watered apples and pears (Lopez et al., 2011). The exposure of fruits to water stress needs to be timing-specific; for example, mango fruits exposed to water stress showed more susceptibility to chilling injury when water stress was applied during cell expansion rather than during cell division or close to harvest (Lechaudel & Joas, 2007).

9.6.1 RESPONSE TOWARDS ABIOTIC STRESSES

Postharvest abiotic stresses influence eating quality, nutritional attributes, shelf life, and susceptibility to pathological and physiological disorders. Postharvest technologies involve the application of stress conditions such as heat, cold, controlled atmospheric conditions, and the addition of chemicals to improve shelf life and product quality. However, in recent times, researchers are concerned about the molecular aspect of the stresses and the use of holistic approaches such as transcriptomic, proteomic, and metabolic strategies, complemented with biochemical analysis and phenotyping to understand stress regulation and its complex regulations at cellular levels. The transition from the vegetative to flowering state is the most important time for plants. To maintain or hasten the plant life cycle, methylation and demethylation are triggered during flowering in response to environmental stresses such as temperature, drought stress, carbon dioxide, and other abiotic/biotic conditions (Shi et al., 2022).

Epigenetic pathways contribute to the phenotypic plasticity of plants and provide the ability to survive under stress conditions (Pikaard et al., 2014). Phenotypic plasticity is the ability of an organism to show changes in its phenotype corresponding to stimuli or environmental stress. Transcript profiling of two similar fruits, peach and nectarine, was conducted by Dagar et al. (2013). It was observed that nectarines had better storability than peaches. They also compared the transcript profiles from both fruits with cold-tolerant varieties of peach and found that the cold-tolerant varieties had a greater number of genes regulating antioxidant pathways, stress proteins related to pathogenesis, amino acid metabolism, and cell wall-modifying enzymes, thus improving storability. Epigenetic modifications to DNA are involved in plants' response to drought stress, as discovered by studying the epigenetic regulation of gene expression in desiccated seeds of common pear (Basdeki & Hagidimitriou, 2019; Michalak et al., 2013). CpHUA1 is part of the papaya sex differentiation gene network. The methylation of CpHUA1 in the CHH context is associated with temperature stress response and sex reversal in papaya (Liu et al., 2018).

A number of miRNAs in banana target stress-related proteins either directly or indirectly, although *miR393* is involved in the regulation of redox in response to stress (Iglesias et al., 2014). According to research published by Muthusamy et al. (2016), miR169 is upregulated in response to drought stress, and it directly targets a nuclear factor Y (transcription factor) to positively regulate the expression of the dehydrin and aquaporin genes. microRNAs (miRNAs) miR156 and miR535 target SPL genes known to regulate heat stress at various stages of banana development, leading to increased tolerance to heat stress (Bhakta et al., 2021; Zhu et al., 2019). Cysteine proteases (CP) are involved in biotic and abiotic stress, plant growth and development, senescence, and cell death. Gong et al. (2022) found 51 CP genes in apples and found that drought affected the expression of different MdCPs in apples, with MdCP37 being the most responsive. Kong et al. (2022) identified 508 differentially methylated regions (DMRs), out of which 59.92% were correlated with transcripts, and 39.43% were associated with methylation levels. 31 DEGs were targeted by methylation and miRNA and were primarily involved in the ROS response and abscisic acid (ABA) signaling.

DNA methylation, histone modifications, and the overexpression of stress-responsive genes are genetic and epigenetic mechanisms that have evolved to help organisms cope with heat and cold stress (Verma et al., 2022). Du et al. (2021) discovered that 305R stimulated histone methylation, histone phosphorylation, DNA methylation, and nucleosome assembly. Specific proteome dynamics have been linked to stress responses and developmental programs (Momo et al., 2022). It is now widely accepted that HATs play a critical role in plant development, growth, and the response to various environmental stresses (Eom et al., 2018). DNA methylation is crucial for the potato's defense against DON stress (Shi et al., 2022). HDACs are proteins that catalyze the removal of acetyl groups from lysine residues in histones, and they are involved in a wide variety of physiological processes, such as stress responses. Around 20 histone deacetylase (HDAC) genes were found in the genome of *Brassica rapa* and these genes fall into three distinct families: RPD3/HDA1, HD2, and SIR2 (Eom et al., 2021). Wang et al. (2019) anticipated 39 lncRNAs as endogenous target mimics (eTMs) for 35 miRNAs; of these, 5 were confirmed to play a role in Chinese cabbage's heat tolerance. Since miRNA entities play a significant role in stress regulation, transgenic approaches may be a useful tool for increasing crop yields while decreasing susceptibility to environmental stresses (Saha et al., 2022).

The climacteric fruits such as apple, tomato, banana, and kiwi exhibit bursts of ethylene at the initiation of ripening (Klee & Giovannoni, 2011). These fruits are typically harvested at a relatively low maturity level. However, the excessive use of ethylene during this period can result in rapid softening and a decline in quality. Modulating the pathways of ethylene biosynthesis holds significant potential for extending shelf life and preserving quality traits post-harvest. The use of inhibitors targeting ethylene-related responses in fresh fruits and vegetables at the molecular level is a promising option with great future prospects. For instance, 1-MCP acts as a competitive inhibitor of ethylene receptors, thereby suppressing ethylene-related responses. Consequently, it has been widely employed to delay ripening in various climacteric fruits (Liu et al., 2015; Zhang et al., 2012). In addition to ripening delay, 1-MCP has been shown to enhance fungal resistance (Jiang et al., 2001; Zhang et al., 2012) and improve the organoleptic qualities of fruits by influencing organic acid dynamics, regulating levels of volatile flavor compounds, and sugar levels (Liu et al., 2016).

9.6.2 RESPONSE TOWARDS BIOTIC STRESS

DNA methylation exercises control over some of the defense genes during normal conditions but when under stress, it may lead to alteration of gene expressions by inducing methylation changes to transcriptional control of specific loci in plant systems (Cercos et al., 2006; An et al., 2011; Frenkel et al., 1977). Such changes may hyperactivate the plant defense system through activation of salicylic acid (SA) signaling pathways imparting pathogen resistance (Li et al., 2018; Qin et al., 2016). Under normal conditions, these pathways remain suppressed until required for defense. Enzymes that deteriorate pectin are involved in the pathogenicity of many different types of plant-harming fungi. These include polygalacturonase (PGs) and pectate lyases (PLs). DNA methylation is involved in the transcription regulation of pectin-depolymerase genes under the catalysis of DNMT, as suggested by a study by Meng et al. (2022), which found that PG and PL activities were up while DNMT activity was down in C. Gloeosporioides and these factors contribute to the spread of the fungus Ceratocystis gloeosporioides in mango fruits. Two different proanthocyanidin-specific R2R3 MYB activators (MaMYBPA1 and MaMYBPA2) and 4 repressors (MaMYBPR1, MaMYBPR2, MaMYBPR3, and MaMYBPR4) were categorized by Rajput et al. (2022) and observed that overexpression of MaMYBPAs alone or with MaMYC in banana fruits induced proanthocyanidin accumulation.

Liu et al. (2022) knocked out the coding gene to study the functions of GCN5 homolog FocGCN5, and they discovered a drastic slowdown of growth and a nearly complete loss of pathogenicity to banana plantlets.

Fungal pathogens are responsible for a large proportion of economic losses in fresh fruits and vegetables. Frequent and long-term application of fungicides poses public health concerns and environmental risks. The overexploitation of fungicides also leads to the development of fungicide resistance in strains, corresponding to economic losses. This has led to the exploration and development of safer alternatives, especially from natural sources. Crop plants also tend to have a memory, a phenomenon in which organisms, including plants, are able to store and recall information from previous experiences or events and have an effect on responses to stress conditions in the future. The mechanisms involved are of epigenetic, metabolic, and electrical origins. One such example of the exploitation of memory is the use of pre-treatments accompanied by low temperatures to acclimatize fruits to freezing conditions and reduce chilling injuries. The new perspective of attributing memory to fruits can be helpful in improving shelf life and post-harvest quality.

9.7 FUTURE PERSPECTIVES

In developing countries like India, addressing the issue of postharvest losses (PHLs) in fruits and vegetables and maintaining quality for extended periods is crucial. The application of biotechnological tools and skills has been instrumental in resolving these challenges. Researchers in plant sciences must delve into gene expression studies and chromatin profiling, as numerous technologies have emerged to facilitate this, enhancing practical skills and knowledge in postharvest handling. It is evident that gene expression can influence fruit quality traits in various tissues. Consequently, new phenotypic and genotypic traits can be modified in the future using transcription factors (TFs). Epigenetic modifications provide scientists with a tool to evaluate the relationship between genotype and phenotype, aiding in reducing postharvest losses. Epigenetic modifications offer phenotypic variation, which can be beneficial for enhancing crops with desirable traits. Further exploration of epigenetic regulations in fruits and vegetables through breeding is essential. Understanding the hierarchy of TFs is crucial for establishing a complete regulatory network in this field. Additionally, in non-climacteric fruits, there is a need for more exploration of the molecular mechanisms of fruit ripening. Epigenetic modifications also contribute to mitigating various biotic and abiotic stresses and improving qualitative and quantitative traits in plants.

Therefore, crops can benefit from epigenetic modifications, necessitating further research in this area.

9.8 CONCLUSION

As the agriculture sector is flourishing day by day and people are concerned about the quality of the food they eat in an effort to improve their health. Understanding the role of epigenetic modifications in fruit and vegetable development, ripening, and senescence will allow for the modification of the epigenome to minimize post-harvest deterioration and breed novel cultivars with improved quality traits without altering the genomic sequence. The epigenetic regulation of fruit and vegetable development before and after harvest is an important and interesting area of research with many hurdles and unresolved issues; thus, more investigation into this topic is required to advance our knowledge of epigenetic regulations in fruits and vegetables for the benefit of mankind. This chapter provides a comprehensive review of the most recent findings regarding epigenetic regulations in fruits and vegetables, both before and after harvest. With the advent of new breeding techniques, modifying a plant's epigenome has emerged as a promising strategy for preserving quality in fruit and vegetables both before and after harvest.

KEYWORDS

- **abiotic stresses**
- **biotic stresses**
- **breeding**
- **epigenome**
- **fungicide resistance**
- **genotypic traits**
- **post-harvest deterioration**
- **postharvest losses**
- **transcription factors**

REFERENCES

Agusti, M., Mesejo, C., Munoz-Fambuena, N., Vera-Sirera, F., De Lucas, M., Martinez-Fuentes, A., & Blazquez, M. A., (2020). Fruit-dependent epigenetic regulation of flowering in *Citrus*. *New Phytologist*, *225*(1), 376–384.

Ahmed, F. F., Hossen, M. I., Sarkar, M. A. R., Konak, J. N., Zohra, F. T., Shoyeb, M., & Mondal, S., (2021). Genome-wide identification of DCL, AGO and RDR gene families and their associated functional regulatory elements analyses in banana (*Musa acuminata*). *PloS One*, *16*(9), e0256873.

An, J. P., Zhang, X. W., You, C. X., Bi, S. Q., Wang, X. F., & Hao, Y. J., (2019). Md WRKY 40 promotes wounding-induced anthocyanin biosynthesis in association with Md MYB 1 and undergoes Md BT 2-mediated degradation. *New Phytologist*, *224*(1), 380–395.

Anwar, S., Brenya, E., Alagoz, Y., & Cazzonelli, C. I., (2021). Epigenetic control of carotenogenesis during plant development. *Critical Reviews in Plant Sciences*, *40*(1), 23–48.

Avramidou, E. V., Ganopoulos, I. V., Doulis, A. G., Tsaftaris, A. S., & Aravanopoulos, F. A., (2015). Beyond population genetics: Natural epigenetic variation in wild cherry (*Prunus avium*). *Tree Genetics & Genomes*, *11*(5), 1–9.

Ay, N., Janack, B., & Humbeck, K., (2014). Epigenetic control of plant senescence and linked processes. *Journal of Experimental Botany*, *65*(14), 3875–3887.

Bai, S., Tuan, P. A., Saito, T., Honda, C., Hatsuyama, Y., Ito, A., & Moriguchi, T., (2016). Epigenetic regulation of MdMYB1 is associated with paper bagging-induced red pigmentation of apples. *Planta*, *244*(3), 573–586.

Basdeki, L., & Hagidimitriou, M., (2019). The role of DNA methylation in perennial plants. *Notulae Scientia Biologicae*, *11*(1), 1–11.

Berr, A., Shafiq, S., & Shen, W. H., (2011). Histone modifications in transcriptional activation during plant development. *Biochimica et Biophysica Acta (BBA)-Gene Regulatory Mechanisms*, *1809*(10), 567–576.

Bhakta, S., Tak, H., & Ganapathi, T. R., (2021). Exploring diverse roles of micro RNAs in banana: Current status, and future prospective. *Physiologia Plantarum*, *173*(4), 1323–1334.

Bilichak, A., Ilnystkyy, Y., Hollunder, J., & Kovalchuk, I., (2012). The progeny of *Arabidopsis thaliana* plants exposed to salt exhibit changes in DNA methylation, histone modifications and gene expression. *PloS One*, *7*(1), e30515.

Bitonti, M. B., Cozza, R., Chiappetta, A., Giannino, D., Castiglione, M. R., Dewitte, W., & Innocenti, A. M., (2002). Distinct nuclear organization, DNA methylation pattern and cytokinin distribution mark juvenile, juvenile-like and adult vegetative apical meristems in peach (*Prunus persica* (L.) Batsch). *Journal of Experimental Botany*, *53*(371), 1047–1054.

Bollati, V., & Baccarelli, A., (2010). Environmental epigenetics. *Heredity*, *105*(1), 105–112.

Bowler, C., & Fluhr, R., (2000). The role of calcium and activated oxygens as signals for controlling cross-tolerance. *Trends in Plant Science*, *5*(6), 241–246.

Bucher, E., Kong, J., Teyssier, E., & Gallusci, P., (2018). Epigenetic regulations of fleshy fruit development and ripening and their potential applications to breeding strategies. In: *Advances in Botanical Research* (Vol. 88, pp. 327–360). Academic Press.

Cao, X., Wei, C., Duan, W., Gao, Y., Kuang, J., Liu, M., & Zhang, B., (2021). Transcriptional and epigenetic analysis reveals that NAC transcription factors regulate fruit flavor ester biosynthesis. *The Plant Journal*, *106*(3), 785–800.

Cercos, M., Soler, G., Iglesias, D. J., Gadea, J., Forment, J., & Talon, M., (2006). Global analysis of gene expression during development and ripening of *Citrus* fruit flesh. A proposed mechanism for citric acid utilization. *Plant Molecular Biology*, *62*(4), 513–527.

Chachar, S., Chachar, M., Riaz, A., Shaikh, A. A., Li, X., Li, X., & Zhang, P., (2022). Epigenetic modification for horticultural plant improvement comes of age. *Scientia Horticulturae*, *292*, 110633.

Chan, S. W. L., Henderson, I. R., & Jacobsen, S. E., (2005). Gardening the genome: DNA methylation in *Arabidopsis thaliana*. *Nature Reviews Genetics*, *6*(5), 351–360.

Chen, T., Ji, D., Zhang, Z., Li, B., Qin, G., & Tian, S., (2021). Advances and strategies for controlling the quality and safety of postharvest fruit. *Engineering*, *7*(8), 1177–1184.

Cheng, J., Wei, G., Zhou, H., Gu, C., Vimolmangkang, S., Liao, L., & Han, Y., (2014). Unraveling the mechanism underlying the glycosylation and methylation of anthocyanins in peach. *Plant Physiology*, *166*(2), 1044–1058.

Cheng, L. Q., Zhang, Z. Q., Chen, H. Z., & Liu, D. P., (2017). Epigenetic regulation in cell senescence. *Journal of Molecular Medicine*, *95*(12), 1257–1268.

Cho, H. J., Kim, G. H., & Choi, C., (2020). Differential gene expression and epigenetic analyses between striped and blushed skinned sports of 'fuji' apple. *Scientia Horticulturae*, *261*, 108944.

Clapier, C. R., & Cairns, B. R., (2012). Regulation of ISWI involves inhibitory modules antagonized by nucleosomal epitopes. *Nature*, *492*(7428), 280–284.

Contreras-De La, R. P. A., Centeotl, A. R., Jaime, A. C. L., Kimberly, F. G. A., & De-la-Peña, C., (2022). Lysine crotonylation: A challenging new player in the epigenetic regulation of plants. *Journal of Proteomics*, 104488.

Dagar, A., Pons, P. C., Marti, I. C., Ziliotto, F., Bonghi, C., Crisosto, C. H., & Granell, A., (2013). Comparative transcript profiling of a peach and its nectarine mutant at harvest reveals differences in gene expression related to storability. *Tree Genetics & Genomes*, *9*(1), 223–235.

Deal, R. B., & Henikoff, S., (2011). Histone variants and modifications in plant gene regulation. *Current Opinion in Plant Biology*, *14*(2), 116–122.

Desta, B., Tena, N., & Amare, G., (2022). Preharvest factors that influence the quality of tomato. *Asian Journal of Research and Review in Agriculture*, 6–12.

Ding, Y., Tao, Y., & Zhu, C., (2013). Emerging roles of microRNAs in the mediation of drought stress response in plants. *Journal of Experimental Botany*, *64*(11), 3077–3086.

Dowen, R. H., Pelizzola, M., Schmitz, R. J., Lister, R., Dowen, J. M., Nery, J. R., & Ecker, J. R., (2012). Widespread dynamic DNA methylation in response to biotic stress. *Proceedings of the National Academy of Sciences*, *109*(32), E2183–E2191.

Du, J. S., Hang, L. F., Hao, Q., Yang, H. T., Ali, S., Badawy, R. S. E., & Lai, Y. S., (2021). The dissection of R genes and locus Pc5. 1 in *Phytophthora capsici* infection provides a novel view of disease resistance in peppers. *BMC Genomics*, *22*(1), 1–16.

Du, W. W., Yang, W., Chen, Y., Wu, Z. K., Foster, F. S., Yang, Z., & Yang, B. B., (2017). Foxo3 circular RNA promotes cardiac senescence by modulating multiple factors associated with stress and senescence responses. *European Heart Journal*, *38*(18), 1402–1412.

Eccleston, A., DeWitt, N., Gunter, C., Marte, B., & Nath, D., (2007). Epigenetics. *Nature*, *447*(7143), 395–396.

Elik, A., Yanik, D. K., Istanbullu, Y., Guzelsoy, N. A., Yavuz, A., & Gogus, F., (2019). Strategies to reduce post-harvest losses for fruits and vegetables. *Strategies*, *5*(3), 29–39.

El-Sharkawy, I., Liang, D., & Xu, K., (2015). Transcriptome analysis of an apple (*Malus* × *domestica*) yellow fruit somatic mutation identifies a gene network module highly associated with anthocyanin and epigenetic regulation. *Journal of Experimental Botany*, *66*(22), 7359–7376.

Eom, S. H., & Hyun, T. K., (2018). Histone acetyltransferases (HATs) in Chinese cabbage: Insights from histone H3 acetylation and expression profiling of HATs in response to abiotic stresses. *Journal of the American Society for Horticultural Science*, *143*(4), 296–303.

Eom, S. H., & Hyun, T. K., (2021). Comprehensive analysis of the histone deacetylase gene family in Chinese cabbage (*Brassica Rapa*): From evolution and expression pattern to functional analysis of BraHDA3. *Agriculture*, *11*(3), 244.

Falavigna, V. D. S., Guitton, B., Costes, E., & Andrés, F., (2019). I want to (bud) break free: The potential role of DAM and SVP-like genes in regulating dormancy cycle in temperate fruit trees. *Frontiers in Plant Science*, *9*, 1990.

FAO, (2011). *Global Food Losses and Food Waste. Extent, Causes and Prevention.* Food and Agriculture Organization of the United Nations, Rome. Available at: http://www.fao.org/docrep/014/mb060e/mb060e00 pdf (accessed on 7 February 2024).

Farinati, S., Rasori, A., Varotto, S., & Bonghi, C., (2017). Rosaceae fruit development, ripening and post-harvest: An epigenetic perspective. *Frontiers in Plant Science*, *8*, 1247.

Fei, Y., Nyikó, T., & Molnar, A., (2021). Non-perfectly matching small RNAs can induce stable and heritable epigenetic modifications and can be used as molecular markers to trace the origin and fate of silencing RNAs. *Nucleic Acids Research*, *49*(4), 1900–1913.

Feschotte, C., Jiang, N., & Wessler, S. R., (2002). Plant transposable elements: Where genetics meets genomics. *Nature Reviews Genetics*, *3*(5), 329–341.

Fortes, A. M., Granell, A., Pezzotti, M., & Bouzayen, M., (2017). Molecular and metabolic mechanisms associated with fleshy fruit quality. *Frontiers in Plant Science*, 1236.

Fraga, M. F., Rodríguez, R., & Cañal, M. J., (2002). Genomic DNA methylation–demethylation during aging and reinvigoration of *Pinus radiata. Tree Physiology*, *22*(11), 813–816.

Fragou, D., Pakkidi, E., Aschner, M., Samanidou, V., & Kovatsi, L., (2019). Smoking and DNA methylation: Correlation of methylation with smoking behavior and association with diseases and fetus development following prenatal exposure. *Food and Chemical Toxicology*, *129*, 312–327.

Francini, A., & Sebastiani, L., (2013). Phenolic compounds in apple (*Malus x domestica* Borkh.): Compounds characterization and stability during postharvest and after processing. *Antioxidants*, *2*(3), 181–193.

Frenkel, C., & Eskin, M., (1977). Ethylene evolution as related to changes in hydroperoxides in ripening tomato fruit. *HortScience*, *12*(6), 552–553.

Fu, C. C., Han, Y. C., Guo, Y. F., Kuang, J. F., Chen, J. Y., Shan, W., & Lu, W. J., (2018). Differential expression of histone deacetylases during banana ripening and identification of MaHDA6 in regulating ripening-associated genes. *Postharvest Biology and Technology*, *141*, 24–32.

Gafni, I., Rai, A. C., Halon, E., Zviran, T., Sisai, I., Samach, A., & Irihimovitch, V., (2022). Expression profiling of four mango FT/TFL1-encoding genes under different fruit load conditions, and their involvement in flowering regulation. *Plants*, *11*(18), 2409.

Gambhir, P., Singh, V., Parida, A., Raghuvanshi, U., Kumar, R., & Sharma, A. K., (2022). Ethylene response factor ERF. D7 activates auxin response factor 2 paralogs to regulate tomato fruit ripening. *Plant Physiology*.

Garcia-Gomez, B. E., Salazar, J. A., Nicolás-Almansa, M., Razi, M., Rubio, M., Ruiz, D., & Martínez-Gómez, P., (2020). Molecular bases of fruit quality in *Prunus* species: An integrated genomic, transcriptomic, and metabolic review with a breeding perspective. *International Journal of Molecular Sciences*, *22*(1), 333.

Gillaspy, G., Ben-David, H., & Gruissem, W., (1993). Fruits: A developmental perspective. *The Plant Cell*, *5*(10), 1439.

Giordano, A., Santo, D. M., Quadrana, L., Pujol, M., Martín-Hernández, A. M., & Garcia-Mas, J., (2022). CRISPR/Cas9 gene editing uncovers the roles of constitutive triple response 1 and repressor of silencing 1 in melon fruit ripening and epigenetic regulation. *Journal of Experimental Botany*.

Gong, X., Deng, W., Wen, B., Xiao, W., & Li, L., (2022). Genome-wide analysis of cysteine proteases (CP) in apple (*Malus domestica)* and functional characterization of MdCP37 in response to drought stress. *Plant Cell, Tissue, and Organ Culture*, 1–11.

Gu, D., Yang, J., Wu, S., Liao, Y., Zeng, L., & Yang, Z., (2021). Epigenetic regulation of the phytohormone abscisic acid accumulation under dehydration stress during postharvest processing of tea (*Camellia sinensis). Journal of Agricultural and Food Chemistry*, *69*(3), 1039–1048.

Gu, T., Ren, S., Wang, Y., Han, Y., & Li, Y., (2016). Characterization of DNA methyltransferase and demethylase genes in *Fragaria vesca. Molecular Genetics and Genomics*, *291*(3), 1333–1345.

Han, Y. C., Kuang, J. F., Chen, J. Y., Liu, X. C., Xiao, Y. Y., Fu, C. C., & Lu, W. J., (2016). Banana transcription factor MaERF11 recruits histone deacetylase MaHDA1 and represses the expression of MaACO1 and *Expansins* during fruit ripening. *Plant Physiology*, *171*(2), 1070–1084.

Hasbun, R., Valledor, L., Berdasco, M., Santamaria, E., Canal, M. J., Rodriguez, R., & Sanchez, M., (2005). *In vitro* proliferation and genome DNA methylation in adult chestnuts. *Acta Horticulturae*, *693*, 333.

He, M., Kong, X., Jiang, Y., Qu, H., & Zhu, H., (2022). MicroRNAs: Emerging regulators in horticultural crops. *Trends in Plant Science*.

Hollister, J. D., & Gaut, B. S., (2009). Epigenetic silencing of transposable elements: A trade-off between reduced transposition and deleterious effects on neighboring gene expression. *Genome Res.*, *19*, 1419–1428.

Huan, C., Jiang, L., An, X., Kang, R., Yu, M., Ma, R., & Yu, Z., (2016). Potential role of glutathione peroxidase gene family in peach fruit ripening under combined postharvest treatment with heat and 1-MCP. *Postharvest Biology and Technology*, *111*, 175–184.

Huang, H., Liu, R., Niu, Q., Tang, K., Zhang, B., Zhang, H., & Lang, Z., (2019). Global increase in DNA methylation during orange fruit development and ripening. *Proceedings of the National Academy of Sciences*, *116*(4), 1430–1436.

Iglesias, M. J., Terrile, M. C., Windels, D., Lombardo, M. C., Bartoli, C. G., Vazquez, F., & Casalongué, C. A., (2014). MiR393 regulation of auxin signaling and redox-related components during acclimation to salinity in *Arabidopsis. PLoS One*, *9*(9), e107678.

Ince, A. G., & Karaca, M., (2021). Tissue and/or developmental stage specific methylation of nrDNA in *Capsicum annuum. Journal of Plant Research*, *134*(4), 841–855.

Jiang, S., Wang, N., Chen, M., Zhang, R., Sun, Q., Xu, H., & Chen, X., (2020). Methylation of MdMYB1 locus mediated by RdDM pathway regulates anthocyanin biosynthesis in apple. *Plant Biotechnology Journal*, *18*(8), 1736–1748.

Jiang, Y., Joyce, D. C., & Terry, L. A., (2001). 1-methylcyclopropene treatment affects strawberry fruit decay. *Postharvest Biology and Technology*, *23*(3), 227–232.

Jones, P. A., (2012). Functions of DNA methylation: Islands, start sites, gene bodies and beyond. *Nature Reviews Genetics*, *13*(7), 484–492.

Jones-Rhoades, M. W., & Bartel, D. P., (2004). Computational identification of plant microRNAs and their targets, including a stress-induced miRNA. *Molecular Cell*, *14*(6), 787–799.

Kitavi, M., Cashell, R., Ferguson, M., Lorenzen, J., Nyine, M., McKeown, P. C., & Spillane, C., (2020). Heritable epigenetic diversity for conservation and utilization of epigenetic germplasm resources of clonal East African Highland banana (EAHB) accessions. *Theoretical and Applied Genetics*, *133*(9), 2605–2625.

Klee, H. J., & Giovannoni, J. J., (2011). Genetics and control of tomato fruit ripening and quality attributes. *Annual Review of Genetics*, *45*(1), 41–59.

Kondhare, K. R., Kumar, A., Patil, N. S., Malankar, N. N., Saha, K., & Banerjee, A. K., (2021). Development of aerial and belowground tubers in potato is governed by photoperiod and epigenetic mechanism. *Plant Physiology*, *187*(3), 1071–1086.

Kong, C., Su, H., Deng, S., Ji, J., Wang, Y., Zhang, Y., & Lv, H., (2022). Global DNA methylation and mRNA-miRNA variations activated by heat shock boost early microspore embryogenesis in cabbage (*Brassica oleracea*). *International Journal of Molecular Sciences*, *23*(9), 5147.

Kumar, A., Kumar, S., Anju, T., & Ramchiary, N., (2018). Genetic, epigenetic, and hormonal regulation of fruit development and ripening in *Capsicum* L. species. *Annual Plant Reviews Online*, 295–356.

Kumar, A., Sharma, D. P., Dogra, R. K., Singh, G., Sharma, R., & Kumar, P., (2022b). Identification and characterization of apple (*Malus* × *domestica* Borkh.) bud sports mutations in the apple growing Northwestern Himalayan region. *Scientia Horticulturae*, *304*, 111308.

Kumar, A., Sharma, D. P., Kumar, P., Sharma, G., & Suprun, I. I., (2022). Comprehensive insights on apple (*Malus* × *domestica* Borkh.) bud sport mutations and epigenetic regulations. *Scientia Horticulturae*, *297*, 110979.

Kumar, G., Rattan, U. K., & Singh, A. K., (2016). Chilling-mediated DNA methylation changes during dormancy and its release reveal the importance of epigenetic regulation during winter dormancy in apple (*Malus x domestica* Borkh.). *PloS One*, *11*(2), e0149934.

Lai, Y. S., Shen, D., Zhang, W., Zhang, X., Qiu, Y., Wang, H., & Li, X., (2018). Temperature and photoperiod changes affect cucumber sex expression by different epigenetic regulations. *BMC Plant Biology*, *18*(1), 1–13.

Lechaudel, M., & Joas, J., (2007). An overview of preharvest factors influencing mango fruit growth, quality and postharvest behavior. *Brazilian Journal of Plant Physiology, 19*, 287–298.

Lempe, J., Flachowsky, H., & Peil, A., (2022). Exploring epigenetic variation for breeding climate resilient apple crops. *Physiologia Plantarum*, e13782.

Lempe, J., Peil, A., & Flachowsky, H., (2021). Time-resolved analysis of candidate gene expression and ambient temperature during bud dormancy in apple. *Frontiers in Plant Science, 12*.

Li, S., Xu, H., Ju, Z., Cao, D., Zhu, H., Fu, D., & Zhu, B., (2018). The RIN-MC fusion of MADS-box transcription factors has transcriptional activity and modulates expression of many ripening genes. *Plant Physiology, 176*(1), 891–909.

Li, X., Wang, X., Zhang, Y., Zhang, A., & You, C. X., (2022). Regulation of fleshy fruit ripening: From transcription factors to epigenetic modifications. *Horticulture Research, 9*.

Liu, C., Long, J., Zhu, K., Liu, L., Yang, W., Zhang, H., & Deng, X., (2016). Characterization of a *Citrus* R2R3-MYB transcription factor that regulates the flavonol and hydroxycinnamic acid biosynthesis. *Scientific Reports, 6*(1), 1–16.

Liu, J., An, B., Luo, H., He, C., & Wang, Q., (2022). The histone acetyltransferase FocGCN5 regulates growth, conidiation, and pathogenicity of the banana wilt disease causal agent *Fusarium oxysporum* f. sp. cubense tropical race 4. *Research in Microbiology, 173*(3), 103902.

Liu, J., Chatham, L., Aryal, R., Yu, Q., & Ming, R., (2018). Differential methylation and expression of HUA1 ortholog in three sex types of papaya. *Plant Science, 272*, 99–106.

Liu, J., Chen, L. Y., Zhou, P., Liao, Z., Lin, H., Yu, Q., & Ming, R., (2021). Sex biased expression of hormone related genes at early stage of sex differentiation in papaya flowers. *Horticulture Research, 8*.

Liu, R., Wang, Y., Qin, G., & Tian, S., (2016). Molecular basis of 1-methylcyclopropene regulating organic acid metabolism in apple fruit during storage. *Postharvest Biology and Technology, 117*, 57–63.

Liu, R., Xu, Y., Li, B., Qin, G., & Tian, S., (2015). The role of 1-methylcyclopropene in lipid peroxidation, anti-oxidant enzyme activities, and ethylene biosynthesis in 'Laiyang' pear (*Pyrus bretschneideri* Rehd.) during fruit ripening. *The Journal of Horticultural Science and Biotechnology, 90*(2), 210–216.

Liu, Y., Tang, M., Liu, M., Su, D., Chen, J., Gao, Y., & Li, Z., (2020). The molecular regulation of ethylene in fruit ripening. *Small Methods, 4*(8), 1900485.

Lopez, G., Larrigaudière, C., Girona, J., Behboudian, M. H., & Marsal, J., (2011). Fruit thinning in 'Conference' pear grown under deficit irrigation: Implications for fruit quality at harvest and after cold storage. *Scientia Horticulturae, 129*(1), 64–70.

Lowe, D., Horvath, S., & Raj, K., (2016). Epigenetic clock analyses of cellular senescence and ageing. *Oncotarget, 7*(8), 8524–8531.

Lu, P., Yu, S., Zhu, N., Chen, Y. R., Zhou, B., Pan, Y., & Zhong, S., (2018). Genome encode analyses reveal the basis of convergent evolution of fleshy fruit ripening. *Nature Plants, 4*(10), 784–791.

Ma, Q., Suo, J., Huber, D. J., Dong, X., Han, Y., Zhang, Z., & Rao, J., (2014). Effect of hot water treatments on chilling injury and expression of a new C-repeat binding factor

(CBF) in 'hongyang' kiwifruit during low temperature storage. *Postharvest Biology and Technology*, *97*, 102–110.

Ma, W., Li, B., Zheng, L., Peng, Y., Tian, R., Yuan, Y., & Ma, B., (2021). Combined profiling of transcriptome and DNA methylome reveal genes involved in accumulation of soluble sugars and organic acid in apple fruits. *Foods*, *10*(9), 2198.

Ma, Y. J., Li, P. T., Sun, L. M., Zhou, H., Zeng, R. F., Ai, X. Y., & Hu, C. G., (2020). HD-ZIP I transcription factor (pthb13) negatively regulates *Citrus* flowering through binding to flowering locus C promoter. *Plants*, *9*(1), 114.

Marfil, C. F., Duarte, P. F., & Masuelli, R. W., (2018). Phenotypic and epigenetic variation induced in newly synthesized allopolyploids and autopolyploids of potato. *Scientia Horticulturae*, *234*, 101–109.

Margueron, R., & Reinberg, D., (2010). Chromatin structure and the inheritance of epigenetic information. *Nature Reviews Genetics*, *11*(4), 285–296.

Meng, L., Wang, J., Li, S., Xu, X., Zhang, Z., Shi, X., & Song, H., (2022). DNA methylation is involved in the regulation of the pectin depolymerase gene of *Colletotrichum gloeosporioides* and accelerates the infection of mango fruit. *Journal of Plant Biochemistry and Biotechnology*, 1–12.

Mesejo, C., Marzal, A., Martínez-Fuentes, A., Reig, C., De Lucas, M., Iglesias, D. J., & Agustí, M., (2022). Reversion of fruit-dependent inhibition of flowering in *Citrus* requires sprouting of buds with epigenetically silenced CcMADS19. *New Phytologist*, *233*(1), 526–533.

Michalak, M., Barciszewska, M. Z., Barciszewski, J., Plitta, B. P., & Chmielarz, P., (2013). Global changes in DNA methylation in seeds and seedlings of Pyrus communis after seed desiccation and storage. *PloS One*, *8*(8), e70693.

Momo, J., Kumar, A., Islam, K., Ahmad, I., Rawoof, A., & Ramchiary, N., (2022). A comprehensive update on *Capsicum* proteomics: Advances and future prospects. *Journal of Proteomics*, 104578.

Murr, R., (2010). Interplay between different epigenetic modifications and mechanisms. *Advances in Genetics*, *70*, 101–141.

Muthusamy, M., Uma, S., Backiyarani, S., & Saraswathi, M. S., (2016). Computational prediction, identification, and expression profiling of microRNAs in banana (*Musa* spp.) during soil moisture deficit stress. *The Journal of Horticultural Science and Biotechnology*, *89*(2), 208–214.

Nasef, I. N., (2018). Short hot water as safe treatment induces chilling tolerance and antioxidant enzymes, prevents decay, and maintains quality of cold-stored cucumbers. *Postharvest Biology and Technology*, *138*, 1–10.

Naser, F., Rabiei, V., Razavi, F., & Khademi, O., (2018). Effect of calcium lactate in combination with hot water treatment on the nutritional quality of persimmon fruit during cold storage. *Scientia Horticulturae*, *233*, 114–123.

Niederhuth, C. E., & Schmitz, R. J., (2014). Covering your bases: Inheritance of DNA methylation in plant genomes. *Molecular Plant*, *7*(3), 472–480.

Olivares, D., García-Rojas, M., Ulloa, P. A., Riveros, A., Pedreschi, R., Campos-Vargas, R., & Defilippi, B. G., (2022). Response mechanisms of "Hass" avocado to sequential 1–methylcyclopropene applications at different maturity stages during cold storage. *Plants*, *11*(13), 1781.

Osorio, S., Scossa, F., & Fernie, A. R., (2013). Molecular regulation of fruit ripening. *Frontiers in Plant Science*, *4*, 198.

Pedreschi, R., & Lurie, S., (2015). Advances and current challenges in understanding postharvest abiotic stresses in perishables. *Postharvest Biology and Technology*, *107*, 77–89.

Peng, M., Ying, P., Liu, X., Li, C., Xia, R., Li, J., & Zhao, M., (2017). Genome-wide identification of histone modifiers and their expression patterns during fruit abscission in litchi. *Frontiers in Plant Science*, *8*, 639.

Pikaard, C. S., & Scheid, O. M., (2014). Epigenetic regulation in plants. *Cold Spring Harbor Perspectives in Biology*, *6*(12), a019315.

Pott, D. M., Vallarino, J. G., & Osorio, S., (2020). Metabolite changes during postharvest storage: Effects on fruit quality traits. *Metabolites*, *10*(5), 187.

Prado, F., Jimeno-González, S., & Reyes, J. C., (2017). Histone availability as a strategy to control gene expression. *RNA Biology*, *14*(3), 281–286.

Prusky, D., (1996). Pathogen quiescence in postharvest diseases. *Annual Review of Phytopathology*, *34*(1), 413–434.

Przybilla, J., Buske, P., Binder, H., & Galle, J., (2013). Histone modifications control DNA methylation profiles during ageing and tumor expansion. *Frontiers in Life Science*, *7*(1, 2), 31–43.

Pu, H., Shan, S., Wang, Z., Duan, W., Tian, J., Zhang, L., & Xu, X., (2020). Dynamic changes of DNA methylation induced by heat treatment were involved in ethylene signal transmission and delayed the postharvest ripening of tomato fruit. *Journal of Agricultural and Food Chemistry*, *68*(33), 8976–8986.

Qin, G., Zhu, Z., Wang, W., Cai, J., Chen, Y., Li, L., & Tian, S., (2016). A tomato vacuolar invertase inhibitor mediates sucrose metabolism and influences fruit ripening. *Plant Physiology*, *172*(3), 1596–1611.

Rajput, R., Naik, J., Stracke, R., & Pandey, A., (2022). *Interplay Between R2R3 MYB-Type Activators and Repressors Regulates Proanthocyanidin Biosynthesis in Banana (Musa acuminata).* bioRxiv.

Ramjan, M. D., & Ansari, M. T., (2018). Factors affecting of fruits, vegetables, and its quality. *J. Med. Plants*, *6*, 16–18.

Rothkegel, K., Sánchez, E., Montes, C., Greve, M., Tapia, S., Bravo, S., & Almeida, A. M., (2017). DNA methylation and small interference RNAs participate in the regulation of MADS-box genes involved in dormancy in sweet cherry (*Prunus avium* L.). *Tree Physiology*, *37*(12), 1739–1751.

Rottach, A., Leonhardt, H., & Spada, F., (2009). DNA methylation-mediated epigenetic control. *Journal of Cellular Biochemistry*, *108*(1), 43–51.

Saha, P., Singh, S., Bhatia, R., Dey, S. S., Das Saha, N., Ghoshal, C., & Kalia, P., (2022). Genomic designing for abiotic stress resistant brassica vegetable crops. In: *Genomic Designing for Abiotic Stress Resistant Vegetable Crops* (pp. 153–185). Springer, Cham.

Sam, C., Figueroa, C. R., & Helen, N., (2014). Movers and shakers in the regulation of fruit ripening: A cross-dissection of climacteric versus non-climacteric fruit. *J. Exp. Bot*, *65*, 4705–4722.

Samantara, K., Shiv, A., De Sousa, L. L., Sandhu, K. S., Priyadarshini, P., & Mohapatra, S. R., (2021). A comprehensive review on epigenetic mechanisms and application of epigenetic modifications for crop improvement. *Environmental and Experimental Botany*, *188*, 104479.

Santamaraa, E., Cañal, M. J., Rodraguez, R., Hasban, R., & Valledor, L., (2006). Dynamics of DNA methylation in chestnut trees development. In: *XXVII International Horticultural Congress-IHC2006: II International Symposium on Plant Genetic Resources of Horticultural* (Vol. 760, pp. 563–566).

Santamaría, M. E., Rodríguez, R., Canal, M. J., & Toorop, P. E., (2011). Transcriptome analysis of chestnut (Castanea sativa) tree buds suggests a putative role for epigenetic control of bud dormancy. *Annals of Botany*, *108*(3), 485–498.

Shi, M., Wang, C., Wang, P., Zhang, M., & Liao, W., (2022). Methylation in DNA, histone, and RNA during flowering under stress condition: A review. *Plant Science*, 111431.

Shi, Y., Qin, Y., Li, F., & Wang, H., (2022). Genome-wide profiling of DNA methylome and transcriptome reveals epigenetic regulation of potato response to DON stress. *Frontiers in Plant Science*, *13*.

Shipman, E. N., Yu, J., Zhou, J., Albornoz, K., & Beckles, D. M., (2021). Can gene editing reduce postharvest waste and loss of fruit, vegetables, and ornamentals? *Horticulture Research*, *8*.

Siddiqui, M. W., (2017). *Preharvest Modulation of Postharvest Fruit and Vegetable Quality*. Academic Press.

Simon, S. A., & Meyers, B. C., (2011). Small RNA-mediated epigenetic modifications in plants. *Current Opinion in Plant Biology*, *14*(2), 148–155.

Sirangelo, T. M., Rogers, H. J., & Spadafora, N. D., (2022). Multi-omic approaches to investigate molecular mechanisms in peach post-harvest ripening. *Agriculture*, *12*(4), 553.

Song, J., Rutjens, B., & Dean, C., (2014). Detecting histone modifications in plants. In: *Plant Epigenetics and Epigenomics* (pp. 165–175). Humana Press, Totowa, NJ.

Sparks, E., Wachsman, G., & Benfey, P. N., (2013). Spatiotemporal signaling in plant development. *Nature Reviews Genetics*, *14*(9), 631–644.

Spiegelman, Z., Golan, G., & Wolf, S., (2013). Don't kill the messenger: Long-distance trafficking of mRNA molecules. *Plant Science*, *213*, 1–8.

Tagliasacchi, A. M., Andreucci, A. C., Giraldi, E., Felici, C., Ruberti, F., & Forino, L. M. C., (2007). Structure, DNA content and DNA methylation of synergids during ovule development in *Malus domestica* Borkh. *Caryologia*, *60*(3), 290–298.

Tang, D., Gallusci, P., & Lang, Z., (2020). Fruit development and epigenetic modifications. *New Phytologist*, *228*(3), 839–844.

Tatsuki, M., Haji, T., & Yamaguchi, M., (2006). The involvement of 1-aminocyclopropane-1-carboxylic acid synthase isogene, Pp-ACS1, in peach fruit softening. *Journal of Experimental Botany*, *57*(6), 1281–1289.

Telias, A., Lin-Wang, K., Stevenson, D. E., Cooney, J. M., Hellens, R. P., Allan, A. C., & Bradeen, J. M., (2011). Apple skin patterning is associated with differential expression of MYB10. *BMC Plant Biology*, *11*(1), 1–15.

Tian, S., Torres, R., Ballester, A. R., Li, B., Vilanova, L., & González-Candelas, L., (2016). Molecular aspects in pathogen-fruit interactions: Virulence and resistance. *Postharvest Biology and Technology*, *122*, 11–21.

Toivonen, P. M., & Hodges, D. M., (2011). Abiotic stress in harvested fruits and vegetables. *Abiotic Stress in Plants: Mechanisms and Adaptations* (pp. 39–58). Croatia: InTech.

Tucker, M. L., & Kim, J., (2015). Abscission research: What we know and what we still need to study. *Stewart Postharvest Rev.*, *11*(7).

Turner, B. M., (2007). Defining an epigenetic code. *Nature Cell Biology*, *9*(1), 2–6.

Verma, N., Giri, S. K., Singh, G., Gill, R., & Kumar, A., (2022). Epigenetic regulation of heat and cold stress responses in crop plants. *Plant Gene*, 100351.

Wang, A., Hu, J., Gao, C., Chen, G., Wang, B., Lin, C., & Zhou, G., (2019). Genome-wide analysis of long non-coding RNAs unveils the regulatory roles in the heat tolerance of Chinese cabbage (*Brassica rapa* ssp. chinensis). *Scientific Reports*, *9*(1), 1–14.

Wei, C., Liu, H., Cao, X., Zhang, M., Li, X., Chen, K., & Zhang, B., (2021). Synthesis of flavour-related linalool is regulated by PpbHLH1 and associated with changes in DNA methylation during peach fruit ripening. *Plant Biotechnology Journal*, *19*(10), 2082–2096.

Xu, X., Yin, L., Ying, Q., Song, H., Xue, D., Lai, T., & Shi, X., (2013). High-throughput sequencing and degradome analysis identify miRNAs and their targets involved in fruit senescence of *Fragaria ananassa*. *PLoS One*, *8*(8), e70959.

Yang, Y. Y., Shan, W., Yang, T. W., Wu, C. J., Liu, X. C., Chen, J. Y., & Kuang, J. F., (2022). MaMYB4 is a negative regulator and a substrate of RING-type E3 ligases MaBRG2/3 in controlling banana fruit ripening. *The Plant Journal*.

Zhang, H., Li, H., Lai, B., Xia, H., Wang, H., & Huang, X., (2016). Morphological characterization and gene expression profiling during bud development in a tropical perennial, litchi chinensis Sonn. *Frontiers in Plant Science*, *7*, 1517.

Zhang, H., Tang, K., Wang, B., Duan, C. G., Lang, Z., & Zhu, J. K., (2014). Protocol: A beginner's guide to the analysis of RNA-directed DNA methylation in plants. *Plant Methods*, *10*(1), 1–9.

Zhang, Y., Yin, X., Xiao, Y., Zhang, Z., Li, S., Liu, X., & Chen, K., (2018). An ETHYLENE RESPONSE FACTOR-MYB transcription complex regulates furaneol biosynthesis by activating QUINONE OXIDOREDUCTASE expression in strawberry. *Plant Physiology*, *178*(1), 189–201.

Zhang, Z., Tian, S., Zhu, Z., Xu, Y., & Qin, G., (2012). Effects of 1-methylcyclopropene (1-MCP) on ripening and resistance of jujube (*Zizyphus jujuba* cv. Huping) fruit against postharvest disease. *LWT-Food Science and Technology*, *45*(1), 13–19.

Zhao, T., Zhan, Z., & Jiang, D., (2019). Histone modifications and their regulatory roles in plant development and environmental memory. *Journal of Genetics and Genomics*, *46*(10), 467–476.

Zhou, J., Chen, B., Albornoz, K., & Beckles, D. M., (2021). Postharvest handling induces changes in fruit DNA methylation status and is associated with alterations in fruit quality in tomato (*Solanum lycopersicum* L.). *Scientia Horticulturae*, *283*, 110090.

Zhou, L., Tian, S., & Qin, G., (2019). RNA methylomes reveal the m6A-mediated regulation of DNA demethylase gene SlDML2 in tomato fruit ripening. *Genome Biology*, *20*(1), 1–23.

Zhou, P., Zhang, X., Fatima, M., Ma, X., Fang, H., Yan, H., & Ming, R., (2020). DNA methylome and transcriptome landscapes revealed differential characteristics of dioecious flowers in papaya. *Horticulture Research*, *7*.

Zhu, H., Chen, P. Y., Zhong, S., Dardick, C., Callahan, A., An, Y. Q., & Liu, Z., (2020). Thermal-responsive genetic and epigenetic regulation of DAM cluster controlling dormancy and chilling requirement in peach floral buds. *Horticulture Research*, *7*.

Zhu, H., Zhang, Y., Tang, R., Qu, H., Duan, X., & Jiang, Y., (2019a). Banana sRNAome and degradome identify microRNAs functioning in differential responses to temperature stress. *BMC Genomics*, *20*(1), 1–16.

Zhu, L. S., Liang, S. M., Chen, L. L., Wu, C. J., Wei, W., Shan, W., & Kuang, J. F., (2019b). Banana MaSPL16 modulates carotenoid biosynthesis during fruit ripening through activating the transcription of lycopene β-cyclase genes. *Journal of Agricultural and Food Chemistry*, *68*(5), 1286–1296.

Ziv, C., & Fallik, E., (2021). Postharvest storage techniques and quality evaluation of fruits and vegetables for reducing food loss. *Agronomy*, *11*(6), 1133.

CHAPTER 10

GENETICALLY MODIFIED LEAFY VEGETABLES

SEKHU ANSARI,[1] VINAY KUMAR,[2] and JUERIA MUSTAQEEM[3]

[1]*Division of Pathology, Cincinnati Children's Hospital Medical Center, Cincinnati, Ohio, USA*

[2]*Pennsylvania State University HMC, Hershey, Pennsylvania, USA*

[3]*Dr. B.R. Ambedkar Center for Biomedical Research, University of Delhi, New Delhi, India*

ABSTRACT

Leafy vegetables are one of the most agronomically and economically important plant groups in the world, raised for their edible leaves or leafy shoots. Spinach, chicory, lettuce, cabbage, etc., are a few of the important leafy vegetables. Leafy vegetables contain all the essential nutrients for a healthy diet and help prevent several lifestyle diseases such as coronary heart diseases, diabetes, and aging. Like other crops, the cultivation and postharvest of leafy vegetables are also threatened by biotic and abiotic stress. Biotechnology has the potential to provide a defense mechanism against these stresses. Recently, several transgenic leafy vegetables have been developed to resist these stresses and reduce post-harvest loss. The nutritional value of leafy vegetables has been enhanced through biotechnological applications. Various transgenic lines of lettuce, spinach, cabbage, etc., have been created.

Bioengineered Fruit and Vegetables: Trends in Postharvest Quality Improvement.
Mohammad Irfan and Pankaj Kumar (Eds.)

10.1 INTRODUCTION

Vegetables are important components of food as they contribute to the staple food of the human diet in almost every part of the world by providing important nutritional requirements of a healthy human diet (Kumari et al., 2022). Vegetables are a good source of vitamins (A, C, B complex, and E), minerals, antioxidants, folates, fibers, etc. Regular consumption of vegetables in the diet helps in the improvement of overall human health and protects the human body against various diseases such as diabetes, some types of cancers, and cardiovascular diseases (Río-Celestino & Font, 2020). It is estimated that the lives of nearly 2.7 million people could be saved per year by sufficient consumption of vegetables (Kanungsukkasem et al., 2009). There are various edible parts in different vegetables, which could be fruit, root, seeds, or their leaves. Usually, the nutrient composition of the vegetable varies with the edible part.

Vegetables are cultivated in almost every part of the world, and a large part of the produce is lost due to various reasons. The major factors for this loss include climate change, biotic and abiotic stresses, and post-harvest losses during handling, maintaining, and storage (Cerda et al., 2017; Guzmán et al., 2022; Kumari et al., 2022). Because of their perishable nature, vegetables are very prone to phytopathogen attacks not only at the postharvest levels but also at pre-harvest (Irfan et al., 2023). The population explosion demands an increase in the amount of vegetable production. Depletion of agricultural land is working as a limiting factor in increased vegetable production. Therefore, the prevention of the loss of vegetables could provide a solution for meeting the challenge of supplying vegetables to the increasing population. Several attempts have been made, including the high use of pesticides and insecticides, but there has been very limited success against them. Besides this, there have always been health and environmental concerns with the overuse of them (Kim et al., 2017). Most vegetables have a shorter shelf life after harvest, soften quickly, and are susceptible to pathogens. It is estimated that nearly 20–35% of vegetables are lost due to decay by pathogens because of inadequate storage during processing and transportation. Vegetables need to be consumed quickly after harvest for optimum nutritional benefits. The use of antimicrobial agents against fungi, bacteria, etc., has brought some success in reducing post-harvest loss, but this reduction is not sufficient. Health and environmental concerns associated with the use of antimicrobial agents have increased demands for the application of biotechnology in vegetable production. Genetic engineering (GE) of vegetable crops has the potential to address these vegetable loss problems faced by farmers, which are not easily addressed by conventional breeding technology (Datta, 2013; Irfan & Datta, 2017; Tayal et al., 2022). The application of biotechnology can help develop vegetables with improved shelf life and nutritional values (Kumar et al., 2016a, b, 2019). The application of biotechnology in vegetable development can help improve the global supply of vegetable crops to the increasing population.

10.2 LEAFY VEGETABLES

There are nearly 400 types of vegetable crops cultivated worldwide. Nearly half of these are leafy vegetables followed by vegetable fruits. Leafy vegetables are a diverse category of agricultural plants raised for their edible leaves or leafy shoots. Due to their multiple health and nutritional benefits, green leafy vegetables (GLV) have lately generated a decent amount of attention (Río-Celestino & Font, 2020). GLV are among the cheap and readily available vegetables and contain all the essential nutrients of a healthy diet. Sufficient daily intake could even help prevent major lifestyle diseases like coronary heart diseases, diabetes, etc. Leafy vegetables are a rich source of fibers, vitamins, minerals, antioxidants, essential amino acids, folates, fatty acids, etc. (Sarkar et al., 2022; Sharma et al., 2022) (Table 10.1). They are the only natural source of folic acid, which is sufficiently high in the leaves of lettuce, asparagus, spinach, etc. Among all the GLVs, spinach (*Spinacia oleracea*), radish (*Raphanus sativus*), *Chenopodium album*, cabbage (*Brassica oleracea*), amaranth (*Amaranthus gangeticus*), fenugreek (*Trigonella foenum*), drumstick (*Moringa oleifera*), etc., are the most widely used GLVs (Sarkar et al., 2022). There are several GLVs which are also used as spices for flavor like bay leaf (*Laurus nobilis*), coriander leaves (*Coriandrum sativum*), basil (*Ocinumbasilicum*), mint (*Mentha spicata*), etc.

TABLE 10.1 Nutrient Constituents of Leafy Vegetables and Their Health Benefits

Vegetable	Nutrients	Health Benefits	References
Lettuce	Carbohydrates, vitamins (A, B, C, K, folate), proteins, minerals (Fe, Ca, Na, Mg), lutein, quercetin, etc.	Vision improvement, bone strengthening, anti-diabetic, Anti-cancerous, weight loss management, maintains skin elasticity, etc.	Shi et al. (2022); Yang et al. (2022)
Spinach	Vitamin K (abundant), vitamin A, B, and C, folates, minerals (Ca, Mg, Na, Fe, Cu, etc.), omega-3 fatty acids, carbohydrates, proteins, nitrates, flavonoids like patuletin, spinacetin, spinatoside, jaceidin, dietary fibers, proteins, etc.	Anti-inflammatory, anti-aging, anti-obesity, anti-cancerous, neural health improvement, maintains blood pressure, vascular health improvement, etc.	Lomnitski et al. (2003); Roberts & Moreau (2016)
Asparagus	Abundance of asparagine, vitamin A and C, minerals (Fe and K), carbohydrates, flavonoids like quercetin, isorhamnetin, proteins, etc.	Weight loss management, healthy pregnancy outcomes, digestive health improvement, lowers blood pressure, etc.	Kohli et al. (2022); Negi et al. (2010)
Cabbage	Vitamins (C, K, and E), minerals (P, S, Cl, Ca, Fe, Sr, K, Cr, Se, and Zn), proteins, fiber, coumarins, flavonoids, anthocyanins terpenes, S-methyl cysteine sulfoxide, glucosinolates, proteins, etc.	Cardiac health improvement, Anti-cancer, antioxidant, anti-inflammatory, etc.	Albert-Puleo (1983); Rokayya et al. (2014)
Chicory	Vitamins, minerals (K, Ca, Mg, Se, Zn), proteins, inulin (starch-like polysaccharide), coumarins, flavonoids like sesquiterpene, lactones (lactucin and lactucopicrin), tannins, alkaloids, etc.	Antioxidant, anti-cancer, anti-inflammatory, anti-parasitic, anti-hepatotoxic, etc.	Nwafor et al. (2017); Puhlmann& Vos (2020)
Coriander	Vitamins (A, K, C), proteins, minerals (like Ca, P, Fe, etc.), fibers, and carbohydrates.	Antibacterial, anti-diabetic, skin health improvement, antifungal antioxidant, etc.	Sahib et al. (2013); Wang et al. (2022)

10.3 GENETICALLY MODIFIED (GM) LEAFY VEGETABLES

Several attempts have been made to develop genetically modified (GM) leafy vegetables to improve their nutritional and postharvest quality attributes using various biotechnological tools. A few examples of these attempts have been discussed in subsections.

10.3.1 LETTUCE

Lettuce is one of the widely consumed leafy vegetables of the Asteraceae family. Efforts have been continuously made to maximize yield, improve nutritional quality, and enhance postharvest durability due to its financial and health importance. Genetic engineering approaches have been utilized to boost the nutritional value of lettuce by overcoming species barriers. The introduction of genes through *Agrobacterium* mediation has enabled the generation of transgenic lettuces. Genetic engineering in lettuce has primarily focused on disease resistance and the development of stress-tolerant varieties. Efforts have also been made to reduce the bitterness of raw lettuce leaves. For instance, the expression of the Miraculin protein from *Richadella dulcifica* berries in the lettuce cultivar "Kaiser" using *Agrobacterium tumefaciens* has successfully eradicated the bitterness of the leaves.

Lettuce leaves contain a moderate amount of nutrients such as iron, zinc, calcium, etc. (Saini et al., 2022). Efforts have been made to biofortify lettuce by increasing the levels of essential micronutrients like iodine, zinc, iron, calcium, and various other nutrients (Saini et al., 2022). Transgenic lettuces have been developed by incorporating genes from other plants. For instance, the ferritin gene from soybean was introduced into lettuce to boost the iron content, the metallothionein mutant β-cDNA from mice was expressed to enhance the zinc content, and the CAX1 gene from *Arabidopsis* was utilized to increase the calcium content (Goto et al., 2000; Park et al., 2009). Similarly, Cho et al. managed to elevate the alpha form (more potent form) of tocopherol in the lettuce cv. Chungchima (Li et al., 2011). They were able to achieve this by expressing the gene for γ-tocopherol methyltransferase from *A. thaliana*(Li et al., 2011). Like several other leafy vegetables, nitrate content is typically higher than the recommended for human consumption. Since lettuce is mostly consumed in its uncooked form, nitrate concentration becomes even more significant (Curtis et al., 1999). Transgenic lettuce plants expressing the nitrate reductase (NR) gene from tobacco help in reducing the nitrate to levels assumed to be safe for human consumption (Curtis et al., 1999; Sobolev et al., 2010).

Since lettuce is consumed for its leaves, which are susceptible to viral attacks and rapid deterioration following harvesting, viral attack on the lettuce leaves results in considerable loss of the lettuce. Lettuce mosaic virus (LMV), tomato spotted wilt virus (TSWV), lettuce big vein associated virus (LBVaV), Mirafiori lettuce virus (MLV), and Tospovirus are among the major viruses involved in the loss (Sobolev et al., 2010). Viral-resistant lettuce plants were generated using the nucleocapsid protein strategy. Transgenic lettuce plants expressing the viral coat proteins provided resistance against viral attack (Sobolev et al., 2010). Transgenic lettuce plants overexpressing the ferritin gene from soybean also improved tolerance to the pathogens. Besides viral attacks on leaves, post-harvest loss of lettuce has been a major concern. Attempts have been made to delay the natural senescence process in lettuce leaves. For example, the ipt gene from *Agrobacterium tumefaciens* was expressed under the senescence-specific promoter SAG12 from *A. thaliana*. The expression of isopentenyl transferase (ipt) in lettuce significantly delayed the senescence of lettuce leaves, thereby increasing the shelf life (McCabe et al., 2001).

Besides biotic stress, the management of abiotic stress has also been a concern addressed by several researchers through transgenic technology. Development of cold and drought-tolerant transgenic lettuce was successful. Transgenic lettuce plants expressing the mutated P5CS gene for delta 1-pyrroline-5-carboxylate synthase provided resistance to cold and drought (Pileggi et al., 2001). Similarly, over-expression of the *Arabidopsis thaliana* ABF3 gene also conferred drought and cold stress (Vanjildorj et al., 2005). The development of herbicide-resistant plants is an important attribute in weed management. For this, the bar gene was introduced into the lettuce cv. Evola using *Agrobacterium* (Mohapatra et al., 1999). In addition to *Agrobacterium*-mediated genetic improvement, CRISPR-Cas has also been used for trait improvement in lettuce. CRISPR-Cas-mediated editing of the upstream open reading frame (uORF) of the genes (LsGGP1 and LsGGP2) involved in vitamin C biosynthesis enhanced the vitamin C content, which in turn also helped in oxidative stress tolerance (Zhang et al., 2018).

10.3.2 BRASSICA CROPS

Brassica crops belong to the diverse Brassicaceae family. Brassica crops are one of the most important vegetable crops which are consumed throughout the world as food and for their health benefits and industrial applications. There are around 39 species of Brassica which are cultivated for their leaves, stems, flowers, buds, edible roots and oilseeds (Poveda et al., 2020). Few important species of the Brassica are *Brassica oleracea* (Cauliflower, cabbage, broccoli, kale, etc.), *Brassica rapa* (Chinese cabbage, turnip, etc.), *Brassica napus* (rapeseed and leaf rape), *Brassica juncea* and *Brassica carinata* (mustards). Among these, Brassica species of cabbage, kale, collards, etc., are leafy vegetables and are consumed for their leaves (Poveda et al., 2020).

Developments of the GE tools created opportunities for the generation of transgenic Brassica crops. Many agronomically important genes were introduced into different Brassica species using Agrobacterium-mediated transformation. The major focus of generating transgenic lines of leafy crops of Brassica has been on biotic and abiotic stress tolerance. It is estimated that nearly 50–60% of Brassica crops are lost due to insects and pests, resulting in substantial economic loss. The high price of pesticides and the health risks associated with them limit their use. Therefore, attempts were made to generate insect resistance, virus resistance, and herbicide resistance transgenics using *Agrobacterium*-mediated transformation. Transgenic cabbage (*Brassica oleracea* var. *capitata*) expressing the cry gene from *Bacillus thuringiensis* were generated which provided the resistance against the diamondback moth (*Plutellaxylostella*) (Mason et al., 2003). Transgenic Chinese cabbage (*Brassica rapa*L. ssp. *Chinensis*) expressing the same cry gene were also resistant to the attack of the insects like *Pieris rapae* and *Trichoplusiani*(Cho et al., 2001). Similarly, CpTI (cowpea trypsin inhibitor) gene was introduced in cabbage (wild cabbage and Chinese cabbage) which provided the resistance to the insect pests like *Pieris rapae L.*, *Erwinia aroidae* and *Heliothisarmigera*(Ma et al., 2017; Poveda et al., 2020). *Xanthomonas campestris* causes the black rot disease in cabbage leaves which is responsible for the considerable loss of the crop. Lee et al. introduced the glucose oxidase gene from *Aspergillus niger* into the cabbage, which significantly reduced the leaf lesions (Lee et al., 2001). Using the nucleocapsid protein strategy, virus-resistant transgenic Chinese cabbage was generated. Transgenic Chinese cabbage plants expressing L-coat protein gene (TMV-L) from Tobacco Mosaic Virus and TuMV-Nib gene from Turnip Mosaic virus were generated (Jun et al., 1995; Zhandong et al., 2007). These plants were resistant to viral attacks of Tobacco Mosaic Virus and Turnip Mosaic virus. All these genetic engineering modifications in plants help prevent post-harvest crop losses.

Development of herbicide-resistant plants is an important attribute in weed management, especially for creeper plants. Chinese cabbage

(*Brassica rapa*L. ssp. *Chinensis*) resistant to the herbicide phosphinotricin (Basta) was generated by introducing the Basta resistance gene bar/ TuMVNIa using Agrobacterium-mediated transformation (Qing et al., 2000; Xu et al., 2008). In addition to biotic stress, tolerance to abiotic stress such as salt, water submergence, and drought is also essential for crops due to the significant crop losses it can cause. Transgenic cabbage with higher tolerance to these abiotic stresses was developed. The bacterial betA gene was introduced in the cabbage (*Brassica oleracea var. capitata*) cultivar 'Golden Acre' through *A. tumefaciens* mediated transformation (Gerszberg, 2018). Overexpression of this gene increased the salt tolerance capacity of the cabbage (Gerszberg, 2018). Similarly, transgenic Chinese cabbage expressing the otsA gene from *E. coli* and LEA gene from *C. annuum* show the remarkable improvement in the tolerance against drought and salt stress (Park et al., 2003, 2005). Transgenic Chinese cabbage expressing the superoxide dismutase and catalase (CAT) genes from either *E. coli* or maize show enhanced tolerance to sulfur dioxide (SO_2) and salt stress (Tseng et al., 2007, 2008). Cabbage is a creeping plant. Therefore, flooding has always been an issue with such crops. *A. tumefaciens* mediated transformation of the cabbage with Vitreoscilla hemoglobin (vhb) gene provided them with greater tolerance to flooding or water submergence (Li et al., 2005). Male sterile plants are required for the generation of hybrid plants. Genetic generation of male sterile plants has substituted the traditional emasculation processes. Lee et al. generated transgenic cabbage expressing cytotoxic diphtheria toxin A-chain (DTxA) gene (Lee et al., 2003). This gene was expressed in the cabbage using the anther-specific promoter of the gene BcA9 from the cabbage. Expression of the DTxA gene in transgenic plants resulted in male sterile cabbage (Lee et al., 2003).

Plants are emerging as a production system for recombinant proteins, including vaccines (Ghosh et al., 2017). Edible plants with a large biomass and the capacity to sustain the expression of foreign proteins without affecting their growth and development are ideal candidates for protein expression. Several crops of the Brassicaceae family meet these criteria and are excellent for expressing recombinant proteins, including vaccines. Collard is one such crop. Transgenic collard plants expressing the smallpox vaccine in their leaves were created (Pogrebnyak et al., 2006). This plant-based expression of pharmaceutical proteins in leafy vegetables of the Brassicaceae family has opened up new possibilities for their use in pharmaceuticals and transgenic biotechnology.

10.3.3 OTHER LEAFY VEGETABLES

There are several other economically important leafy vegetables like spinach, asparagus, and Chinese kale, etc., which are widely consumed across the globe (Sarkar et al., 2022). Attempts have been made for the genetic modifications of these crops (Table 10.2). Spinach is a nutrient-rich plant. A considerable amount of spinach is lost due to the viral attack. Transgenic spinach plants were generated using the *Agrobacterium*-mediated transformation. Cucumber mosaic virus (CMV) coat protein was expressed in spinach for the defense against the viral attack of the CMV (Yang et al., 1997). Pat gene from *Steptomyceshygroscopicus* was also expressed in Spinach which provided the resistance against the herbicide glufosinate ammonium (Block et al., 1987).

Color modification of leaves has been used as a market strategy for the better economics of the vegetables. Chinese kale was genetically modified for their color (Sun et al., 2020). Genes involved in the color development were modified using the CRISPR/Cas9 system (Sun et al., 2020). Genetic editing of genes BoaCRTISO or BaPDS1/BaPDS2 results in the change in color to yellow and albino, respectively (Sun et al., 2020).

Asparagus (*Asparagus officinalis L.*) is a nutrient-rich popular vegetable, also known as the king of vegetables. Stem wilt disease of asparagus is responsible for considerable economic loss in the asparagus industry (Encina & Regalado, 2022; Treuren et al., 2012). Transgenic asparagus plants expressing the hevein-like gene were generated using Agrobacterium-mediated transformation (Chen et al., 2019). These transgenic plants have resistance against stem wilt diseases.

10.4 HEALTH BENEFITS OF LEAFY VEGETABLES

Consumption of GLV has increased in recent times owing to the increase in awareness about their nutritional benefits (Figure 10.1). Due to their health benefits, the U.S. Departments of Health and Human Services (HHS) and Agriculture (USDA) have recommended one serving of leafy vegetables per day out of the recommended five servings of vegetables. GLVs provide ample amounts of essential vitamins (A, B2, B9, C, etc.), fibers, proteins, and minerals (Ca, Fe, Na, Zn, P, Cu, etc.) for humans and help in the prevention of several human diseases (Morris et al., 2018; Sarkar et al., 2022; Wallace et al., 2020).

TABLE 10.2 Review of the Few Genetically Engineered Leafy Vegetables

Character	Transgene	Aim	References
		Lettuce	
Taste enhancement	Miraculin gene from berries of *Richadelladulcifica.*	Taste improvement	Sun et al. (2006)
Biofortification	Ferritin gene from soybean.	Increase the iron content.	Goto et al. (2000)
–	Metallothionein mutant β-cDNA from mice.	Increase the Zn content.	Park et al. (2009)
–	CAX1 gene from *Arabidopsis.*	Increase the calcium content.	Zuo et al. (2002)
–	γ-tocopherol methyltransferase from *A. thaliana.*	Increase the alpha form (more potent form) of tocopherol.	Li et al. (2011)
Metabolic engineering	Nitrate reductase gene from tobacco.	Bring down the nitrate level to safer level of human consumption.	Curtis et al. (1999); Sobolev et al. (2010)
Virus resistance	LMV gene, TSWV gene, LBVaV gene and MLV genes from lettuce mosaic virus (LMV), tomato spotted wilt virus (TSWV), lettuce big vein associated virus (LBVaV) and Mirafiori lettuce virus (MLV), respectively.	Resistance to viral attack of lettuce mosaic virus, tomato spotted wilt virus, lettuce big vein associated virus and Mirafiori lettuce virus.	Curtis (2006); Davey et al. (n.d.-a, b, 2007)
Delayed senescence	*ipt*gene from *A. tumefaciens.*	Increase the shelf life.	McCabe et al. (2001)
Abiotic stress	Mutated P5CS gene for delta 1-pyrroline-5-carboxylate synthase.	Cold and drought tolerant.	Pileggi et al. (2001)
Herbicide resistance	*ABF3* gene from of the *Arabidopsis thaliana bar* gene.	Herbicide resistance and cold and drought tolerance.	Vanjildorj et al. (2005)
		Cabbage	
Insect resistance	Cry gene from *Bacillus thuringiensis.* CpTI (cowpea trypsin inhibitor) gene.	Resistance against *Plutellaxylostella*), *Pieris rapae, Trichoplusiani.* Resistance against *Pieris rapae* L., *Erwinia aroidae*, and *Heliothisarmigera*	Cho et al. (2001); Mason et al. (2003); Ma et al. (2017); Poveda et al. (2020)
Bacterial resistance	Glucose oxidase gene from *Aspergillus niger.*	*Xanthomonas campestris* resistance.	Lee et al. (2001)
Virus resistance	L-coat protein gene (TMV-L) from tobacco mosaic virus and TuMV-Nib gene from turnip mosaic virus.	Resistant to tobacco mosaic virus and turnip mosaic virus.	Jun et al. (1995); Zhandong et al. (2007)
Herbicide resistance	Bar/TuMVNIa	Phosphinotricin (Basta) resistance	Qing et al. (2000); Xu et al. (2008)

TABLE 10.2 *(Continued)*

Character	Transgene	Aim	References
Abiotic stress tolerance	betA gene The otsA gene from *E. coli* LEA gene from *C. annuum* Superoxide dismutase and catalase genes from either *E. coli* or maize. VHb gene.	Salt tolerance. Drought and salt stress tolerance. Enhanced tolerance to sulfur dioxide and salt stress. Enhanced toleration to the flooding or water submergence.	Gerszberg (2018); Park et al. (2003, 2005); Tseng et al. (2007, 2008); Li et al. (2005)
Male sterility	Cytotoxic diphtheria toxin A-chain (DTxA) gene.	Male sterility	Lee et al. (2003)
		Spinach	
Virus resistance	CMV gene from cucumber mosaic virus (CMV).	Resistance to CMV	Yang et al. (1997)
Herbicide resistance	Pat gene from *Steptomyceshygroscopicus.*	Herbicide (glufosinate ammonium) resistance.	Block et al. (1987)
		Asparagus	
Fungal resistance	Hevein like gene	Resistance to stem wilt disease.	Treuren et al. (2012)

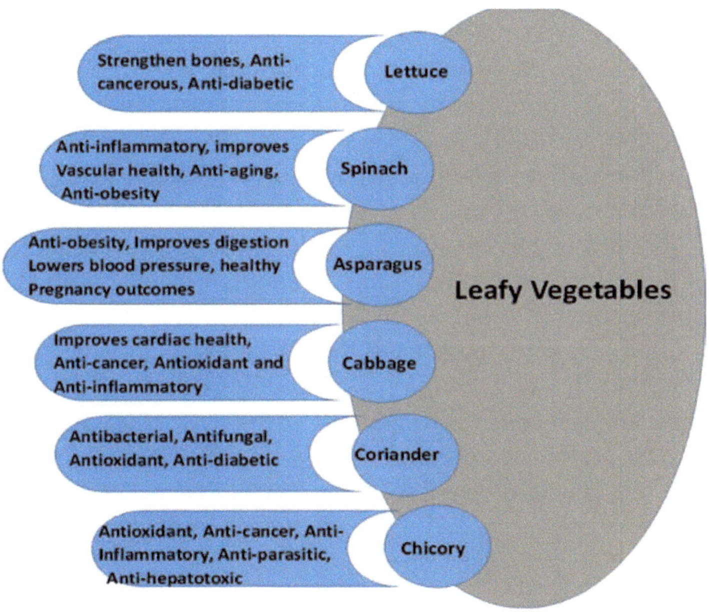

FIGURE 10.1 Health benefits of leafy vegetables: Leafy vegetables such as spinach, chicory, cabbage, lettuce, etc., offer numerous health benefits such as anti-aging, anti-inflammatory properties, digestive health improvement, bone strengthening, anti-parasitic effects, cardiac health improvement, etc.

10.5 CONCLUSION

Leafy vegetables are a diverse category of agricultural plants that have been consumed by humans since ancient times due to their immense health benefits. Leafy vegetables are rich in essential nutrients required for human beings. Lettuce, cabbage, spinach, and chicory are among the principal leafy vegetables (Brouwer-Brolsma et al., 2020; Treuren et al., 2012). Biotic and abiotic stress, along with post-harvest loss, are responsible for a considerable amount of vegetable loss. Genetic engineering of plants could solve this problem. The application of biotechnology in the development of vegetables can help improve the global supply of vegetable crops to the increasing population (Kumari et al., 2022; Irfan et al., 2023). Biotechnological applications in plants have expanded dramatically since their introduction (Irfan & Datta, 2017). Transgenic technology has been adopted in vegetables as well. Transgenic leafy vegetables resistant to phytopathogens and having the delayed senescence phenotype help reduce post-harvest loss (Ghosh et al., 2013). Transgenic field trials for a large number of vegetables, including leafy vegetables, have been documented. However, only a handful of transgenic vegetables are approved for human consumption globally. The common man is afraid of the supposed risks associated with the consumption of GM vegetables. This issue needs to be addressed through knowledge impartation and technological improvements such as marker-free transformation of plants. It can be presumed that in the future, transgenic vegetables will be increasingly used to address the effects of global climate change, biotic and abiotic stress. It is expected that in the near future, the abundance of data suggesting the minimal environmental and health impact of transgenics will compel authorities to approve transgenic vegetables. Currently, transgenic lettuces, spinach, cabbages, and many other leafy vegetables are primarily tailored based on nutritional value, biotic or abiotic stress. In the future, we can anticipate the approval of these vegetables for human consumption.

KEYWORDS

- **abiotic stress**
- **biotic stress**
- **genetically modified**
- **leafy vegetables**
- **salt tolerance**
- **transgenic vegetables**
- **virus resistance**
- **vitamins**

REFERENCES

Albert-Puleo, M., (1983). Physiological effects of cabbage with reference to its potential as a dietary cancer-inhibitor and its use in ancient medicine. *Journal of Ethnopharmacology*, *9*(2, 3), 261–272. https://doi.org/10.1016/0378-8741(83)90035-1.

Block, M. D., Botterman, J., Vandewiele, M., Dockx, J., Thoen, C., Gossele, V., Movva, N. R., et al., (1987). Engineering herbicide resistance in plants by expression of a detoxifying enzyme. *The EMBO Journal*, *6*(9), 2513. https://doi.org/10.1002/J.1460-2075.1987.TB02537.X.

Brouwer-Brolsma, E. M., Brandl, B., Buso, M. E. C., Skurk, T., & Manach, C., (2020). Food intake biomarkers for green leafy vegetables, bulb vegetables, and stem vegetables: A review. *Genes & Nutrition*, *15*, 7. https://doi.org/10.1186/s12263-020-00667-z.

Cerda, R., Avelino, J., Gary, C., Tixier, P., Lechevallier, E., & Allinne, C., (2017). Primary and secondary yield losses caused by pests and diseases: Assessment and modeling in coffee. *PloS One*, *12*(1). https://doi.org/10.1371/JOURNAL.PONE.0169133.

Chen, H., Guo, A., Lu, Z., Tan, S., Wang, J., Gao, J., Zhang, S., et al., (2019). *Agrobacterium tumefaciens*-mediated transformation of a hevein-like gene into asparagus leads to stem wilt resistance. *PLoS One*, *14*(10). https://doi.org/10.1371/journal.pone.0223331.

Cho, H. S., Cao, J., Ren, J. P., & Earle, E. D., (2001). Control of lepidopteran insect pests in transgenic Chinese cabbage (*Brassica rapa* ssp. Pekinensis) transformed with a synthetic Bacillus thuringiensis cry1C gene. *Plant Cell Reports*, *20*(1), 1–7. https://doi.org/10.1007/S002990000278.

Curtis, I. S., (2006). Lettuce (*Lactuca sativa* L.). *Methods in Molecular Biology (Clifton, N.J.)*, *343*, 449–458. https://doi.org/10.1385/1-59745-130-4:449.

Curtis, I. S., Power, J. B., Laat, A. M. M. D., Caboche, M., & Davey, M. R., (1999). Expression of a chimeric nitrate reductase gene in transgenic lettuce reduces nitrate in leaves. *Plant Cell Reports*, *18*(11), 889–896. https://doi.org/10.1007/S002990050680/METRICS.

Datta, A., (2013). Genetic engineering for improving quality and productivity of crops. *Agriculture & Food Security*, *2*(1), 15. https://doi.org/10.1186/2048-7010-2-15.

Davey, M. R., Anthony, P., Van, H. P., Power, J. B., & Lowe, K. C., (2007). Lettuce. In: Pua, E. C., & Davey, M. R., (eds.), *Transgenic Crops IV* (pp. 221–249). Springer Berlin Heidelberg. https://doi.org/10.1007/978-3-540-36752-9_12.

Encina, C. L., & Regalado, J. J., (2022). Aspects of *in vitro* plant tissue culture and breeding of asparagus: A review. *Horticulturae, 8*(5), 439. https://doi.org/10.3390/HORTICULTURAE8050439.

Gerszberg, A., (2018). Tissue culture and genetic transformation of cabbage (*Brassica oleracea* var. *Capitata*): An overview. *Planta, 248*(5), 1037. https://doi.org/10.1007/S00425-018-2961-3.

Ghosh, S., Irfan, M., & Datta, A., (2017). Applications of RNA silencing in improving plant traits for industrial use. In: Dalmay, T., (ed.), *Plant Gene Silencing: Mechanisms and Applications* (pp.128–146). CAB International 2017, Biotechnology Series-5. ISBN: 1780647670, 9781780647678, https://doi.org/10.1079/9781780647678.0128.

Ghosh, S., Singh, U. K., Meli, V. S., Kumar, V., Kumar, A., Irfan, M., Chakraborty, N., et al., (2013). Induction of senescence and identification of differentially expressed genes in tomato in response to monoterpene. *PLoS One, 8*(9), e76029. https://doi.org/10.1371/journal.pone.0076029.

Goto, F., Yoshihara, T., & Saiki, H., (2000). Iron accumulation and enhanced growth in transgenic lettuce plants expressing the iron- binding protein ferritin. *Theoretical and Applied Genetics*, *100*(5), 658–664. https://doi.org/10.1007/S001220051336/METRICS.

Guzmán, M. G., Cellini, F., Fotopoulos, V., Balestrini, R., & Arbona, V., (2022). New approaches to improve crop tolerance to biotic and abiotic stresses. *Physiologia Plantarum*, *174*(1), e13547. https://doi.org/10.1111/PPL.13547.

Irfan, M., & Datta, A., (2017). Improving food nutritional quality and productivity through genetic engineering. *International Journal of Cell Science & Molecular Biology, 2*(1), 555576. https://doi.org/10.19080/IJCSMB.2017.02.555576.

Irfan, M., Kumar, P., Ahmad, M. F., & Siddiqui, M. W., (2023). Biotechnological interventions in reducing losses of tropical fruits and vegetables. *Current Opinion in Biotechnology, 79*, 102850. doi.org/10.1016/j.copbio.2022.102850.

Jun, S., Kwon, S. Y., Pack, K. Y., & Paek, K. H., (1995). *Agrobacterium*-mediated transformation and regeneration of fertile transgenic plants of Chinese cabbage (*Brassica campestris* ssp. Pekinensis cv. 'spring flavor'). *Plant Cell Reports*, *14*(10), 620–625. https://doi.org/10.1007/BF00232725.

Kanungsukkasem, U., Ng, N., Minh, H. V., Razzaque, A., Ashraf, A., Juvekar, S., Ahmed, S. M., & Bich, T. H., (2009). Fruit and vegetable consumption in rural adults population in INDEPTH HDSS sites in Asia. *Global Health Action*, *2*(1), 35–43. https://doi.org/10.3402/gha.v2i0.1988.

Kim, K. H., Kabir, E., & Jahan, S. A., (2017). Exposure to pesticides and the associated human health effects. *Science of The Total Environment*, *575*, 525–535. https://doi.org/10.1016/J.SCITOTENV.2016.09.009.

Kohli, D., Champawat, P. S., & Mudgal, V. D., (2022). Asparagus (*Asparagus racemosus* L.) roots: Nutritional profile, medicinal profile, preservation, and value addition. *Journal of the Science of Food and Agriculture*. https://doi.org/10.1002/JSFA.12358.

Kumar, V., Chattopadhyay, A., Ghosh, S., Irfan, M., Chakraborty, N., Chakraborty, S., & Datta, A., (2016a). Improving nutritional quality and fungal tolerance in soybean and grass pea by expressing an oxalate decarboxylase. *Plant Biotechnology Journal, 14*, 1394–1405. https://doi.org/10.1111/pbi.12503.

Kumar, V., Irfan, M., & Datta, A., (2019). Manipulation of oxalate metabolism in plants for improving food quality and productivity. *Phytochemistry, 158*, 103–109. doi: 10.1016/j.phytochem.2018.10.029. Epub 2018 Nov 28. PMID: 30500595.

Kumar, V., Irfan, M., Ghosh, S., Chakraborty, N., Chakraborty, S., & Datta, A., (2016b). Fruit ripening mutants reveal cell metabolism and redox state during ripening. *Protoplasma, 253*, 581–594. https://doi.org/10.1007/s00709-015-0836-z.

Kumari, C., Sharma, M., Kumar, V., Sharma, R., Kumar, V., Sharma, P., Kumar, P., & Irfan, M., (2022). Genome editing technology for genetic amelioration of fruits and vegetables for alleviating post-harvest loss. *Bioengineering 9*(4), 176. https://doi.org/10.3390/bioengineering9040176.

Lee, Y. H., Chung, K. H., Kim, H. U., Jin, Y. M., Kim, H. I., & Park, B. S., (2003). Induction of male sterile cabbage using a tapetum-specific promoter from *Brassica campestris* L. ssp. Pekinensis. *Plant Cell Reports*, *22*(4), 268–273. https://doi.org/10.1007/S00299-003-0688-4/METRICS.

Lee, Y. H., Yoon, I. S., Suh, S. C., & Kim, H. I., (2001). Enhanced disease resistance in transgenic cabbage and tobacco expressing a glucose oxidase gene from *Aspergillus Niger*. *Plant Cell Reports*, *20*(9), 857–863. https://doi.org/10.1007/S00299-001-0416-X.

Li, X., Peng, R. H., Fan, H. Q., Xiong, A. S., Yao, Q. H., Cheng, Z. M., & Li, Y., (2005). Vitreoscilla hemoglobin overexpression increases submergence tolerance in cabbage. *Plant Cell Reports*, *23*(10, 11), 710–715. https://doi.org/10.1007/S00299-004-0872-1/METRICS.

Li, Y., Wang, G., Hou, R., Zhou, Y., Gong, R., Sun, X., & Tang, K., (2011). Engineering tocopherol biosynthetic pathway in lettuce. *Biologia Plantarum*, *55*(3), 453–460. https://doi.org/10.1007/S10535-011-0110-Y/METRICS.

Lomnitski, L., Bergman, M., Nyska, A., Ben-Shaul, V., & Grossman, S., (2003). Composition, efficacy, and safety of spinach extracts. *Nutrition and Cancer*, *46*(2), 222–231. https://doi.org/10.1207/S15327914NC4602_16.

Ma, X., Zhu, Z., Li, Y., Yang, G., & Pei, Y., (2017). Expressing a modified cowpea trypsin inhibitor gene to increase insect tolerance against *Pieris rapae* in Chinese cabbage. *Horticulture Environment and Biotechnology*, *58*(2), 195–202. https://doi.org/10.1007/S13580-017-0188-Z/METRICS.

Mason, P., Braun, L., Warwick, S. I., Zhu, B., & Stewart, C. N., (2003). Transgenic Bt-producing *Brassica napus*: *Plutella xylostella* selection pressure and fitness of weedy relatives. *Environmental Biosafety Research*, *2*(4), 263–276. https://doi.org/10.1051/EBR:2003016.

Mccabe, M. S., Garratt, L. C., Schepers, F., Jordi, W. J. R. M., Stoopen, G. M., Davelaar, E., Hans, J., et al., (2001). *Effects of P SAG12-IPT Gene Expression on Development and Senescence in Transgenic Lettuce, 1*. https://doi.org/10.1104/pp.010244.

Mohapatra, U., McCabe, M. S., Power, J. B., Schepers, F., Arend, A. V. D., & Davey, M. R., (1999). Expression of the bar gene confers herbicide resistance in transgenic lettuce. *Transgenic Research*, *8*(1), 33–44. https://doi.org/10.1023/A:1008891216134/METRICS.

Morris, M. C., Wang, Y., Barnes, L. L., Bennett, D. A., Dawson-Hughes, B., & Booth, S. L., (2018). Nutrients and bioactives in green leafy vegetables and cognitive decline: Prospective study. *Neurology*, *90*(3), E214–E222. https://doi.org/10.1212/WNL.0000000000004815.

Negi, J., Singh, P., Joshi, G., Rawat, M., & Bisht, V., (2010). Chemical constituents of *Asparagus*. *Pharmacognosy Reviews*, *4*(8), 215. https://doi.org/10.4103/0973-7847.70921.

Nwafor, I. C., Shale, K., & Achilonu, M. C., (2017). Chemical composition and nutritive benefits of chicory (*Cichorium intybus*) as an ideal complementary and/or alternative livestock feed supplement. *The Scientific World Journal, 2017*. https://doi.org/10.1155/2017/7343928.

Park, B. J., Liu, Z., Kanno, A., & Kameya, T., (2005). Genetic improvement of Chinese cabbage for salt and drought tolerance by constitutive expression of a *B. napus* LEA gene. *Plant Science*, *169*(3), 553–558. https://doi.org/10.1016/J.PLANTSCI.2005.05.008.

Park, S. H., Jun, S. S., An, G., Hong, Y. N., & Park, M. C., (2003). A comparative study on the protective role of trehalose and LEA proteins against abiotic stresses in transgenic Chinese cabbage (*Brassica campestris*) overexpressing CaLEA orotsA. *Journal of Plant Biology, 46*(4), 277–286. https://doi.org/10.1007/BF03030375.

Park, S., Elless, M. P., Park, J., Jenkins, A., Lim, W., IV, E. C., & Hirschi, K. D., (2009). Sensory analysis of calcium-biofortified lettuce. *Plant Biotechnology Journal*, *7*(1), 106–117. https://doi.org/10.1111/J.1467-7652.2008.00379.X.

Pileggi, M., Pereira, A. A. M., Suva, J. D. S., Pileggi, S. A. V., & Verma, D. P. S., (2001). An improved method for transformation of lettuce by agrobacterium tumefaciens with a gene that confers freezing resistance. *Brazilian Archives of Biology and Technology*, *44*(2), 191–196. https://doi.org/10.1590/S1516-89132001000200013.

Pogrebnyak, N., Markley, K., Smirnov, Y., Brodzik, R., Bandurska, K., Koprowski, H., & Golovkin, M., (2006). Collard and cauliflower as a base for production of recombinant antigens. *Plant Science*, *171*(6), 677. https://doi.org/10.1016/J.PLANTSCI.2006.06.017.

Poveda, J., Francisco, M., Cartea, M. E., & Velasco, P., (2020). Development of transgenic brassica crops against biotic stresses caused by pathogens and arthropod pests. *Plants*, *9*(12), 1664. https://doi.org/10.3390/plants9121664.

Puhlmann, M. L., & Vos, W. M. De., (2020). Back to the roots: Revisiting the use of the fiber-rich *Cichorium intybus* L. taproots. *Advances in Nutrition (Bethesda, Md.)*, *11*(4), 878–889. https://doi.org/10.1093/ADVANCES/NMAA025.

Qing, C. M., Fan, L., Lei, Y., Bouchez, D., Tourneur, C., Yan, L., & Robaglia, C., (2000). Transformation of Pakchoi (*Brassica rapa* L. ssp. Chinensis) by agrobacterium infiltration. *Molecular Breeding*, *6*(1), 67–72. https://doi.org/10.1023/A:1009658128964/METRICS.

Río-Celestino, M. D., & Font, R., (2020). *The Health Benefits of Fruits and Vegetables*. *9*, 369. https://doi.org/10.3390/foods9030369.

Roberts, J. L., & Moreau, R., (2016). Functional properties of spinach (*Spinacia oleracea* L.) phytochemicals and bioactives. *Food & Function*, *7*(8), 3337–3353. https://doi.org/10.1039/C6FO00051G.

Rokayya, S., Li, C. J., Zhao, Y., Li, Y., & Sun, C. H., (2014). Cabbage (*Brassica oleracea* L. var. capitata) phytochemicals with antioxidant and anti-inflammatory potential. *Asian Pacific Journal of Cancer Prevention: APJCP*, *14*(11), 6657–6662. https://doi.org/10.7314/APJCP.2013.14.11.6657.

Sahib, N. G., Anwar, F., Gilani, A. H., Hamid, A. A., Saari, N., & Alkharfy, K. M., (2013). Coriander (*Coriandrum sativum* L.): A potential source of high-value components for functional foods and nutraceuticals–a review. *Phytotherapy Research: PTR*, *27*(10), 1439–1456. https://doi.org/10.1002/PTR.4897.

Saini, R. K., Song, M. H., Yu, J. W., Lee, J. H., Ahn, H. Y., Keum, Y. S., & Lee, J. H., (2022). Profiling of nutritionally vital bioactive compounds in emerging green leafy vegetables: A comparative study. *Foods*, *11*(23), 3867. https://doi.org/10.3390/FOODS11233867/S1.

Sarkar, T., Salauddin, M., Roy, S., Chakraborty, R., Rebezov, M., Shariati, M. A., Thiruvengadam, M., & Rengasamy, K. R. R., (2022). Underutilized green leafy vegetables: Frontier in fortified food development and nutrition. *Critical Reviews in Food Science and Nutrition*. https://doi.org/10.1080/10408398.2022.2095555.

Sharma, S., Padhi, S., Kumari, M., Patnaik, S., & Sahoo, D., (2022). Antioxidant potential of selected wild edible leafy vegetables of Sikkim Himalayan Region: Effects of cooking methods and gastrointestinal digestion on activity. *Frontiers in Nutrition*, *9*. https://doi.org/10.3389/fnut.2022.861347.

Shi, M., Gu, J., Wu, H., Rauf, A., Emran, T. B., Khan, Z., Mitra, S., et al., (2022). Phytochemicals, nutrition, metabolism, bioavailability, and health benefits in lettuce—A comprehensive review. *Antioxidants*, *11*(6). https://doi.org/10.3390/ANTIOX11061158.

Sobolev, A. P., Testone, G., Santoro, F., Nicolodi, C., Iannelli, M. A., Amato, M. E., Ianniello, A., et al., (2010). Quality traits of conventional and transgenic lettuce (*Lactuca sativa* L.) at harvesting by NMR metabolic profiling. *Journal of Agricultural and Food Chemistry*, *58*(11), 6928–6936. https://doi.org/10.1021/jf904439y.

Sun, B., Jiang, M., Zheng, H., Jian, Y., Huang, W. L., Yuan, Q., Zheng, A. H., et al., (2020). Color-related chlorophyll and carotenoid concentrations of Chinese kale can be altered through CRISPR/Cas9 targeted editing of the carotenoid isomerase gene BoaCRTISO. *Horticulture Research*, *7*(1). https://doi.org/10.1038/s41438-020-00379-w.

Sun, H. J., Cui, M. L., Ma, B., & Ezura, H., (2006). Functional expression of the taste-modifying protein, miraculin, in transgenic lettuce. *FEBS Letters*, *580*(2), 620–626. https://doi.org/10.1016/j.febslet.2005.12.080.

Tayal, R., Kumar, V., & Irfan, M., (2022). Harnessing the power of hydrogen sulphide (H_2S) for improving fruit quality traits. *Plant Biol (Stuttg)*, *24*(4), 594–601. doi: 10.1111/plb.13372. Epub 2021 Dec 5. PMID: 34866296.

Treuren, R. V., Coquin, P., & Lohwasser, U., (2012). Genetic resources collections of leafy vegetables (lettuce, spinach, chicory, artichoke, asparagus, lamb's lettuce, rhubarb, and rocket salad): Composition and gaps. *Genetic Resources and Crop Evolution*, *59*(6), 981–997. https://doi.org/10.1007/s10722-011-9738-x.

Tseng, M. J., Liu, C. W., & Yiu, J. C., (2007). Enhanced tolerance to sulfur dioxide and salt stress of transgenic Chinese cabbage plants expressing both superoxide dismutase and catalase in chloroplasts. *Plant Physiology and Biochemistry*, *45*(10, 11), 822–833. https://doi.org/10.1016/J.PLAPHY.2007.07.011.

Tseng, M. J., Liu, C. W., & Yiu, J. C., (2008). Tolerance to sulfur dioxide in transgenic Chinese cabbage transformed with both the superoxide dismutase containing manganese and catalase genes of *Escherichia coli*. *Scientia Horticulturae, 115*(2), 101–110. https://doi.org/10.1016/J.SCIENTA.2007.08.008.

Vanjildorj, E., Bae, T. W., Riu, K. Z., Kim, S. Y., & Lee, H. Y., (2005). Overexpression of *Arabidopsis* ABF3 gene enhances tolerance to drought and cold in transgenic lettuce (*Lactuca sativa*). *Plant Cell, Tissue, and Organ Culture, 83*(1), 41–50. https://doi.org/10.1007/S11240-005-3800-3/METRICS.

Wallace, T. C., Bailey, R. L., Blumberg, J. B., Burton-Freeman, B., Chen, C. Y. O., Crowe-White, K. M., Drewnowski, A., et al., (2020). Fruits, vegetables, and health: A comprehensive narrative, umbrella review of the science and recommendations for enhanced public policy to improve intake. *Critical Reviews in Food Science and Nutrition, 60*(13), 2174–2211. https://doi.org/10.1080/10408398.2019.1632258.

Wang, X., Liu, Y., Wang, Y., Dong, X., Wang, Y., Yang, X., Tian, H., & Li, T., (2022). Protective effect of coriander (*Coriandrum sativum* L.) on high-fructose and high-salt diet-induced hypertension: Relevant to improvement of renal and intestinal function. *Journal of Agricultural and Food Chemistry, 70*(12), 3730–3744. https://doi.org/10.1021/ACS.JAFC.2C00267.

Xu, H., Wang, X., Zhao, H., & Liu, F., (2008). An intensive understanding of vacuum infiltration transformation of pakchoi (*Brassica rapa* ssp. Chinensis). *Plant Cell Reports, 27*(8), 1369–1376. https://doi.org/10.1007/S00299-008-0564-3/METRICS.

Yang, X., Gil, M. I., Yang, Q., & Tomás-Barberán, F. A., (2022). Bioactive compounds in lettuce: Highlighting the benefits to human health and impacts of preharvest and postharvest practices. *Comprehensive Reviews in Food Science and Food Safety, 21*(1), 4–45. https://doi.org/10.1111/1541-4337.12877.

Yang, Y., Al-Khayri, J. M., & Anderson, E. J., (1997). Transgenic spinach plants expressing the coat protein of cucumber mosaic virus. *In Vitro Cellular and Developmental Biology – Plant, 33*(3), 200–204. https://doi.org/10.1007/S11627-997-0022-X/METRICS.

Zhandong, Y., Shuangyi, Z., & Qiwei, H., (2007). High level resistance to turnip mosaic virus in Chinese cabbage (*Brassica campestris* ssp. Pekinensis (Lour) Olsson) transformed with the antisense NIb gene using marker-free *Agrobacterium tumefaciens* infiltration. *Plant Science, 172*(5), 920–929. https://doi.org/10.1016/J.PLANTSCI.2006.12.018.

Zhang, H., Si, X., Ji, X., Fan, R., Liu, J., Chen, K., Wang, D., & Gao, C., (2018). Genome editing of upstream open reading frames enables translational control in plants. *Nature Biotechnology, 36*(9), 894–899.

Zuo, X., Zhang, Y., Wu, B., Chang, X., & Ru, B., (2002). Expression of the mouse metallothionein mutant ββ-cDNA in the lettuces (*Lactuca sativa* L.). *Chinese Science Bulletin, 47*(7), 558–562. https://doi.org/10.1360/02TB9128/METRICS.

CHAPTER 11

GENETIC AND GENOME ENGINEERING OF UNDERUTILIZED FRUIT AND VEGETABLE CROPS: CHALLENGES AND OPPORTUNITIES FOR POSTHARVEST QUALITY IMPROVEMENT

MEGHA,[1] NITESH SINGH,[2] NISHA SINGH,[3] and JITENDER KUMAR[4]

[1]*Amity University, Manesar, Gurugram, Haryana, India*

[2]*Department of Biosciences, UIBT, Chandigarh University, India*

[3]*Gujarat Biotechnology University, Gandhinagar, Gujarat, India*

[4]*Assistant Professor, Department of Plant Science, Central University of Himachal Pradesh, India*

ABSTRACT

Underutilized fruit and vegetable crops offer immense potential for food security and improved nutrition, but their postharvest quality often hinders wider adoption. This chapter explores how genetic and genome engineering tools can address these challenges. We discuss the vast genetic diversity of underutilized crops in India and highlight the growing importance of plant genome editing/engineering tools. We showcase various applications of genome editing systems in enhancing underutilized crop resilience, yield, and nutritional quality. While acknowledging the significant impact

Bioengineered Fruit and Vegetables: Trends in Postharvest Quality Improvement.
Mohammad Irfan and Pankaj Kumar (Eds.)

of these technologies, this chapter also discusses the challenges associated with public acceptance and regulatory hurdles. Looking ahead, we emphasize the future prospects of genome editing in unlocking the potential of underutilized crops, contributing to sustainable food systems, and achieving global food security goals.

11.1 INTRODUCTION

One of the most well-known fields of life science, systems biology, is rapidly developing due to the integration of bioengineering principles with fundamental biological science knowledge. Researchers have been interested in crop enhancement by foreign gene transfer over the past 30 years (Stephens & Barakate, 2017). However, gene transfer technologies are always a contentious issue, particularly when it comes to the safety and risk of edible crops. To address the many difficulties, synergistic interactions between enhanced technologies, institutional supports, favorable government policies, and farmer awakening are required (Sahoo et al., 2021). Despite the fact that different countries' agro-climatic circumstances allow them to grow a higher percentage of cultivated and a lower percentage of recognized vegetable and fruit crops, little attention has been paid to known underutilized crops (Figure 11.1). Underutilized crops/plant species are those having untapped potential to contribute to food security, health (nutritional/medicinal), revenue generation, and environmental services (Jena et al., 2018). Several crops with high nutritional qualities were once widely consumed by humans. Approximately 80,000 plant species have been directly used by humans for food, fodder, fiber, medicinal, and industrial purposes, according to various studies on the evaluation of mankind. More than 25,000 of these are edible, and over 7,000 have either been domesticated or at some point taken from the wild for nutrition (Muthamilarasan et al., 2019). Self-reliance on food may not always imply nutritional sufficiency. In fact, significant inequalities in the availability of calories, proteins, minerals, and vitamins persist. Furthermore, malnutrition is a major concern in many countries. Increased availability of fruits and vegetables could significantly reduce these inadequacies (Tiozon et al., 2021). Underutilized food crops are classified into five groups: pseudo cereals and millets, grain legumes, root and tubers, green vegetables, and fruits. The majority of pseudo cereals and millets are of Chinese or African origin and have a lengthy history of cultivation. Before rice or any other crops, broomcorn (*Panicum miliaceum*) and foxtail millet (*Setaria italica*) were the first domesticated plants. The Bambara groundnut (*Vigna subterranea*), an indigenous African legume, is recognized for its adaptability in poor soils of hot and arid conditions where other crops fail to grow (Bernard et al., 2018). These underutilized crops were not mainstreamed due to low yields and a lack of national policy to link with the value chain in order to promote agro-business. Due to their neglect, genetic diversity and unique gene pools from agricultural areas have been overlooked (Robbeets et al., 2021).

FIGURE 11.1 Different underutilized crops, significant contributor to growing food and nutritional insecurity.

The availability of genetic resources is critical in such a process. Farmers continue to cultivate and reproduce local varieties (landraces) based on their adaptation to specific settings and the availability of important features. New genomes and phenomics methods enable comprehensive characterization of such resources, which could improve their application in breeding (Silva, 2014). New plant breeding techniques, particularly cis-genic gene transfer and next-generation precision genome engineering based on genome editing technologies, could be used to access the genetic resources, also exploiting them in functional investigations and streamlined breeding strategies. Such techniques, however, necessitate rapid transformation protocols, substantial genetic resources, and precise understanding before they can be effectively used in real breeding operations (Cardi, 2016). Next-generation breeding technology (NGS), often known as genome engineering/editing (GE), is promising and increasing in popularity for improving economically essential features in crops. TALENs, CRISPR, mega-nucleases (MNs), and zinc-finger nucleases (ZFNs), are all important GE tools. The most recent of these methods, CRISPR, is revolutionizing agriculture. The core CRISPR/Cas system has improved, making it a more accurate, precise, and adaptable tool. A range of genome-engineered crops are currently accessible for production in various parts of the world (Aglawe et al., 2022). Moreover, gene-editing tools are being used with more sophisticated delivery mechanisms and increased precision. This chapter discusses sophisticated gene-editing technologies for controlling gene expression in plants in order to boost crop yield. In this chapter, we focused on available genome-edited underutilized food crops and current genome engineering research on food crops. We also provided an overview of genome engineering method advancements/tools, regulation, and the scope of genome engineering in future crops.

11.2 GENETIC DIVERSITY OF UNDERUTILIZED CROP PRESENT IN INDIA

Plant genetic resources must be conserved and used to enhance agricultural crops. The significance of these resources has increased in recent years due to changes in ownership and regulatory frameworks related to biodiversity. India is home to many tropical fruit species, which play a crucial role in supporting the livelihoods of rural populations (Diengngan & Hasan, 2015). Vegetables are gaining popularity due to their nutritional and medicinal value. However, the gap between vegetable supply and demand is widening over time. India's diverse agro-ecologies provide ample opportunities for cultivating a wide range of uncultivated or alternative vegetables compared to other commercially popular vegetables. India boasts a variety of vegetable species with broad applications, including leguminous varieties (Kumar et al., 2018). Tribal communities in India's

Western Ghats, Maharashtra, and Northeastern States have traditionally depended on non-timber forest products and specific agri-horti-forestry practices for their livelihoods. Native fruit species in India include tamarind (*Tamarindus indica*), jackfruit (*Artocarpus heterophyllus*), Indian gooseberry (*Emblica officinalis*), ber (*Zizyphus mauritiana*), Citrus, Musa, Pyrus, and Malus (Table 11.1).

The benefits of India's underutilized fruits are widely known, but sadly, few of these fruit crops have made it into farmers' orchards. Minor fruits have tremendous nutritional and therapeutic significance, despite being referred to as "lesser-known fruits" in popular culture (Shukla & Jain, 2019). In spite of India's abundant germplasm, there were few standard varieties created. Due to their high level of tolerance and wide range of adaptability, they can survive under harsh climatic and environmental situations. These fruits have promise for use in sustainable agriculture as well. Therefore, it is important to take into account research and development efforts, farmer awareness, and the viability of cultivating these lesser-known fruits.

11.3 PLANT GENOME EDITING/ENGINEERING TOOLS AND TECHNOLOGY

Underutilized crops can help meet human energy, health, and food security needs. Therefore, it is crucial to promote their cultivation and preservation. Most native fruits are limited in variety, so it is advisable to focus on their advancement and the development of optimal propagation methods and agro-techniques (Figure 11.2). The use of genome engineering to enhance crop resilience and yield is expected to be significant. In recent years, several new genome editing systems have been developed and are discussed in subsections.

11.3.1 CRISPR/CAS

Analysis of the genes involved in the isozyme conversion of alkaline phosphatase in Escherichia coli led to the discovery of the CRISPR-Cas system (Ishino et al., 1987). In plants, genome editing using the CRISPR/Cas9 system was first reported by Li et al. (2013). In many applications,

TABLE 11.1 Different Underutilized Fruits and Vegetables in India with Their Benefits

Sl. No.	Common Name	Scientific Name	Significance	References
1.	Amla	*Emblica officinalis*	It includes iron, calcium, and phosphorus. It is cost concentrated form of vitamin C. It treats constipation, scurvy, and sleeplessness. Additionally, it helps with hemorrhage, leucorrhea, and uterine bleeding. It lowers a blood sugar level.	Hasan et al. (2016)
2.	Bael/stone apple	*Aegle marmelos*	Moisture makes up 61.5% of its composition, followed by protein (18%), fat (0.3%), minerals (1.7%), fiber (2.9%), and carbs (31.8%). It contains calcium, phosphorus, iron, carotene, thiamine, riboflavin, niacin, and vitamin C among other vitamins and minerals. The fruit is used to treat colds, diarrhea, and eye dryness.	Vijayalakshmi & Venkatalakshmi (2017)
3.	Jackfruit	*Artocarpus heterophyllus*	The fruit has varying amounts of carbohydrate, protein, fiber, fat, calcium, iron, phosphorus, carotene, thiamine, riboflavin, niacin, and vitamin C. The latex promotes the healing of glandular swellings, abscesses, and snakebites. A good source of potassium is also present in jackfruit, which helps to control high blood pressure and support cardiovascular health.	Khan et al. (2021)
4.	Wood apple	*Feronia limonia*	18.1% of the pulp is composed of carbohydrates, 7.1% of protein, 3.7% of fat, 5.0% of fiber, and 1.9% of minerals. Calcium, phosphorus, iron, and vitamins including carotene, riboflavin, niacin, and thiamine are all abundant in the pulp. When unripe, the fruit is used to treat diarrhea and dysentery and as a liver and heart tonic.	Srivastava et al. (2019)
5.	Jamun	*Syzygiumcumini*	The tree is best suited for planting along roadsides and as a windbreak. Fruits are rich in fiber, protein, iron, calcium, phosphorus, and glucose. It is used as an antidote for strychnine poisoning as well as for diabetes, diarrhea, eczema, and blood purification. It reduces blood pressure and has stomachic, carminative, and diuretic properties.	Saleem et al. (2016)
6.	Ber/Indian Jujube	*Zyziphusmauritiana*	It contains a lot of vitamin B complex and A. For diarrhea and dysentery, root, and bark infusion are beneficial, while leaf can be used as a gargle for sore throats and bleeding gums. The seeds have aphrodisiac properties. Ber root powder helpful in treatment of wounds, fever, and ulcers.	Abdallah et al. (2016)

TABLE 11.1 *(Continued)*

Sl. No.	Common Name	Scientific Name	Significance	References
7.	Elephant foot yam	*Amorphophallus paeoniifolius*	Triacotane, lupeol, betulinicacid, and stigma sterol are all present. Additionally, it has blood purifying qualities and is used to treat piles, tumors, asthma, and rheumatism.	Navya et al. (2017)
8.	Pointed gourd/ parwal	*Trichosanthes dioica*	It is a good source of minerals, vitamins A and C, and carbs. Its juice is utilized in subacute cases of liver enlargement, baldness, and as a tonic.	Khatua et al. (2016)

CRISPR/Cas9 is regarded as versatile, affordable, and adaptable. It has drawn a lot of interest due to its great efficiency, simple alteration, and exploitation in gene mutation and transcriptional regulation in plants (Song et al., 2016). Many economically important crops have benefited from CRISPR research, including rice (*Oryza sativa*), wheat (*Triticum aestivum*), maize (*Zea mays*), tomato (*Solanum lycopersicum*), potato (*Solanum tuberosum*), tobacco (*Nicotiana tabacum*), and cotton (*Gossypium* spp.) (Chen et al., 2019). With the advancements in CRISPR-Cas, significant methods have been used to improve crops through metabolic engineering and host gene regulation. A number of new CRISPR systems, such as Cas12 (Zaidi et al., 2017; Merker et al., 2020), Cas13a (Aman et al., 2018; Abudayyeh et al., 2017), and Cas13b (Smargon et al., 2017), are currently in the pipeline. Another intriguing recent advance is crop chromosome engineering, which allows for resequencing the plant genomes, genetic linkages, and somatic chromosome engineering (Beying et al., 2020; Schmidt et al., 2020). These advancements might make it easier to transmit resistance genes to elite cultivars.

11.3.2 ZINC FINGER NUCLEASES (ZFNS)

Zinc finger nucleases (ZFNs) were the first generation of chimerically built nuclease-based genome editing tools that were created after the functioning Cys2-His2 zinc finger (ZF) domain was discovered (Kim et al., 1996). ZFNs are specialized DNA cleavage proteins that can be targeted to cut DNA sequences at particular locations (Martínez-Fortún et al., 2017). With the development of ZFPs, it is easy to target a particular

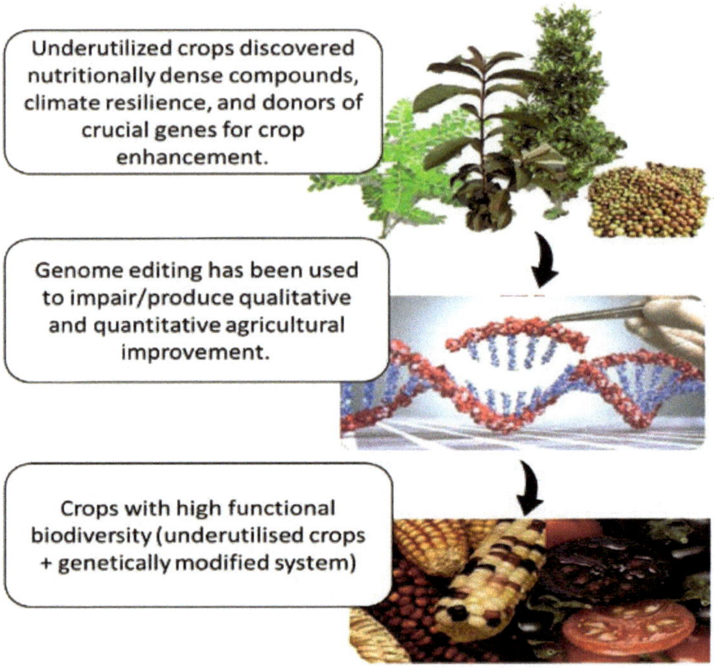

FIGURE 11.2 Underutilized crops have been enhanced to address a range of agricultural challenges through the application of science-based technology.

DNA sequence in genome-editing. Arabidopsis, Nicotiana, maize, petunia, soybean, rapeseed, rice, apple, and fig have all been altered using ZFNs till now (Ran et al., 2017). Trait stacking in maize (Ainley et al., 2013) and rice (Cantos et al., 2014) also done with the help of ZFN-mediated targeted transgene integration. Also used for assembled a number of important traits, to improve the crop production and constraint the climatic issues.

11.3.3 TRANSCRIPTION ACTIVATOR-LIKE EFFECTOR NUCLEASES (TALENS)

A new technique known as the TALEN (transcription activator-like effector nucleases) system was created in 2011 to improve the effectiveness, safety, and accessibility of genome editing. The transcription activator-like effectors (TALES) made by the phytopathogenic *Xanthomonas* genus of bacteria gave rise to the TALEN system (Boch & Bonas, 2010;

Urnov et al., 2010). Numerous plants, including Arabidopsis, Nicotiana, Brachypodium, barley, potatoes, tomatoes, sugarcane, flax, rapeseed, soybean, rice, maize, and wheat, have shown that TALENs can modify the genome (Martínez-Fortún et al., 2017; Ran et al., 2017). Incorporating TALENs and donor DNA into *gemini* virus replicons greatly boosted their copy number, which in turn increased the effectiveness of homologous recombination (HR) and is a good example of how TALEN-mediated gene insertion might improve crops. This was also performed in tomato (Cermak et al., 2015); purple tomatoes with high anthocyanin levels were produced by inserting a potent promoter upstream of the gene responsible for anthocyanin production. These illustrations highlight the TALEN technology's enormous potential for crops' trait enhancement (Saurabh, 2021).

Utilizing more recent research and technology, genome editing enables the modification of a specific region of the genome, improving the accuracy of the correction or insertion, minimizing cell toxicity, and providing perfect reproducibility for a variety of desirable applications in crops (Voytas & Gao, 2014). Other methods, in addition to ZFN, CRISPR/Cas9, and TALENs, have been employed to gather data on gene function. To target gene expression, for instance, HR is employed (Gaj et al., 2013). The use of this approach has been limited by low efficiency, prolonged investigations, mutagenesis consequences, and off-target effects. In contrast to HR, targeted knocking down of gene expression using RNAi technology has proven to be a quick and affordable way to inhibit gene expression. For genome editing to be effective, plant transformation and plant regeneration are essential (Abdallah et al., 2015).

11.4 APPLICATION OF GENOME ENGINEERING SYSTEM FOR UNDERUTILIZED CROPS

Pests, diseases, and changing climatic conditions are among the biotic and abiotic challenges threatening the agriculture sector around the world. Despite such hurdles and risks, the globe has supplied a number of resources and technology that could be used to create a strategy for ensuring long-term food security and eradicating hunger and malnutrition. To overcome such issues and feed the world's rapidly rising population, will and policies aimed at improving underutilized crops with favorable features are required. Traditional crop enhancement methods have been used for a long time (Sreenivasulu & Fernie, 2022).

1. Heritable traits such as blooming period, plant height, and root shape are examples of qualities that can be engineered to be altered or controlled in crops. In sorghum, flowering time (FT) was modified through heritable targeted mutagenesis by the CRISPR/Cas9-mediated deletion of two related genes, FT and gibberellin 2-oxidase 5 (Ga2ox5) (Char et al., 2020). Another study on creating quantitative trait variation in tomatoes revealed that 40% of the variations were within neighboring genes or structural variants, 80% were in regulatory regions, and 10% were associated with changes in expression. The data show that structural changes in cis-regulatory regions disrupt expression variation (Alonge et al., 2020).
2. Unexpected genetic changes have been made with the support of CRISPR-Cas9 to enhance metabolic pathways, tolerance to biotic (pathogenic fungi, bacteria, or viruses) or abiotic stresses (cold, drought, salt), nutritional content, yield, and grain quality, as well as to produce haploid seeds and increase herbicide resistance (Li et al., 2017, 2018; Okada et al., 2019). There is great promise for treating diseases like tomato brown rugose fruit virus and maize lethal necrosis disease by engineering genetic resistance to viruses and other infections (Luria et al., 2017; Garcia-Ruiz, 2018; Wamaitha et al., 2018).
3. Using CRISPR/Cas to investigate insect-plant interactions in order to enhance and domesticate underutilized species such as groundcherry and goldenberry. Metabolite profiling was performed on leaf, fruit, and husks, with an emphasis on the steroidal chemical class; *withanolides* exhibit insect-feeding deterrent characteristics. CRISPR/Cas is used to create mutant plants that target the promoters or coding sequences of genes involved in *withanolide* synthesis in order to better understand the interaction between *withanolides* and plants (Cable et al., 2021).
4. Biology has been transformed by genome editing. Genome editing has the potential to enhance traits of heterozygous, polyploid, and vegetatively reproduced crops such as the cultivated potato, *Solanum tuberosum* Group. *Tuberosum* L. has increased cold-induced sweetening resistance, processing efficiency, herbicide tolerance, starch quality, and self-incompatibility in diploid and tetraploid clones using CRISPR/Cas9 and TALEN reagents (Nadakuduti et al., 2018).

11.5 IMPACT AND CHALLENGES OF GENOME ENGINEERING SYSTEM FOR UNDERUTILIZED CROPS

Gene editing, which has been widely demonstrated and used in plant science, enables precise, heritable genome mutagenesis without permanent transgenesis. In addition to developing underutilized crops for climate resilience features, special emphasis must be paid to boosting crop quality (El-Mounadi et al., 2020). Sweet potato's nutritious quality, for example, is determined by its carotene content, whereas grass pea's ODAP content should be eliminated. Similarly, phytic acid and other antinutritional components found in millet grains must be decreased in order to improve mineral and vitamin absorption (Muthamilarasan et al., 2019). To date, CRISPR/Cas has been applied in many underutilized crops with diverse genetic origins, resulting in novel alleles and positive phenotypes for breeders, producers, and consumers, as well as contributing to food production for the world's growing population (Shan et al., 2020). Crops including finger millet (Hema et al., 2014), foxtail millet (Ceasar et al., 2017; Santos et al., 2020) and pigeon pea (Ghosh et al., 2017) are increasingly being developed with transformation-capable technologies such as finger millet, foxtail millet, and pigeon pea. In quinoa, cotyledon-nod with hypocotyl, cotyledons, and hypocotyl fragments developed a quick transformation system with hairy roots with a transformation efficiency of 32%–68% (Wang et al., 2021).

Despite more than three decades of scientific development, most crops remain difficult to alter and regenerate. Hence, genome editing in agriculture needs to be enhanced to accelerate their potential in terms of yielding and disease resistance (Altpeter et al., 2016). Recent research has found that the best phytohormone regimens and development regulator (DR) genes can increase the efficiency of plant transformation. The expression of particular DR gene combinations in plant somatic cells results in the development of meristematic tissue capable of renewing shoots. However, beyond the few crops examined, further advancement and widespread adoption of the technique for crop plants remains difficult (Maher et al., 2020). Aside from improving crop quality, it is also critical to raise public knowledge about the significance of underutilized crops. Additionally, it would be challenging to add value and connect farmers growing underutilized commodities with the market. But if the right policies are created and put into practice to popularize and regulate the commercialization of these crops, they will be elevated to the "major" category, providing long-term food and nutrition security.

11.6 FUTURE PROSPECTS

Genetic engineering (GE) is an efficient, practical, and cost-effective approach for biofortifying important crops with deficient vitamins or minerals. Furthermore, because no foreign DNA is inserted and there are fewer regulatory constraints, the products of this technology are easily accepted by society. For the synthesis and production of commercially relevant plant products like flavors, perfumes, essential oils, medications, and pigments, for instance, improved pharmaceutical compound manufacturing, gene editing may be a promising metabolic engineering strategy. As a result of impending climate change, more genetically diverse underutilized crops must be discovered and developed; these activities must be framed within a breeding pipeline approach and should be taken into account during the product design process. Considering nutritional concerns and rising population, some underused crops have been modified utilizing genome editing methods to improve nutrition, quality, and yield. CRISPR technology, for example, for underutilized crops, has the potential to help meet the 2030 goal of eradicating hunger, food insecurity, and all types of human malnutrition.

KEYWORDS

- **breeding**
- **CRISPER/Cas**
- **disease resistance**
- **food insecurity**
- **genetic engineering**
- **human malnutrition**
- **pharmaceutical compound**
- **transcription activator-like effector nucleases**
- **underutilized crops**

REFERENCES

Abdallah, E. M., Elsharkawy, E. R., & Ed-Dra, A., (2016). Biological activities of methanolic leaf extract of *Ziziphus mauritiana. Pharm. Commun. Biosci. Biotech. Res. Comm. Thomson Reuters ISI ESC Crossref Index; J. NAAS J. Score, 9*(4), 605–614.

Abdallah, N. A., Prakash, C. S., & McHughen, A. G., (2015). Genome editing for crop improvement: Challenges and opportunities. *GM Crops & Food, 6*(4), 183–205.

Abudayyeh, O. O., Gootenberg, J. S., Essletzbichler, P., Han, S., Joung, J., Belanto, J. J., & Zhang, F., (2017). RNA targeting with CRISPR–Cas13. *Nature, 550*(7675), 280–284.

Aglawe, S. B., Magar, N. D., Dhawane, Y., Bhamare, D., Shah, P., Devi, S. J. S., & Barbadikar, K. M., (2022). Genome editing crops in food and futuristic crops. In: *Recent Advances in Food Biotechnology* (pp. 401–445). Springer, Singapore.

Ainley, W. M., Sastry-Dent, L., Welter, M. E., Murray, M. G., Zeitler, B., Amora, R., & Petolino, J. F., (2013). Trait stacking via targeted genome editing. *Plant Biotechnology Journal, 11*(9), 1126–1134.

Alonge, M., Wang, X., Benoit, M., Soyk, S., Pereira, L., Zhang, L., & Lippman, Z. B., (2020). Major impacts of widespread structural variation on gene expression and crop improvement in tomato. *Cell, 182*(1), 145–161.

Aman, R., Ali, Z., Butt, H., Mahas, A., Aljedaani, F., Khan, M. Z., & Mahfouz, M., (2018). RNA virus interference via CRISPR/Cas13a system in plants. *Genome Biology, 19*(1), 1–9.

Bernard, N., Losologolo, M., Batlang, U., et al., (2018). Symbiotic performance of grain and wild herbaceous legumes in the Okavango Delta and Tswapong region of Botswana. *Symbiosis, 75*(3), 179–188.

Beying, N., Schmidt, C., Pacher, M., Houben, A., & Puchta, H., (2020). CRISPR–Cas9-mediated induction of heritable chromosomal translocations in *Arabidopsis. Nature Plants, 6*(6), 638–645.

Boch, J., & Bonas, U., (2010). Xanthomonas AvrBs3 family-type III effectors: Discovery and function. *Annual Review of Phytopathology, 48*, 419–436.

Cable, J., Ronald, P. C., Voytas, D., Zhang, F., Levy, A. A., Takatsuka, A., & Bagchi, R., (2021). Plant genome engineering from lab to field—A keystone symposia report. *Annals of the New York Academy of Sciences, 1506*(1), 35–54.

Cantos, C., Francisco, P., Trijatmiko, K. R., Slamet-Loedin, I., & Chadha-Mohanty, P. K., (2014). Identification of "safe harbor" loci in indica rice genome by harnessing the property of zinc-finger nucleases to induce DNA damage and repair. *Frontiers in Plant Science, 5*, 302.

Cardi, T., (2016). Cisgenesis and genome editing: Combining concepts and efforts for a smarter use of genetic resources in crop breeding. *Plant Breeding, 135*(2), 139–147.

Ceasar, S. A., Baker, A., & Ignacimuthu, S., (2017). Functional characterization of the PHT1 family transporters of foxtail millet with development of a novel agrobacterium-mediated transformation procedure. *Scientific Reports, 7*(1), 1–16.

Cermak, T., Baltes, N. J., Čegan, R., Zhang, Y., & Voytas, D. F., (2015). High-frequency, precise modification of the tomato genome. *Genome Biology, 16*(1), 1–15.

Chen, K., Wang, Y., Zhang, R., Zhang, H., & Gao, C., (2019). CRISPR/Cas genome editing and precision plant breeding in agriculture. *Annu. Rev. Plant Biol., 70*(1), 667–697.

Da Silva, D. J. C., (2014). Guiding strategies for breeding vegetable cultivars. *Agricultural Sciences*, *5*(1), 9.

Diengngan, S., & Hasan, M. A., (2015). Genetic diversity of underutilized fruits in India for environmental sustainability. *Adv. Plants Agric. Res.*, *2*(7), 00076.

El-Mounadi, K., Morales-Floriano, M. L., & Garcia-Ruiz, H., (2020). Principles, applications, and biosafety of plant genome editing using CRISPR-Cas9. *Frontiers in Plant Science*, *11*, 56.

Gaj, T., Gersbach, C. A., & Barbas, III. C. F., (2013). ZFN, TALEN, and CRISPR/Cas-based methods for genome engineering. *Trends in Biotechnology*, *31*(7), 397–405.

Garcia, R. M. T., Knapp, A. N., & Garcia-Ruiz, H., (2018). Profile of genetically modified plants authorized in Mexico. *GM Crops & Food*, *9*(3), 152–168.

Ghosh, G., Ganguly, S., Purohit, A., Chaudhuri, R. K., Das, S., & Chakraborti, D., (2017). Transgenic pigeonpea events expressing Cry1Ac and Cry2Aa exhibit resistance to *Helicoverpa armigera*. *Plant Cell Reports*, *36*(7), 1037–1051.

Hasan, M. R., Islam, M. N., & Islam, M. R., (2016). Phytochemistry, pharmacological activities and traditional uses of *Emblica officinalis*: A review. *International Current Pharmaceutical Journal*, *5*(2), 14–21.

Hema, R., Vemanna, R. S., Sreeramulu, S., Reddy, C. P., Senthil-Kumar, M., & Udayakumar, M., (2014). Stable expression of mtlD gene imparts multiple stress tolerance in finger millet. *PloS One*, *9*(6), e99110.

Ishino, Y., Shinagawa, H., Makino, K., Amemura, M., & Nakata, A., (1987). Nucleotide sequence of the iap gene, responsible for alkaline phosphatase isozyme conversion in *Escherichia coli*, and identification of the gene product. *Journal of Bacteriology*, *169*(12), 5429–5433.

Jena, A. K., Deuri, R., Sharma, P., & Singh, S. P., (2018). Underutilized vegetable crops and their importance. *Journal of Pharmacognosy and Phytochemistry*, *7*(5), 402–407.

Khan, A. U., Ema, I. J., Faruk, M., Tarapder, S. A., Khan, A. U., Noreen, S., & Adnan, M., (2021). A review on importance of *Artocarpus heterophyllus* L. (Jackfruit). *Journal of Multidisciplinary Applied Natural Science*.

Khatua, S., Pandey, A., & Biswas, S. J., (2016). Phytochemical evaluation and antimicrobial properties of *Trichosanthes dioica* root extract. *Journal of Pharmacognosy and Phytochemistry*, *5*(5), 410.

Kim, Y. G., Cha, J., & Chandrasegaran, S., (1996). Hybrid restriction enzymes: Zinc finger fusions to Fok I cleavage domain. *Proceedings of the National Academy of Sciences*, *93*(3), 1156–1160.

Kumar, R., Rajasree, V., Sagar, L., Tripura, U., Karthick, K., Karthick, R., & Prasath, G., (2018). Underutilized vegetables as rich source of medicinal value: A boon for the pharmaceutical industries and farmer income. *International Journal of Chemical Studies*, *6*(4), 3320–3323.

Li, A., Jia, S., Yobi, A., Ge, Z., Sato, S. J., Zhang, C., & Holding, D. R., (2018). Editing of an alpha-kafirin gene family increases, digestibility, and protein quality in sorghum. *Plant Physiology*, *177*(4), 1425–1438.

Li, J. F., Norville, J. E., Aach, J., McCormack, M., Zhang, D., Bush, J., & Sheen, J., (2013). Multiplex and homologous recombination–mediated genome editing in *Arabidopsis* and *Nicotiana benthamiana* using guide RNA and Cas9. *Nature Biotechnology*, *31*(8), 688–691.

Li, J., Zhang, H., Si, X., Tian, Y., Chen, K., Liu, J., & Gao, C., (2017). Generation of thermosensitive male-sterile maize by targeted knockout of the ZmTMS5 gene. *Journal of Genetics and Genomics= Yi Chuan Xue Bao*, *44*(9), 465–468.

Luria, N., Smith, E., Reingold, V., Bekelman, I., Lapidot, M., Levin, I., et al., (2017). A new Israeli tobamovirus isolate infects tomato plants harboring Tm-22 resistance genes. *PloS One, 12*, e0170429.

Maher, M. F., Nasti, R. A., Vollbrecht, M., Starker, C. G., Clark, M. D., & Voytas, D. F., (2020). Plant gene editing through de novo induction of meristems. *Nature Biotechnology*, *38*(1), 84–89.

Martínez-Fortún, J., Phillips, D. W., & Jones, H. D., (2017). Potential impact of genome editing in world agriculture. *Emerging Topics in Life Sciences*, *1*(2), 117–133.

Merker, L., Schindele, P., Huang, T. K., Wolter, F., & Puchta, H., (2020). Enhancing in planta gene targeting efficiencies in *Arabidopsis* using temperature-tolerant CRISPR/LbCas12a. *Plant Biotechnology Journal*, *18*(12), 2382.

Muthamilarasan, M., Singh, N. K., & Prasad, M., (2019). Multi-omics approaches for strategic improvement of stress tolerance in underutilized crop species: A climate change perspective. *Advances in Genetics*, *103*, 1–38.

Nadakuduti, S. S., Buell, C. R., Voytas, D. F., Starker, C. G., & Douches, D. S., (2018). Genome editing for crop improvement–applications in clonally propagated polyploids with a focus on potato (*Solanum tuberosum* L.). *Frontiers in Plant Science*, *9*, 1607.

Navya, K., Desai, K. D., Tandel, Y. N., & Sheth, S. G., (2017). Effect of integrated nutrient management on growth, yield, and quality of elephant foot yam [*Amorphophallus paeoniifolius* (Dennst.) Nicolson]. *Int. J. Chem. Stud.*, *5*(4), 1766–1769.

Okada, A., Arndell, T., Borisjuk, N., Sharma, N., Watson-Haigh, N. S., Tucker, E. J., & Whitford, R., (2019). CRISPR/Cas9-mediated knockout of Ms1 enables the rapid generation of male-sterile hexaploid wheat lines for use in hybrid seed production. *Plant Biotechnology Journal*, *17*(10), 1905–1913.

Ran, Y., Liang, Z., & Gao, C., (2017). Current and future editing reagent delivery systems for plant genome editing. *Science China Life Sciences*, *60*(5), 490–505.

Robbeets, M., Bouckaert, R., Conte, M., Savelyev, A., Li, T., An, D. I., & Ning, C., (2021). Triangulation supports agricultural spread of the Transeurasian languages. *Nature*, *599*(7886), 616–621.

Sahoo, G., Swamy, S., Rout, S., Wani, A., & Mishra, A., (2021). Exploitation of wild leafy vegetables and under-utilized fruits: Consequences for food and nutritional security. *Annals of the Romanian Society for Cell Biology*, *25*(6), 5656–5668.

Saleem, U., Ali, N., & Ahmad, B., (2016). Does *Syzygium cumini* possess significant pharmacological effects an overview. *Pharmacologyonline*, *2*, 26–29.

Santos, C. M., Romeiro, D., Silva, J. P., Basso, M. F., Molinari, H. B. C., & Centeno, D. C., (2020). An improved protocol for efficient transformation and regeneration of *Setaria italica*. *Plant Cell Reports*, *39*(4), 501–510.

Saurabh, S., (2021). Genome editing: Revolutionizing the crop improvement. *Plant Molecular Biology Reporter*, 1–21.

Schmidt, C., Fransz, P., Rönspies, M., Dreissig, S., Fuchs, J., Heckmann, S., & Puchta, H., (2020). Changing local recombination patterns in *Arabidopsis* by CRISPR/Cas mediated chromosome engineering. *Nature Communications*, *11*(1), 1–8.

Shan, S., Soltis, P. S., Soltis, D. E., & Yang, B., (2020). Considerations in adapting CRISPR/Cas9 in nongenetic model plant systems. *Applications in Plant Sciences*, *8*(1), e11314.

Shukla, S., & Jain, S., (2019). Underutilized fruits: Lost nutritional treasure. *Think India Journal*, *22*(34), 1339–1353.

Smargon, A. A., Cox, D. B., Pyzocha, N. K., Zheng, K., Slaymaker, I. M., Gootenberg, J. S., & Zhang, F., (2017). Cas13b is a type VI-B CRISPR-associated RNA-guided RNase differentially regulated by accessory proteins Csx27 and Csx28. *Molecular Cell*, *65*(4), 618–630.

Song, G., Jia, M., Chen, K., Kong, X., Khattak, B., Xie, C., & Mao, L., (2016). CRISPR/Cas9: A powerful tool for crop genome editing. *The Crop Journal*, *4*(2), 75–82.

Sreenivasulu, N., & Fernie, A. R., (2022). Diversity: Current and prospective secondary metabolites for nutrition and medicine. *Current Opinion in Biotechnology*, *74*, 164–170.

Srivastava, R., Mishra, N., Agarwal, S., & Mishra, N., (2019). Pharmacological and phytochemical properties of kaitha (*Feronia limonia* L.): A review. *Plant Archives*, *19*, 608–611.

Stephens, J., & Barakate, A., (2017). In: Thomas, B., Murray, B G., &Murphy, D. J., (eds.), *Gene Editing Technologies–ZFNs, TALENs, and CRISPR/Cas9, Encyclopedia of Applied Plant Sciences* (2nd edn., pp. 157–161). Cambridge, MA: Academic.

Tiozon, R. J. N., Sartagoda, K. J. D., Fernie, A. R., & Sreenivasulu, N., (2021). The nutritional profile and human health benefit of pigmented rice and the impact of post-harvest processes and product development on the nutritional components: A review. *Critical Reviews in Food Science and Nutrition*, 1–28.

Urnov, F. D., Rebar, E. J., Holmes, M. C., Zhang, H. S., & Gregory, P. D., (2010). Genome editing with engineered zinc finger nucleases. *Nature Reviews Genetics*, *11*(9), 636–646.

Vijayalakshmi, S., & Venkatalakshmi, P., (2017). An updated review on the phytopharmacological significance of *Aegle marmelos* (L.). *World Journal of Pharmaceutical Research*, *6*(16), 1390–1422.

Voytas, D. F., & Gao, C., (2014). Precision genome engineering and agriculture: Opportunities and regulatory challenges. *PLoS Biology*, *12*(6), e1001877.

Wamaitha, M. J., Nigam, D., Maina, S., Stomeo, F., Wangai, A., Njuguna, J. N., & Garcia-Ruiz, H., (2018). Metagenomic analysis of viruses associated with maize lethal necrosis in Kenya. *Virology Journal*, *15*(1), 1–19.

Wang, Y., Zhang, Y., Dai, C., Ma, J., Zhou, Y., Jaikishun, S., et al., (2021). The establishment of two efficient transformation systems in quinoa. *Research Square*. doi: 10.21203/rs.3.rs-364280/v1.

Zaidi, S. S. E. A., Mahfouz, M. M., & Mansoor, S., (2017). CRISPR-Cpf1: A new tool for plant genome editing. *Trends in Plant Science*, *22*(7), 550–553.

Zhang, Y., Massel, K., Godwin, I. D., & Gao, C., (2018). Applications and potential of genome editing in crop improvement. *Genome Biology*, *19*(1), 1–11.

CHAPTER 12

BIOSENSORS: RECENT ADVANCEMENTS IN POSTHARVEST MANAGEMENT OF FRUITS

ANKIT KUMAR,[1] RAJAT CHANDEL,[2] RAMANDEEP KAUR,[2] SATISH KUMAR,[3] NEHA GAUTAM,[4] and VIKAS KUMAR[2]

[1]*Center of Food Science and Technology, CCS-HAU, Hisar, Haryana, India*

[2]*Department of Food Science and Technology, Punjab Agricultural University, Ludhiana, Punjab, India*

[3]*Department of Food Science and Technology, Dr. Y. S. Parmar University of Horticulture and Forestry, Nauni, Solan, Himachal Pradesh, India*

[4]*Department of Basic Sciences, Dr. Y. S. Parmar University of Horticulture and Forestry, Nauni, Solan, Himachal Pradesh, India*

ABSTRACT

With one-third of produced food being wasted annually, there's an urgent need for effective post-harvest management strategies. Spoiled food not only contributes to economic losses but also poses serious health risks, including foodborne illnesses responsible for millions of deaths each year. In this context, traditional analytical methods for assessing fruit quality are limited by their time-consuming nature, expense, and reliance on laboratory settings. This chapter explores recent advancements in biosensor

Bioengineered Fruit and Vegetables: Trends in Postharvest Quality Improvement.
Mohammad Irfan and Pankaj Kumar (Eds.)

technology as a promising solution for real-time, on-field monitoring of fruit quality. Biosensors, leveraging bio-based sensing components, offer rapid and accurate detection of analytes, enabling continuous monitoring throughout the post-harvest process. The chapter discusses the evolution of biosensors, from early enzyme-based designs to modern transduction-based systems incorporating diverse biological and transduction elements. Various types of biosensors are highlighted, emphasizing their potential applications in food safety, including pathogen and toxin recognition, early infection diagnosis, and detection of pathogens during quiescent stages. Furthermore, the chapter explores the role of biosensors in fruit ripening, post-harvest detection of gases, and identification of quality characteristics, underscoring their significance in mitigating post-harvest losses and ensuring food security. While biosensors offer promising solutions, future research must address challenges such as reusability, simplicity, speed, and validation against existing standards to realize their full potential in revolutionizing post-harvest management practices.

12.1 INTRODUCTION

Worldwide, one-third of the total food produced is wasted completely in the previous three decades, generating malvaceous situations of food shortage which ultimately lead to food quality and safety issues among consumers (Jedermann et al., 2014). In this sequence, the WHO has revealed that spoiled food generates around 6.5 billion instances of disease and 4.2 million deaths annually. Additionally, diarrhea, food poisoning, and other diseases/infections are serious human ailments due to the ingestion of rotten fruit items. The devastating part concerning this deterioration is related to inappropriate management throughout production as well as post-harvesting procedures. Besides this, economic progress as well as improvement in standards of living have boosted healthy and fresh fruit consumption throughout all seasons.

Fruit loss during transit is a major concern due to improper handling and management of the regulated environment as well as long transportation. The quality of fruits is determined by various factors such as consumer standards, fruit health, nutritional status, storage stability, etc. (Yousefi et al., 2019). End users primarily focus on the appearance of packaging, flavor, ripeness, aroma, and packaging date indicated on the label when evaluating packaged fruit products. Various methods are available to assess the quality attributes of fruit and its products. Chromatography, spectroscopy, and other laboratory procedures are commonly used in the food sector to evaluate fruit product quality. These traditional optical technologies provide chemical, spatial, and physical information about fruit quality (Wang et al., 2015). However, these traditional analytical methods are laboratory-based, time-consuming, expensive, and require specialized equipment and expertise to yield accurate results. Therefore, there is a need for a fast and highly reliable monitoring system to evaluate fruit quality, enabling on-field and real-time assessment of relevant commodities. Recently, several portable and wireless sensing platforms have emerged, gaining popularity on a global scale (Vigneault et al., 2009). These devices are specially integrated into production systems to monitor soil, crops, diseases, and potential pesticide misuse. On the other hand, wireless sensor and smart packaging systems offer viable alternatives for real-time fruit and product assessment. On-field sensors are used to analyze substances in the field and post-harvest operations, while smart packaging is employed for real-time monitoring of physicochemical changes (Zhai et al., 2022). This chapter will discuss the recent developments in sensing platforms and biosensor technology for monitoring fruit quality.

12.2 BIOSENSOR

The word "biosensor" is a powerful and novel analytical tool that incorporates bio-based sensing components for recognition, with a wide variety of applications such as diagnostics, healthcare, food safety, biomedicine, defense, and environmental monitoring. It operates on the premise of chemical and biological processes, where output signals are proportional to the analyte concentration (Gu & Liu, 2020). The primary objective of biosensors is to provide rapid, real-time, reliable, and accurate detection of the concerned analyte. In ideal conditions, biosensors can monitor reversibly and continuously. A biosensor is an incredibly tiny device capable of sensing and responding to physical stimuli within a range of 1 billionth of a meter (Neme et al., 2021). Dr. Leland C Clark, known as the father of biosensors, introduced the notion of employing a biological sensing element for the detection of diverse analytes. He created an 'enzyme electrode' in 1960 for monitoring glucose levels using an immobilized glucose oxidase enzyme. This paved the way for a wide array of sensors to detect biological components utilizing a range of enzymes, such as urea detection using urease, NADH using NAD+, glutamate dehydrogenase, and lactate dehydrogenase. While these sensors were based on a relatively basic concept, the next generation of sensors includes more sensitive identification components and complex detection techniques. These include appropriate receptors like enzymes, oligonucleotides, antibodies, nanoparticles, cells, and imprinting polymers, along with temperature and pH changes (Mehrotra, 2016). The changes in these elements upon perceiving a signal are monitored through optical, electrochemical, calorimetric, acoustic, piezoelectric, magnetic, and micromechanical transducers (Dhakane et al., 2018).

12.2.1 TYPES OF BIOSENSORS

Biosensors are categorized based on their biological elements (enzymes, antibodies, microorganisms, biological tissue, and organelles) or their transduction element. The change in the physicochemical composition is the basis of transduction, which creates the sensing. Therefore, different types of biosensors are available based on the transducer and biological elements as depicted in Table 12.1, along with the benefits and limitations of biosensors in Table 12.2.

12.2.2 CONSTRUCTION AND WORKING

Biosensors consist of three main components:

i. Bioreceptor: Which recognizes the target analyte.
ii. Transducer: For converting the recognition event into a measurable electrical signal.
iii. Processing device.

The bioreceptor comprises physiologically sensitive elements, i.e., receptors, enzymes, antibodies, nucleic acids, microbes, and organelles. It receives signals originating from the analyte and then sends them to the transducer. The transducer measures the changes happening as a result of the response to the biological component and converts the energy into measurable electrical output. Transducers are categorized as mass (piezoelectric crystals), electrochemical (electrodes), optical (optrodes), and thermistors. The signals from the transducer are captured in the detector

TABLE 12.1 Biosensors: Different Types with Respective Characteristics

Type	Subtype	Characteristics
Immunosensors	Enzyme-based biosensors	Immobilization of enzyme is the basis for their creation, i.e., reproducibility, sensitivity, stability, storage, quick response as well as high selectivity. Analytes are identified by labeling the antibodies with respective enzymes.
Optical biosensors	Fluorescence resonance energy transfer (FRET)-based biosensors.	It is used to quantify the protein-protein interactions as well as bio-molecular dynamics between DNA and protein. It also analyzes the conformational changes in proteins.
–	Surface plasmon resonance biosensors.	In this, reflected light with a variety of refractive indices produces surface plasmon resonance which alters the matrix turbidity due to altering refractive index on another side of the metal film in which bio-element is settled in an immobilized state.
–	Colorimetric biosensors	It is an appealing optical method, enables easy recognition of the pathogens by changing color.
Electrochemical biosensors	Amperometric biosensors	It is a universal method used for pathogen detection which utilizes the electrochemical reactions to analyze along with continuously monitoring the current shift.
–	Potentiometric biosensors	It persuades the use of electrodes (ion-selective) to transcript the biological reaction induced by pathogens and convert them into an electrical signal which is then transduced.
–	Impedimetric biosensors	Due to metabolism of microbes, ionic compounds and inert substrates of acidic products, i.e., lactate, amino acids, and acetates are which is then sensed by produced impedimetric biosensor due to alterations in conductivity of microbial environment.
Mass-sensitive biosensors	Piezoelectric biosensors	The sensor's surface is coated with a selective binding agent (e.g., antibodies) in which the solution having bacteria is mounted. Bacteria deceases the oscillation frequency by binding with antibodies.
–	Magnetoelastic biosensors	The increase in mass of analyte induces a modification in the resonance frequency when a target analyte is in direct contact with the alloy sensor surface.

Source: Mathushika (2021).

TABLE 12.2 Benefits and Limitations of Biosensors

Benefits	Limitations
• Selection and use of bioreceptor is of key importance for identification.	• Low repeatability and its accuracy decrease with time.
• Highly effective in estimation of food constituents.	• Heat sterilization is not possible.
• Empower growers to undertake their own quality control, rather than out-sourcing.	• Stability of biological material varies with environmental conditions.
• Faster, cheaper, and reliable.	• Intoxication of cells due to molecules diffusing.

Source: Dhakane et al. (2018).

and then transmitted to a microprocessor. Subsequently, the data obtained is transferred to the output where it is displayed or stored (Swarnapriya & Gurve, 2020).

The basic biosensor framework includes a substrate, i.e., silicon, glass, and polymers (polydimethyl siloxane, polymethyl methacrylate, etc.) coated with a conductive layer, i.e., silicon dioxide, polysilicon, metal like gold, silicon nitride, and metal oxides. Highly sensitive sensors (e.g., thickness shear mode and immunosensor) may be produced employing piezoelectric materials, i.e., quartz crystal, lithium niobate, potassium sodium tartrate, etc., as a substrate, linked with electromechanical detectors (Dhakane et al., 2018). As mentioned in the previous section, biosensors rely on biologically active elements that are very selective and there is a possibility of customizing the precise interaction between the sensor and the analyte (Strano et al., 2022). It is vital to display rapid reactions and real-time assessment of the target molecule to work well. The qualities of a prospective biosensor own to its repeatability, reliability, and linearity. False-negative findings will not be appreciated and there should also be negligible false positive outcomes. Good biosensors are mainly automatic devices that involve minimum input and possess a simple interface. It should be comfortable to use, low-priced, portable, and small in size (Mathushika, 2021). Nowadays, the Use of a typical smartphone for the detection of analytes inside the food that controls reporting criteria and food recalls is very popular in European countries. Platform tools are emerging for linking consumers to relevant technology, although these initiatives are not yet widespread in other parts of the world (McLamore et al., 2021).

12.2.2.1 NANOMATERIALS AS BIOSENSORS

Recently, nanomaterial-based biosensors have been engineered because of their exceptional physical and chemical features. Apart from the typical bio-elements, nanomaterial-modified bio-elements can greatly increase the stability, specificity, and sensitivity of biosensors. On the other hand, the huge surface area of a nanomaterial can encapsulate more biomolecules to increase the number of reaction sites (Xu et al., 2009). Moreover, the use of nanomaterials for biosensors provides variable detection limits based on the samples to be studied and helps the tweaking of the level of sensitivity according to the needs. For instance, silver nanoparticles (AuNPs) display several intriguing features, including the promotion of the electron transfer after modification of the electrode and the production of diverse ultraviolet-visible (UV-Vis) absorption spectra in its aggregation state (Barthelmebs et al., 2010). In consideration of another instance of the spoiling of onions, the origin of the sensing resides in the alteration of optical characteristics of innovative silver NPs owing to the integration of volatile sulfur. When spoilage is exaggerated, the organosulfur compounds (unstable) in the onion are exposed and rearranged to create a highly stable complex with AgNPs that finally leads to color changes. Additionally, the propensity of AgNP's trustworthiness in terms of antibacterial action will undoubtedly be proven as a better alternative for food safety. Consequently, nanomaterial-based biosensors possess vital scope in the detection of post-harvest rotting in onions (Sachdev et al., 2016). Therefore, the utilization of nanomaterials to produce new biosensors can provide a substantial way for their applications in foodborne pathogen detection (Kaittanis et al., 2010). Detection and management of pathogens in food are critically crucial during every phase in the farm-to-table continuum. Hence, Nanomaterials should be carefully selected according to the intended effects (Zhang et al., 2020).

12.2.3 E-TONGUE AND E-NOSE

In the present scenario, e-nose and e-tongue have been employed in the food sector, which is an artificial way of olfaction. Technically, these are the collaborative approach to electronics, biochemistry, and artificial intelligence (AI) (Wasilewski et al., 2019). The e-nose is simply a gas multisensory machine that assesses the complete gas mixture instead of specific volatiles.

The sensing array (optical and electrochemical sensors), signal processing unit, data collecting system along with pattern recognition unit are the basic units of e-nose. The mechanism of e-nose is defined as when any volatile chemicals (from microbial, post-harvest, etc.) react with the biorecognition component on the multisensory array of e-nose, and thereby, the pattern recognition unit assesses the particular chemical concerning their database (Karakaya et al., 2020). Recently, nanoparticles, i.e., carbon nanotubes (CNTs), tin oxide, silicon, and titanium nanowires, etc., are being utilized to engineer e-nose sensing arrays as they possess high surface/volume ratio, microelectronic processing, and high sensitivity toward the analyte (Chauhan et al., 2020). However, the functionality of the e-nose technology relies on construction material, specifically, the part which is used to predict the alterations in the voltage signals under the effect of the target analyte. However, the calculations and maintenance of customized gas chambers possess a range of operational problems which often result in flawed output. Therefore, there is an urge to formulate cost-effective as well as compact techniques as sensing tools (Sachdev et al., 2016).

Similarly, e-tongue is also a piece of multimodal equipment for the quick qualitative analysis of a varied range of liquid foods which is based on technology similar to e-nose (Yadollahi et al., 2010). Among multiple uses of e-tongue, it offers significant promise to evaluate the shelf stability of a wide range of products. It also provides a rapid as well as objective assessment system for assessing the fruit quality, particularly, it can differentiate various ripening stages and can also trace internal quality changes (i.e., TSS, ascorbic acid, total acid, and sugars) (Strano et al., 2022).

12.3 APPLICATIONS

12.3.1 FOOD SAFETY

According to the WHO, over 1.5 million fatalities are reported due to diarrheal illnesses yearly. These fatalities are related to the ingestion of infected food. In this sequence, a substantial amount of the budget of most underdeveloped countries is wasted due to food-borne illnesses. From the above instance, it is clear that food safety and security are visualized as important issues that need urgent assessment in developing nations to overcome social and economic erroneous conditions (Von Sonnenburg et al., 2000). Technically, any food-borne illness arises when contaminated food (with the pathogen of their respective toxin) is consumed by an individual. The toxin/pathogen possesses the potential of inducing a disease known as a 'hazard.' The identification and characterization of hazards are important steps in the process of microbiological food safety and risk assessment. There are several molecular approaches for the identification and characterization of microbes, such as PCR, Fluorescence microscopy, hybridization, etc. Almost all studies concerning microbial ecology are based on ribosomal DNA/RNA sequences. Specifically, rRNA detection relies solely on pure bacterial cultures, making it incompatible in all situations. Therefore, rapid detection and real-time approaches are essential for improved quality assessment of fruits and fruit products. The main objectives of these approaches include enhancing detection specificity, broad-scale application, reducing analysis time, and lowering resource requirements compared to molecular methods. Automated approaches play a significant role in industrial microbial investigations by enabling high-throughput analysis of numerous samples (Table 12.3) (Maukonen et al., 2003). Therefore, future research should focus on developing biosensors capable of operating in real-time scenarios (Aggarwal et al., 2021).

The measures in food security can be classified into two categories: postharvest loss and food biosecurity. Food biosecurity denotes food contamination and destruction by natural, unjust economic gain, warfare, or exacting retribution. On the other side, postharvest loss emphasizes the nutrients and edible conditions in food that need to be conserved by technologies between the harvest time and the consuming moment. Since the duration spans from minutes to years, the technologies concentrating on postharvest loss reduction are crucial in conserving and lowering the loss (Neethirajan et al., 2018). Therefore, reliable and rapid analytical methods are increasingly needed in the food business for the determination of particular chemical components in food items. The demand derives from growing regulatory action and heightened consumer awareness about food composition and safety. The food and drink industry needs speedy and economical solutions for quality control as existing wet chemistry and analytical practices are time-consuming and may need highly specialized people and expensive equipment. Also, the rising need for online assessment of food components leads research and business to hunt for practical solutions, and biosensors may be a potential.

TABLE 12.3 Application of Biosensors

Product	Analyte	Conventional Approach	Biosensors	Benefits	References
Fruits	Formalin	HPLC, GC-MS, NIR, e-tongue, E-nose, spectroscopy.	ATP bioluminescence biosensor.	Cheap Simple, and reliable technique, doesn't require sample purification.	Aini et al. (2016)
Vegetables	MSG	HPLC	Amperometric glutamate biosensor	Good response reproducibility and l stability.	Soldatkina et al. (2017)
Juices and nectar	Glucose	HPLC	Conductometric glucose oxidase-based biosensor.	Good storage stability and reproducibility.	Dudchenko et al. (2016)
Climacteric fruits	Artificial ripening agents	GC-MS, HPLC	Electrochemical, Pt/CeO_2/AChE bioelectrode.	Good storage stability, reproducibility, and sensitivity.	Janssen et al. (2014)
Fermented beverages	Authentication	GC	Optical ethanol biosensor.	Good reproducibility as well as storage stability.	Kuswandi et al. (2014)
Fruits and vegetables	Organophosphates and carbamates	GC, HPLC	Colorimetric biosensing method using acetylcholinesterase (AChE).	Good reproducibility	Sun et al. (2017)
Fruits and vegetables	Genetically modified crops	PCR and DNA extraction methods.	Colorimetric biosensor	Label-free detection of bacterial, rapid analysis, sensitive simple, and onsite testing.	Jung et al. (2015)
Apple	Quality assessment of packaged juices.	Spread plate method, HPLC.	Colorimetric biosensor.	Non-invasive as well as fast technique.	Rai et al. (2019)
Citrus	Flavonoids	HPLC	Amperometric DNA biosensor	Rapid response time, precise, good sensitivity.	Ensafi et al. (2016)

12.3.1.1 BIOSENSORS FOR PATHOGEN OR TOXIN RECOGNITION

Apart from pathogens directly, toxins produced by them cause severe harm to the shelf life of fruits and fruit products. Various forms of biosensors are highly prevalent and are used in recognition of an assortment of potential toxins created by them. To be sure about safety as well as product quality in the food processing industry in a rapid, precise, and easy way, biosensors are recognized as vital alternatives to traditional approaches. The incorporation of biosensing elements in packaging material is extremely prevalent. In the majority of situations, antibodies are bonded to the packaging surface to detect pathogens or toxins; as a consequence, when antibodies come into contact with pathogens, a visual signal is shown to inform the customer (Joshi et al., 2021). There are several more uses of nano-biosensors which include virus identification employing antibody sensor arrays on self-assembled nanoscale block copolymer patterns, and the detection of food-borne toxins using multifunctional nanoparticles. It is also utilized in the detection of pore-forming toxins, and molecularly imprinted polymers for plant and insect virus identification (Kumar Upadhyay et al., 2022). However, this method requires further study to become commercially accessible. The development of biosensor platforms for fungal diagnoses in crops might change the postharvest management of fruits and could result in the introduction of tiny, cost-effective, and sensitive equipment for on-site usage (Harpaz et al., 2022).

12.3.1.2 EARLY DIAGNOSIS OF INFECTIONS

Deploying spectroscopy (NIR, Raman spectroscopy, NMR, etc.) and image-based techniques coupled with image processing subunits facilitates correlating the optical properties of the analyte with various quality parameters (Singh et al., 2012). Considering these aspects, early diagnosis of a pathological condition offers several advantages, such as early detection for planning efforts to limit colonization, ultimately reducing postharvest losses (PHLs). In recent decades, numerous breakthroughs have been reported regarding the swift and accurate identification of concealed fungal colonization in fruit under field conditions, during handling, and subsequent storage. These breakthroughs have driven the development of biosensors for detecting fungal attacks before visible signs of infection appear in fruits (Vinet & Zhedanov, 2011).

12.3.1.3 DETECTION OF PATHOGENS AT THE QUIESCENT STAGE

Fungi from numerous genera are responsible for the major postharvest deterioration of fruits. After entering the young fruit or its parts, these fungi can stay in a quiescent state, and then they transition to the pathogenic state during storage. These quiescent attacks are extremely small and cannot be identified visually during post-harvest processes (Prusky et al., 2013). Keeping these things in view, the demand for sensitive, economical, and reliable devices is significant, which will also assess the presence of microbes during the quiescent period. For instance, *Colletotrichum gloeosporioides*, the most predominant disease agent in harvested fruit, possesses three lifecycle stages after its conidial germination, i.e., penetration, latent, and finally the necrosis stage, which ultimately causes deterioration. In the initial stage, it penetrates the cuticle of the fruit and maintains itself in a quiescent condition until ripening. After ripening, it shifts to the necrosis stage, which ultimately results in the anthracnose of the fruit. During these different phases, the fungus considerably up regulates its stage-specific genes (Alkan et al., 2015). The mRNA transcript sequences of this fungus are utilized as indicators of pathogenicity. Hence, a biosensor based on RNA detection can be an easier, cheaper, faster, and approach to assessing the diseases in fruit crops (Zamir et al., 2020).

12.3.2 BIOSENSORS IN POST-HARVEST MANAGEMENT OF FRUITS

Post-harvest handling is a complex combination of processes that involve various operations from harvesting to packaging. The demand for accurate and cost-effective technologies to evaluate product quality is continuously increasing. In the food industry, product quality is assessed based on both internal and external characteristics. Internal quality is influenced by factors such as nutrient profile, TSS, acidity, pH, enzymes, etc., while external quality is determined by attributes like smell, color, texture, etc.

Internal quality features are crucial factors that customers still prioritize. Biosensors offer a promising solution to meet this demand. Biosensors have been developed to address specific requirements related to fresh and processed fruits (Kuswandi et al., 2014).

12.3.2.1 BIOSENSORS IN FRUIT RIPENING

Fruits and vegetables are very perishable commodities due to their poor shelf life. Climacteric fruits, such as mangoes, tomatoes, bananas, and avocados, are harvested just before ripening begins because shipping ripe fruits is not feasible. In contrast, non-climacteric fruits do not ripen after harvesting. Therefore, they must be carefully harvested to ensure optimal quality. This knowledge helps producers with long-term storage and transportation of produce for proper marketing and consumption. Often, to meet market demands and reap benefits, fruits are picked prematurely. However, the quality of such fruits is unsatisfactory due to inconsistent ripening and unpleasant flavor. Typically, these products are ripened using various artificial chemicals, leading to deception among consumers. To earn more benefits, farmers and marketers usually cater the fruits with the injudicious concentration of artificial ripening chemicals. This negligence leads to the deposition of hazardous chemicals as well as their residues in fruits, which are highly detrimental to human health (Ahmad & Siddiqui, 2015). For the evaluation of the health impacts of these compounds, numerous research has been undertaken on the mechanism as well as their effects on fruit quality. Artificial ripening mediators, i.e., calcium carbide, form acetylene by reacting with water. This acetylene further enters the fruits through the diffusion process; it majorly functions as ethylene acts in natural ripening. The industrial-grade calcium carbide also contains some traces of phosphorous hydrides and arsenic, which are highly dangerous to the workers associated with this operation (Islam et al., 2016). On the other hand, the use of ethylene is considered safer for human consumption; however, they are not budget-friendly. Therefore, pocket-friendly chemicals, i.e., ethephon, calcium carbide, and ethylene glycol, are utilized for the artificial ripening process, despite their potential harms. Besides this, various other chemicals, i.e., oxytocin, are used in fruit markets to instantly increase the size of fruits in the field.

Calcium carbide is a potent carcinogenic chemical and is banned in most countries except for industrial use. It poses various health risks, such as inflammation, allergies, kidney failure, and abnormalities in hematological markers (lymphocytes, WBC, and granulocytes) (Asif, 2012). Due to its alkaline nature, calcium carbide is a mucosal irritant in the abdominal region and causes irritation in the eyes, nose, mouth, throat, lungs, and pulmonary edema (Seo et al., 2012). On the other hand, Ethephon is a plant growth regulator not recommended for ripening various fruits (Cunha et al., 2018). Similarly, oxytocin, a veterinary drug, is not advised for fruits (Chandravathi et al., 2016). Therefore, the evaluation of artificial ripening chemicals is crucial as volatile residues in some cases make detection challenging. Current methods like HPLC and GC-MS provide standard measurements but are time-consuming and highly specific to certain compounds. The need for rapid, specialized, and highly sensitive techniques for this issue can be met through biosensors. Researchers have developed a biosensor to detect calcium carbide in mangoes by inhibiting AChE using a combination of calcium carbide and ethylene, which is detected by an amperometric sensor (Ramachandra et al., 2016).

Naturally ripened fruits differ significantly from artificial ones in terms of nutrients, titratable acidity (TA), TSS, and sensory qualities (Dudchenko et al., 2016). Residues left on the fruit surface during artificial ripening can also be evaluated as an indication of ripening. Biosensors can be integrated to detect the gases released during the decomposition of chemicals in the ripening process. Therefore, extensive research is required to develop various methods to maintain fruit quality and extend the shelf life of perishable fruits using different biosensors.

12.3.2.2 POST-HARVEST DETECTION OF GASES IN FRUITS

To minimize fruit rotting during transit and maintain marketable quality, fruit goods are delivered in specialized preservation chambers to delay gaseous release and subsequent ripening. In these specialized structures, fruits are subjected to a controlled level of oxygen (2–3%), high humidity, low temperature, and moderate CO_2 levels (Jochum et al., 2016). Since ethylene enhances the respiration rate and causes several degradative changes (Gao et al., 2020), its concentration needs to be negligible. Monitoring ammonia is also a major concern due to health issues and its potential harm to stored fruits (Yang & Jiao, 2017). Recent research has shown infrared thermal emission-based sensing devices for ethylene detection in banana fruit samples. Integration of IR emitters with silicon temperature sensors and housing the system in a gas test cell are providing synergistic results. When released ethylene is placed in the test cell, the target molecules absorb IR light, which lowers the temperature and is subsequently sensed by appropriate sensors (Wei et al., 2015).

Some researchers have employed the dielectric property of SnO_2 nanoparticles to detect ethylene in fruits. The SnO_2 with Pd NPs displayed greater conductivity with the analyte and altered the depletion area. This platform demonstrated a reduction in capacitance due to an increase in ethylene concentration, which can be employed for RFID tags to assess the quality of fruits (Agarwal et al., 2012). In another work, using polystyrene and polyhexylthiophene organic field-effect transistors (OFET) were created to detect ammonia in fruit storage (Han et al., 2016). In this platform, the lone pair of ammonia (diffused) reacts with the sensors' platform, resulting in a decrement in current which lowers the net positive charge. Therefore, a reduction in current was achieved due to enhanced ammonia concentration (Du et al., 2019).

12.3.3 RECOGNITION OF QUALITY CHARACTERISTICS OF FRUITS

A dish is considered of superior quality if its internal, as well as exterior constituents, are examined adequately. Varied kinds of food possess different features to be assessed. In fruits, shape, size, scent, texture, taste, brightness, juice content, etc., are major sensory qualities. Determining the internal quality characteristics demands sophisticated chemical techniques. With the development of biosensing platforms for the detection of various quality characteristics, various post-harvest operations have become easier. In continuation of this, numerous research investigations are being undertaken to assess fruit quality indicators, i.e., flavonoid content of citrus, sugar content in juices, etc. (Dudchenko et al., 2016). Among all, freshness is considered the major quality indicator which decides its value in the marketplace. Therefore, biosensors are necessary to detect quality swiftly, in real-time. *In situ* and with better precision (Kaur, Kumar, & Verma, 2014). Therefore, a variety of biosensors needed to be designed for the assessment of various quality attributes with great stability, selectivity, and reproducibility (Soldatkin et al., 2013).

12.3.3.1 DETERMINATION OF PH AND TITRATABLE ACIDITY (TA)

The titratable acidity (TA) and pH are recognized as crucial quality features for fruits, where pH progressively increases in due course of ripening, and TA decreases notably. This acidity in fruits is the function of reducing sugars and organic acids (Batista-Silva et al., 2018). Malic acid is found in pear, apple, loquat, etc., whereas citrus fruits are rich in citric acid (Etienne et al., 2013). In the biological system, intermediate products of the TCA cycle play a key role in fruit ripening which alters the level of different substrates in fruits, thus, enzyme-based sensors are key devices for assessment of the qualitative changes in fruits. Researchers have engineered an enzyme-based electrochemical sensor to detect ascorbic acid, D-glucose, etc., in fruits (Jawaheer et al., 2003). The researchers further employed a membrane made of cellulose acetate to design an enzymatic sensor where hydrogen peroxide (H_2O_2) and cellulose acetate were produced with the virtue of enzymatic reactions. In successive stages, they developed a paper-based microfluidic chip device (μPAD) to assess the glucose in fruits (Li et al., 2017). The researchers have also devised an e-tongue to assess the quality characters of apricots (Kantor et al., 2008). They examined the influence of 1-MCP apricot ripening under storage conditions.

According to studies, as the hardness, TSS, and pH of the kiwifruit under cold storage conditions were lowered, the loss of aromatic compounds of volatile nature was also observed. In this sequence, the Chinese scientists' team used PEN2-based e-nose integrated with a metal oxide sensor to assess the quality attributes of strawberry juice (Qiu et al., 2015). E-tongue arrays were also employed in the experiment to examine the acidity as well as other quality attributes of strawberry juice. Similarly, metal oxide sensing devices were used to formulate the parameters for quality assessment in stored bananas (Sanaeifar et al., 2016). These devices were used to discriminate the banana fruits based on pH, TSS, etc., at different post-harvest stages. For real-time monitoring of the quality of grapes, the Indonesian researchers' team developed a colorimetric pH-based sensor (Kuswandi & Murdyaningsih, 2017). In this, when grapes are subjected to chlorophenol red, then, the sensitive chlorophenol red corroborated the changes in pH, TSS, etc., and color change was immediately noticed in my eyes.

12.3.3.2 DETERMINATION OF CARBOHYDRATES

Carbohydrates are important and diverse biomolecules involved in a wide range of physiological processes, i.e., growth and development, protein trafficking, cell differentiation, cell communication, and signal transduction (Chevolot et al., 2010). Several microbial biosensors, e.g., employing *E. coli* K12 mutants, *S. cerevisiae*, and *Gluconobacteroxydans* have been designed for measuring CHO in fruit samples. A microbial sensor based on *Pseudomonas fluorescens* and an oxygen electrode was employed for glucose measurement. The biosensor was produced for the detection of lactose and sucrose utilizing glucose oxidase connected to the microbial cell surface by concanavalin A or polyethyleneimine (D'Souza, 1989). A technique was devised to protect immobilized glucose oxidase against low pH deactivation (pH_3) using a cellulose acetate membrane coated with Tween-80, thereby expanding the range of glucose detection in fresh fruits including citrus fruits. Carbohydrates are found within the cell, on the cell surface, and in the extracellular matrix. Any irregularity in the glycosylation pattern may be exploited as target analytes for the detection of various disorders too, including Alzheimer's disease and malignancies (Zhao et al., 2020).

12.3.3.3 RECOGNITION OF AROMA IN FRUITS

Aroma/flavor is the sensation of aromatic molecules by olfaction or taste buds. In this process, aldehydes, phenolics, and triterpenoids offer bitterness while sugar (soluble) and acids (organic) are responsible for sweet as well as sour tastes (Zhu et al., 2020). Sensing of aroma compounds is challenging to examine due to the mixing of different volatile molecules as well as various perceived chemical sensations. For real-time and on-field sensing, a Malaysian research team designed an MIP-based QCM sensor to detect the volatile compound in mango during ripening (Hawari et al., 2013). The team assessed the fruit freshness with better sensitivity and higher selectivity concerning target volatile analytes. They also established an MIP-based trimethoxy silane QCM sensor to selectively detect furaneol in preserved mango (Ghatak et al., 2019). This sensor displayed higher selectivity and better resolution for furaneol. In another experiment, workers created a highly sensitive e-nose for the detection of 1-MCP along with calcium coated on apple fruit (Kantor et al., 2008). 1-MCP and calcium are used in combination to lower fruit ripening. Recent studies combine e-nose and sensors to increase the sensitivity and gain reliability in the detection of aroma compounds (Pudake & Jain, 2021).

12.3.4 RECOGNITION OF GENETICALLY MODIFIED (GM) CROPS

Recently, genetically engineered products are also entering the marketplace with a promise of greater quality and production. This places an additional duty on producers to maintain food safety. Various solutions have been provided by workers to address problems of food safety arising from GM foods, with biosensors playing an important role in the screening of GM foods. GM foods can be distinguished based on their diverse biochemical characteristics using multiple approaches. Among these, a colorimetric biosensor was recently developed to detect GM food using a rapid and sensitive approach (Jung et al., 2015). Portable biosensing devices, such as biosensors, have been proven as an ideal solution for fruit assessment to eliminate detrimental substances from fruits. Biosensors are anticipated as the best technology to monitor important parameters crucial for the modern food business and to maintain the supply chain of fruits.

12.3.5 MIDENTIFICATION OF ALLERGENS AND TOXINS

Food poisoning is a sensitive aspect concerning health. All around the world, many cases of toxicity can be observed, such as cyanide from cassava, allergies from peanuts, aflatoxin, dioxin, etc. Acute food toxicity and contamination have proven to be serious problems for human health, leading to long-lasting impairment and potentially death. In cases of food contamination, sewage and polluted air are possible reasons for the easy contamination of fruits and fruit products. PCR and plate count approaches have been considered viable means to assess contamination and toxicity. However, DNA-based biosensors have surpassed traditional approaches due to their quick, easy, and precise detection procedures. Recently, the quality of infant food contaminated with *Bacillus Cereus* was assessed using a DNA-based biosensor (Izadi et al., 2016). On the other hand, a biosensor designed to detect *E. coli* in drinking water was found to be quicker and highly sensitive, allowing for convenience in field investigations (Ettenauer et al., 2015).

12.3.6 BIOSENSORS IN FRUIT PRODUCTS

12.3.6.1 FERMENTED FOODS AND ALCOHOLIC BEVERAGES

Fermentation of food is a frequent means of its preservation, development of new types, and augmentation of its nutritional contents (Barthelmebs et al., 2010). Production of diverse fermented foodstuffs and drinks demands a carefully conducted fermentation operation system, therefore highlighting a requirement for sensitive, precise, and economical techniques for quality control of these goods and managing the fermentation process (Kim et al., 2009). Moreover, there exists a considerable demand for quick and economical detection of numerous components of drinks together with the food-borne and water-borne diseases, toxins, and pesticide residues with high specificity. In this regard, biosensing technology provides a novel quick, and efficient analytical strategy for monitoring food quality and safety in the food and bioprocessing sectors. Inherent features of immunosensors, i.e., sensitivity, specificity, speed, ease, and on-site analysis can be used for various applications. Enzyme and microbial-based biosensors are commonly employed in the beverage industry. Much research attention has been paid to the development of biosensors for ethanol because of its importance in food fermentation process as a main or by-product and for its toxicity. For measurement of ethanol, a reagent-less alcohol dehydrogenase (ADH2)-based biosensors and microbial biosensors based on immobilized cells of *Candida tropicalis/Saccharomyces ellipsoideus* were created. A potentiometric oxygen electrode based on immobilized yeast cells (*Saccharomyces ellipsoideus*) was successfully used for the selective determination of ethanol in alcoholic beverages. The response for ethanol takes 7 minutes for the steady-state method and 2 minutes for kinetic measurements. A linear calibration was achieved over more than two decades of ethanol concentration. Determination of the total and natural antioxidant capacity of the red and white wines was carried out using a superoxide dismutase immobilized on the H_2O_2 electrode. Researchers also measured polyphenol, sulfite, and ascorbic acid contents of samples using a tyrosinase biosensor, a sulfite oxidase biosensor, and an ascorbate oxidase biosensor, respectively (Akyilmaz & Dinçkaya, 2005).

12.3.6.2 FRUIT JUICES, FRESH FRUITS, AND VEGETABLES

Isocitrate analysis of fruit juices was conducted using a CO_3^{2-} selective electrode and isocitric dehydrogenase enzyme immobilization in flow injection analysis. The isocitrate biosensor showed a linear relationship between the potential difference and logarithmic isocitrate concentration. The interfering effect of major sugars and organic acids on the sensor system was below 5%. There was no significant difference in isocitrate concentrations of the fruit juices measured by the biosensor and GC. Detection and quantification of pyruvate in juice from macerated onion tissue were examined using an amperometric biosensor to assess pungency in onions. A strong correlation was observed between the biosensor response and known pyruvate concentrations in onion as determined by the modified colorimetric assay (Dhakane et al., 2018).

12.3.6.3 BIOSENSORS ON RECOGNITION OF FOOD ADDITIVES

The major difficulty in post-harvest of fruits is their short shelf life, which requires careful management as it is the major source of repeated post-harvest fruit losses. The limited shelf life of fruit is associated with their higher water activity and appropriate nutrient profile for microbial proliferation. Therefore, several food additives are utilized to maintain the quality characteristics of fruits and prevent the microbial degradation of fruits, ultimately enhancing the shelf life. They are also essential in packaging and shipping operations. Food additives serve various purposes, such as prolonging the shelf life of food, enhancing flavor, and imparting an attractive appearance to foods. It is important to use them within the maximum allowable limits set by the relevant authorities (Verbeke & Pieniak, 2006). Improper use of these chemicals can pose a serious hazard to human health. Flavor enhancers like glutamate (MSG) are commonly used in the food processing industry. However, it is crucial to use these additives within prescribed limits for good health, as overuse may lead to significant health risks. Therefore, biosensors are employed to detect excessive usage of preservatives, glutamate, and other agents in food (Ramachandra et al., 2016).

12.3.7 FOOD PACKAGING AND STORAGE

Management of fruit gluts is also very challenging in the marketplace concerning food sustainability. The overproduce must be processed or stored to curtail food waste. The statistics for fruit wastage range from 20% to 50%. The packaging of stored and processed food plays a major role in reducing post-harvest losses to a great extent. Traditionally, farmers used to store the produce in their fields before sale. Therefore, the microenvironment of these structures needed to be regulated and monitored continuously for the prolonged shelf life of produce (Kuswandi et al., 2011).

In this scenario, biosensors are proven to be an essential part of controlled storage as well as intelligent food packaging systems. Package optimization is a must in the present scenario to reduce contamination, avoid moisture loss, and eventually enhance shelf life. Intelligent packaging with a regulated inner gaseous environment has turned towards clever packaging. The ultimate aim of these packages is not only to prolong the shelf life of fruits but also to track the continuously changing internal microenvironment with time. For instance, Intelligent tags are placed within the package to assess the degradation of goods inside the package. These tags can be directly connected with biosensing elements that provide visual suggestive information to the consumer immediately (Saini et al., 2016). These tags particularly sense through a polymer film which is then connected to the biosensor and radio frequency identification components (RFID). The inner environment of the food packet consists of various gases which are detected by a molecular recognizing material. RFID tags can be connected directly to the food packages for regular assessment of packaged food. A transponder, an antenna for receiving signals, a microchip for data storage, and a reader are part of an RFID chip device.

Biosensors in food packages operate within specific conditions that are limited to the microenvironment of the package. Despite their various applications, they have certain limitations related to practical considerations. These devices need to be designed to function across different temperature ranges, light exposure levels, humidity levels, and concentrations of gases. Moreover, they must provide fast, reliable responses with high throughput to enable real-time identification of deteriorating packages. Several biosensors have been developed for this purpose, but none have been commercialized due to their limited reversibility and sensitivity. In addition to packaging, intelligent food logistics is gaining traction in global markets for long-distance shipments. Currently, only a few basic cold chain systems offer real-time monitoring of the microenvironment during fruit transportation. Monitoring ethylene levels, an indicator of ripening, and assessing mold infections by determining volatile components are also essential in food logistics during shipments (Kundu et al., 2019).

12.4 CONCLUSIONS AND FUTURE PERSPECTIVES

Conventional sensing techniques assess the quality of fruits in bulk. Therefore, intrinsic quality variables were non-negotiable for assessment. Biosensors have emerged in the global market to bridge the technology gaps in conventional and modern approaches. However, biosensor research is still lacking in several aspects that need to be addressed in the near future. A self-cleaning system must be developed for reusability and to extend the useful life of biosensors. Simpler construction with faster component determination is crucial. Research on the applications of microfluidics technology in biosensors for real-time, label-free, highly sensitive, and multiplexed biomolecular detection in a fully automated form should be conducted. The results from new generation biosensors must be thoroughly validated against existing standards to obtain commercial approval.

KEYWORDS

- **biosensors**
- **conventional sensing**
- **ethylene**
- **global market**
- **long-distance shipments**
- **microenvironment**
- **real-time monitoring**

REFERENCES

Agarwal, M., Balachandran, M. D., Shrestha, S., & Varahramyan, K., (2012). SnO_2 nanoparticle-based passive capacitive sensor for ethylene detection. *Journal of Nanomaterials, 2012*. https://doi.org/10.1155/2012/145406.

Aggarwal, S., Sehgal, S., & Tandon, A. W., (2021). Biosensors-types and application in food processing industry. *Journal of Postharvest Technology*, *2021*(1), 1–19. www.jpht.in (accessed on 7 February 2024).

Ahmad, M. S., & Siddiqui, M. W., (2015). Factors affecting postharvest quality of fresh fruits. *Postharvest Quality Assurance of Fruits*, 7–32. https://doi.org/10.1007/978-3-319-21197-8_2.

Aini, B. N., Siddiquee, S., & Ampon, K., (2016). Development of formaldehyde biosensor for determination of formalin in fish samples; Malabar red snapper (*Lutjanus malabaricus*) and longtail tuna (*Thunnus tonggol*). *Biosensors*, *6*(3), 32. https://doi.org/10.3390/bios6030032.

Akyilmaz, E., & Dinçkaya, E., (2005). An amperometric microbial biosensor development based on *candida tropicalis* yeast cells for sensitive determination of ethanol. *Biosensors and Bioelectronics*, *20*(7), 1263–1269. https://doi.org/10.1016/j.bios.2004.04.010.

Alkan, N., Friedlander, G., Ment, D., Prusky, D., & Fluhr, R., (2015). Simultaneous transcriptome analysis of *Colletotrichum gloeosporioides* and tomato fruit pathosystem reveals novel fungal pathogenicity and fruit defense strategies. *New Phytologist*, *205*(2), 801–815. https://doi.org/10.1111/nph.13087.

Asif, M., (2012). Physico-chemical properties and toxic effect of fruit-ripening agent calcium carbide. *Annals of Tropical Medicine and Public Health*, *5*(3), 150. https://doi.org/10.4103/1755-6783.98602.

Barthelmebs, L., Calas-Blanchard, C., Istamboulie, G., Marty, J. L., & Noguer, T., (2010). Biosensors as analytical tools in food fermentation industry. *Advances in Experimental Medicine and Biology*, *698*, 293–307. https://doi.org/10.1007/978-1-4419-7347-4_22.

Batista-Silva, W., Nascimento, V. L., Medeiros, D. B., Nunes-Nesi, A., Ribeiro, D. M., Zsögön, A., & Araújo, W. L., (2018). Modifications in organic acid profiles during fruit development and ripening: Correlation or causation? *Frontiers in Plant Science*, *871*, 1689. https://doi.org/10.3389/fpls.2018.01689.

Chandravathi, C., Mahadevan, K., Sheela, S. K. P. K., & Hashir, A. R., (2016). Detection of oxytocin in fruits and vegetables using wireless sensor. *Indian Journal of Science and Technology*, *9*(16), 10. https://doi.org/10.17485/ijst/2016/v9i16/92211.

Chauhan, N., Soni, S., Agrawal, P., Balhara, Y. P. S., & Jain, U., (2020). Recent advancement in nanosensors for neurotransmitters detection: Present and future perspective. *Process Biochemistry*, *91*, 241–259. https://doi.org/10.1016/j.procbio.2019.12.016.

Chevolot, Y., Vidal, S., Laurenceau, E., Morvan, F., Vasseur, J. J., & Souteyrand, E., (2010). Carbohydrates as recognition receptors in biosensing applications. *Recognition Receptors in Biosensors*, 275–341. https://doi.org/10.1007/978-1-4419-0919-0_7.

Cunha, F. A., Cunha, M. Da. C. S. O., Da Frota, S. M., Mallmann, E. J. J., Freire, T. M., Costa, L. S., Paula, A. J., et al., (2018). Biogenic synthesis of multifunctional silver nanoparticles from *Rhodotorula glutinis* and *Rhodotorula mucilaginosa*: Antifungal, catalytic and cytotoxicity activities. *World Journal of Microbiology and Biotechnology*, *34*(9), 1–15. https://doi.org/10.1007/s11274-018-2514-8.

D'souza, S. F., (1989). Potentials of co-immobilizates in biochemical processing: The current state of the art. *Journal of Microbial Biotechnology*, *4*(1), 63–73.

Dhakane, J. P., Lad, M. M., Patil, S. S., & Dawange, S. P., (2018). Applications of biosensors in food analysis. *Beverage and Food World, 45*(9), 27, 28.

Du, D., Wang, J., Wang, B., Zhu, L., & Hong, X., (2019). Ripeness prediction of postharvest Kiwifruit using a MOS E-nose combined with chemometrics. *Sensors (Switzerland)*, *19*(2), 419. https://doi.org/10.3390/s19020419.

Dudchenko, O. Y., Pyeshkova, V. M., Soldatkin, O. O., Akata, B., Kasap, B. O., Soldatkin, A. P., & Dzyadevych, S. V., (2016). Development of silicalite/glucose oxidase-based biosensor and its application for glucose determination in juices and nectars. *Nanoscale Research Letters*, *11*(1), 1–7. https://doi.org/10.1186/s11671-016-1275-2.

Ensafi, A. A., Karbalaei, S., Heydari-Bafrooei, E., & Rezaei, B., (2016). Biosensing of naringin in marketed fruits and juices based on its interaction with DNA. *Journal of the Iranian Chemical Society*, *13*(1), 19–27. https://doi.org/10.1007/s13738-015-0707-8.

Etienne, A., Génard, M., Lobit, P., Mbeguié-A-Mbéguié, D., & Bugaud, C., (2013). What controls fleshy fruit acidity? A review of malate and citrate accumulation in fruit cells. *Journal of Experimental Botany*, *64*(6), 1451–1469. https://doi.org/10.1093/jxb/ert035.

Ettenauer, J., Zuser, K., Kellner, K., Posnicek, T., & Brandl, M., (2015). Development of an automated biosensor for rapid detection and quantification of *E. coli* in water. *Procedia Engineering*, *120*, 376–379. https://doi.org/10.1016/j.proeng.2015.08.643.

Gao, J., Zhang, Y., Li, Z., & Liu, M., (2020). Role of ethylene response factors (ERFs) in fruit ripening. *Food Quality and Safety*, *4*(1), 15–20. https://doi.org/10.1093/fqsafe/fyz042.

Ghatak, B., Naskar, H., Ali, S. B., Tudu, B., Pramanik, P., Mukherji, S., & Bandyopadhyay, R., (2019). Development of furaneol imprinted polymer based QCM sensor for discrimination of artificially and naturally ripened mango. *ISOEN 2019 – 18th International Symposium on Olfaction and Electronic Nose, Proceedings*. https://doi.org/10.1109/ISOEN.2019.8823431.

Gu, N., & Liu, S., (2020). Introduction to biosensors. *Journal of Materials Chemistry B*, *8*(16), 3168–3170. https://doi.org/10.1039/d0tb90051f.

Han, S., Zhuang, X., Shi, W., Yang, X., Li, L., & Yu, J., (2016). Poly(3-hexylthiophene)/polystyrene (P3HT/PS) blends based organic field-effect transistor ammonia gas sensor. *Sensors and Actuators, B: Chemical*, *225*, 10–15. https://doi.org/10.1016/j.snb.2015.11.005.

Harpaz, D., Duanis-Assaf, D., Alkan, N., & Eltzov, E., (2022). Detection of postharvest pathogenic fungi by RNA markers in high-throughput platform on streptavidin plate. *Postharvest Biology and Technology*, *183*, 111728. https://doi.org/10.1016/j.postharvbio.2021.111728.

Hawari, H. F., Samsudin, N. M., Shakaff, A. Y. M., Wahab, Y., Hashim, U., Zakaria, A., Ghani, S. A., & Ahmad, M. N., (2013). Highly selective molecular imprinted polymer (MIP) based sensor array using interdigitated electrode (IDE) platform for detection of mango ripeness. *Sensors and Actuators, B: Chemical*, *187*, 434–444. https://doi.org/10.1016/j.snb.2013.01.045.

Islam, M. N., Mursalat, M., & Khan, M. S., (2016). A review on the legislative aspect of artificial fruit ripening. *Agriculture and Food Security*, *5*(1), 1–10. https://doi.org/10.1186/s40066-016-0057-5.

Izadi, Z., Sheikh-Zeinoddin, M., Ensafi, A. A., & Soleimanian-Zad, S., (2016). Fabrication of an electrochemical DNA-based biosensor for *Bacillus cereus* detection in milk and infant formula. *Biosensors and Bioelectronics*, *80*, 582–589. https://doi.org/10.1016/j.bios.2016.02.032.

Janssen, S., Schmitt, K., Blanke, M., Bauersfeld, M. L., Wöllenstein, J., & Lang, W., (2014). Ethylene detection in fruit supply chains. *Philosophical Transactions of the Royal Society A: Mathematical, Physical and Engineering Sciences*, *372*(2017). https://doi.org/10.1098/rsta.2013.0311.

Jawaheer, S., White, S. F., Rughooputh, S. D. D. V., & Cullen, D. C., (2003). Development of a common biosensor format for an enzyme based biosensor array to monitor fruit quality. *Biosensors and Bioelectronics*, *18*(12), 1429–1437. https://doi.org/10.1016/S0956-5663(03)00073-3.

Jedermann, R., Pötsch, T., & Lloyd, C., (2014). Communication techniques and challenges for wireless food quality monitoring. *Philosophical Transactions of the Royal Society A: Mathematical, Physical and Engineering Sciences*, *372*(2017). https://doi.org/10.1098/rsta.2013.0304.

Jochum, T., Rahal, L., Suckert, R. J., Popp, J., & Frosch, T., (2016). All-in-one: A versatile gas sensor based on fiber enhanced Raman spectroscopy for monitoring postharvest fruit conservation and ripening. *Analyst*, *141*(6), 2023–2029. https://doi.org/10.1039/c5an02120k.

Joshi, U., Bisht, T. S., Rawat, L., & Mamgain, A., (2021). Smart packaging: Modern way for reducing post-harvest losses of horticultural produce. *International Journal of Agricultural Sciences*, *17*(AAEBSSD), 331–339. https://doi.org/10.15740/has/ijas/17-aaebssd/331-339.

Jung, S. H., Jang, H., Lim, M. C., Kim, J. H., Shin, K. S., Kim, S. M., Kim, H. Y., et al., (2015). Chromatic biosensor for detection of phosphinothricin acetyltransferase by use of polydiacetylene vesicles encapsulated within automatically generated immunohydrogel beads. *Analytical Chemistry*, *87*(4), 2072–2078. https://doi.org/10.1021/ac501795x.

Kaittanis, C., Santra, S., & Perez, J. M., (2010). Emerging nanotechnology-based strategies for the identification of microbial pathogenesis. *Advanced Drug Delivery Reviews*, *62*(4, 5), 408–423. https://doi.org/10.1016/j.addr.2009.11.013.

Kantor, D. B., Hitka, G., Fekete, A., & Balla, C., (2008). Electronic tongue for sensing taste changes with apricots during storage. *Sensors and Actuators, B: Chemical*, *131*(1), 43–47. https://doi.org/10.1016/j.snb.2007.12.003.

Karakaya, D., Ulucan, O., & Turkan, M., (2020). Electronic nose and its applications: A survey. *International Journal of Automation and Computing*, *17*(2), 179–209. https://doi.org/10.1007/s11633-019-1212-9.

Kim, Y. H., Park, J. S., & Jung, H. I., (2009). An impedimetric biosensor for real-time monitoring of bacterial growth in a microbial fermentor. *Sensors and Actuators, B: Chemical*, *138*(1), 270–277. https://doi.org/10.1016/j.snb.2009.01.034.

Kumar, U. T., Varun, K. V. S., Baran, S. A., Upadhye, V. J., Khan, F., Pandey, P., Amjad, K. M., et al., (2022). Nanotechnology-based advancements in postharvest management of horticultural crops. *Phyton*, *91*(3), 471–487. https://doi.org/10.32604/phyton.2022.017258.

Kundu, M., Krishnan, P., Kotnala, R. K., & Sumana, G., (2019). Recent developments in biosensors to combat agricultural challenges and their future prospects. *Trends in Food Science and Technology*, *88*, 157–178. https://doi.org/10.1016/j.tifs.2019.03.024.

Kuswandi, B., & Murdyaningsih, E. A., (2017). Simple on package indicator label for monitoring of grape ripening process using colorimetric pH sensor. *Journal of Food*

Measurement and Characterization, *11*(4), 2180–2194. https://doi.org/10.1007/s11694-017-9603-5.

Kuswandi, B., Irmawati, T., Hidayat, M. A., Jayus, & Ahmad, M., (2014). A simple visual ethanol biosensor based on alcohol oxidase immobilized onto polyaniline film for halal verification of fermented beverage samples. *Sensors (Switzerland)*, *14*(2), 2135–2149. https://doi.org/10.3390/s140202135.

Kuswandi, B., Wicaksono, Y., Jayus, Abdullah, A., Heng, L. Y., & Ahmad, M., (2011). Smart packaging: Sensors for monitoring of food quality and safety. *Sensing and Instrumentation for Food Quality and Safety*, *5*(3, 4), 137–146. https://doi.org/10.1007/s11694-011-9120-x.

Li, L., Zhao, C., Zhang, Y., Yao, J., Yang, W., Hu, Q., Wang, C., & Cao, C., (2017). Effect of stable antimicrobial nano-silver packaging on inhibiting mildew and in storage of rice. *Food Chemistry*, *215*, 477–482. https://doi.org/10.1016/j.foodchem.2016.08.013.

Mathushika, J. M., (2021). Emerging concepts and practices in post-harvest management of horticultural crops revisited. *International Journal of Current Science Research and Review*, *4*(8), 859–876. https://doi.org/10.47191/ijcsrr/v4-i8-04.

Maukonen, J., Mättö, J., Wirtanen, G., Raaska, L., Mattila-Sandholm, T., & Saarela, M., (2003). Methodologies for the characterization of microbes in industrial environments: A review. *Journal of Industrial Microbiology and Biotechnology*, *30*(6), 327–356. https://doi.org/10.1007/s10295-003-0056-y.

McLamore, E. S., Alocilja, E., Gomes, C., Gunasekaran, S., Jenkins, D., Datta, S. P. A., Li, Y., et al., (2021). FEAST of biosensors: Food, environmental and agricultural sensing technologies (FEAST) in North America. *Biosensors and Bioelectronics*, *178*. https://doi.org/10.1016/j.bios.2021.113011.

Mehrotra, P., (2016). Biosensors and their applications: A review. *Journal of Oral Biology and Craniofacial Research*, *6*(2), 153–159. https://doi.org/10.1016/j.jobcr.2015.12.002.

Neethirajan, S., Ragavan, V., Weng, X., & Chand, R., (2018). Biosensors for sustainable food engineering: Challenges and perspectives. *Biosensors*, *8*(1). https://doi.org/10.3390/bios8010023.

Neme, K., Nafady, A., Uddin, S., & Tola, Y. B., (2021). Application of nanotechnology in agriculture, postharvest loss reduction and food processing: Food security implication and challenges. *Heliyon*, *7*(12), e08539. https://doi.org/10.1016/j.heliyon.2021.e08539.

Prusky, D., Alkan, N., Mengiste, T., & Fluhr, R., (2013). Quiescent and necrotrophic lifestyle choice during postharvest disease development. *Annual Review of Phytopathology*, *51*, 155–176. https://doi.org/10.1146/annurev-phyto-082712-102349.

Pudake, R. N., & Jain, U., (2021). *Biosensors in Agriculture: Recent Trends and Future Perspectives*. http://www.springer.com/series/16076 (accessed on 7 February 2024).

Qiu, S., Wang, J., & Gao, L., (2015). Qualification and quantization of processed strawberry juice based on electronic nose and tongue. *LWT*, *60*(1), 115–123. https://doi.org/10.1016/j.lwt.2014.08.041.

Rai, M., Ingle, A. P., Gupta, I., Pandit, R., Paralikar, P., Gade, A., Chaud, M. V., & Dos, S. C. A., (2019). Smart nanopackaging for the enhancement of food shelf life. *Environmental Chemistry Letters*, *17*(1), 277–290. https://doi.org/10.1007/s10311-018-0794-8.

Ramachandra, B. L., Gumpu, M. B., Nesakumar, N., Krishnan, U. M., & Rayappan, J. B. B., (2016). Calcium carbide in mangoes: An electrochemical way for detection. *Analytical Methods*, *8*(23), 4590–4599. https://doi.org/10.1039/c6ay01314g.

Sachdev, D., Kumar, V., Maheshwari, P. H., Pasricha, R., Deepthi, & Baghel, N., (2016). Silver based nanomaterial, as a selective colorimetric sensor for visual detection of post-harvest spoilage in onion. *Sensors and Actuators, B: Chemical*, *228*, 471–479. https://doi.org/10.1016/j.snb.2016.01.049.

Saini, S., Sillard, C., Naceur, B. M., & Bras, J., (2016). Nisin anchored cellulose nanofibers for long term antimicrobial active food packaging. *RSC Advances*, *6*(15), 12422–12430. https://doi.org/10.1039/c5ra22748h.

Sanaeifar, A., Mohtasebi, S. S., Ghasemi-Varnamkhasti, M., & Ahmadi, H., (2016). Application of MOS based electronic nose for the prediction of banana quality properties. *Measurement: Journal of the International Measurement Confederation*, *82*, 105–114. https://doi.org/10.1016/j.measurement.2015.12.041.

Seo, J. W., Lee, J. H., Son, I. S., Kim, Y. J., Kim, D. Y., Hwang, Y., Chung, H. A., et al., (2012). Acute oxalate nephropathy caused by ethylene glycol poisoning. *Kidney Research and Clinical Practice*, *31*(4), 249–252. https://doi.org/10.1016/j.krcp.2012.09.007.

Singh, D. R., Singh, S., Salim, K. M., & Srivastava, R. C., (2012). Estimation of phytochemicals and antioxidant activity of underutilized fruits of Andaman Islands (India). *International Journal of Food Sciences and Nutrition*, *63*(4), 446–452. https://doi.org/10.3109/09637486.2011.634788.

Soldatkin, O. O., Peshkova, V. M., Saiapina, O. Y., Kucherenko, I. S., Dudchenko, O. Y., Melnyk, V. G., Vasylenko, O. D., et al., (2013). Development of conductometric biosensor array for simultaneous determination of maltose, lactose, sucrose, and glucose. *Talanta*, *115*, 200–207. https://doi.org/10.1016/j.talanta.2013.04.065.

Soldatkina, O. V., Soldatkin, O. O., Kasap, B. O., Kucherenko, D. Y., Kucherenko, I. S., Kurc, B. A., & Dzyadevych, S. V., (2017). A novel amperometric glutamate biosensor based on glutamate oxidase adsorbed on silicalite. *Nanoscale Research Letters*, *12*(1), 1–8. https://doi.org/10.1186/s11671-017-2026-8.

Strano, M. C., Altieri, G., Allegra, M., Renzo, G. C. Di., Paterna, G., Matera, A., & Genovese, F., (2022). Postharvest technologies of fresh *Citrus* fruit: Advances and recent developments for the loss reduction during handling and storage. *Horticulturae*, *8*(7). https://doi.org/10.3390/horticulturae8070612.

Sun, Z., Tian, L., Guo, M., Xu, X., Li, Q., & Weng, H., (2017). A double-film screening card for rapid detection of organophosphate and carbamate pesticide residues by one step in vegetables and fruits. *Food Control*, *81*, 23–29. https://doi.org/10.1016/j.foodcont.2017.05.012.

Swarnapriya, R., & Gurve, V., (2020). *Role of Nanomaterials in Vegetable Crops*. E Scientia Letters.

Verbeke, W., & Pieniak, Z., (2006). Benefit beliefs, attitudes, and behavior towards fresh vegetable consumption in Poland and Belgium. *Acta Alimentaria*, *35*(1), 5–16. https://doi.org/10.1556/AAlim.35.2006.1.2.

Vigneault, C., Bourgeois, G., Toussaint, V., Charlebois, D., & LeBlanc, D. I., (2009). Potential for modelling postharvest quality of fresh fruit and vegetables. *Fresh Produce*, *3*(1), 12–22.

Vinet, L., & Zhedanov, A., (2011). A "missing" family of classical orthogonal polynomials. *Journal of Physics A: Mathematical and Theoretical*, *44*(8), 1–15. https://doi.org/10.1088/1751-8113/44/8/085201.

Von, S. F., Tornieporth, N., Waiyaki, P., Lowe, B., Feruski, L. F., Dupont, H. L., Mathewson, J. J., & Steffen, R., (2000). Risk and aetiology of diarrhoea at various tourist destinations. *Lancet*, *356*(9224), 133–134. https://doi.org/10.1016/S0140-6736(00)02451-X.

Wang, H., Peng, J., Xie, C., Bao, Y., & He, Y., (2015). Fruit quality evaluation using spectroscopy technology: A review. *Sensors (Switzerland)*, *15*(5), 11889–11927. https://doi.org/10.3390/s150511889.

Wasilewski, T., Migoń, D., Gębicki, J., & Kamysz, W., (2019). Critical review of electronic nose and tongue instruments prospects in pharmaceutical analysis. *Analytica Chimica Acta*, *1077*, 14–29. https://doi.org/10.1016/j.aca.2019.05.024.

Wei, Y., Chang, J., Lian, J., & Liu, T., (2015). A coal mine multi-point fiber ethylene gas concentration sensor. *Photonic Sensors*, *5*(1), 67–71. https://doi.org/10.1007/s13320-014-0206-z.

Xu, K., Huang, J., Ye, Z., Ying, Y., & Li, Y., (2009). Recent development of nano-materials used in DNA biosensors. *Sensors*, *9*(7), 5534–5557. https://doi.org/10.3390/s90705534.

Yadollahi, A., Arzani, K., & Khoshghalb, H., (2010). The role of nanotechnology in horticultural crops postharvest management. *Acta Horticulturae*, *875*, 49–56. https://doi.org/10.17660/ActaHortic.2010.875.4.

Yang, W. Z., & Jiao, Y., (2017). The international research progress of ammonia (NH_3) emissions and emissions reduction technology in farmland ecosystem. *IOP Conference Series: Earth and Environmental Science*, *59*(1), 12024. https://doi.org/10.1088/1755-1315/59/1/012024.

Yousefi, H., Su, H. M., Imani, S. M., Alkhaldi, K., Filipe, C. D., & Didar, T. F., (2019). Intelligent food packaging: A review of smart sensing technologies for monitoring food quality. *ACS Sensors*, *4*(4), 808–821. https://doi.org/10.1021/acssensors.9b00440.

Zamir, D., Galsurker, O., Alkan, N., & Eltzov, E., (2020). Detection of quiescent fungi in harvested fruit using CMOS biosensor: A proof of concept study. *Talanta*, *217*, 120994. https://doi.org/10.1016/j.talanta.2020.120994.

Zhai, X., Zhou, S., Zhang, R., Wang, W., & Hou, H., (2022). Antimicrobial starch/poly(butylene adipate-co-terephthalate) nanocomposite films loaded with a combination of silver and zinc oxide nanoparticles for food packaging. *International Journal of Biological Macromolecules*, *206*, 298–305. https://doi.org/10.1016/j.ijbiomac.2022.02.158.

Zhang, R., Belwal, T., Li, L., Lin, X., Xu, Y., & Luo, Z., (2020). Nanomaterial-based biosensors for sensing key foodborne pathogens: Advances from recent decades. *Comprehensive Reviews in Food Science and Food Safety*, *19*(4), 1465–1487. https://doi.org/10.1111/1541-4337.12576.

Zhao, L., Kong, D., Wu, Z., Liu, G., Gao, Y., Yan, X., Liu, F., et al., (2020). Interface interaction of MoS_2 nanosheets with DNA based aptameric biosensor for carbohydrate antigen 15–3 detection. *Microchemical Journal*, *155*, 104675. https://doi.org/10.1016/j.microc.2020.104675.

Zhu, C., Lu, Q., Zhou, X., Li, J., Yue, J., Wang, Z., & Pan, S., (2020). Metabolic variations of organic acids, amino acids, fatty acids, and aroma compounds in the pulp of different pummelo varieties. *LWT*, *130*, 109445. https://doi.org/10.1016/j.lwt.2020.109445.

CHAPTER 13

BIOENGINEERING OF FRUITS AND VEGETABLES USING EMERGING TOOLS: ROLE OF H_2S, NO, H_2O_2 IN POSTHARVEST QUALITY IMPROVEMENT

ANURADHA TYAGI,[1] DHARMENDRA NATH BHATT,[2] SEKHU ANSARI,[3] and VINAY KUMAR[4]

[1]*Department of cBRN, Institute of Nuclear Medicine and Allied Sciences, Brig S.K. Mazumdar, Delhi, India*

[2]*Laboratory of Immunogenetics, National Institute of Allergy and Infectious Diseases, National Institutes of Health, Bethesda, Maryland, USA*

[3]*Division of Pathology, Cincinnati Children's Hospital Medical Center, Cincinnati, Ohio, USA*

[4]*Pennsylvania State University HMC, Hershey, Pennsylvania, USA*

ABSTRACT

Poor postharvest management can lead to significant reductions in quality and shelf life, resulting in economic losses. Fruit ripening, senescence, and physiological changes are regulated by various metabolic pathways influenced by endogenous signaling molecules like NO, H_2S, and H_2O_2. Despite once being considered harmful, these molecules are now

Bioengineered Fruit and Vegetables: Trends in Postharvest Quality Improvement.
Mohammad Irfan and Pankaj Kumar (Eds.)

recognized for their beneficial signaling functions, including extending shelf life and inducing disease resistance. The chapter elucidates how genetic engineering targeting biosynthesis and communication pathways of H_2S, NO, and H_2O_2 can mitigate senescence, reduce post-harvest losses, and enhance antioxidant enzyme activity, ultimately prolonging the quality and shelf life of fruits and vegetables. Furthermore, the chapter explores the collaborative or antagonistic interactions of these molecules with other compounds such as ethylene, shedding light on their complex roles in regulating postharvest processes. Also, the importance of understanding the crosstalk between these pathways highlights the need for further research to decode their complex mechanisms, which could revolutionize postharvest preservation strategies and contribute to addressing global food shortages.

13.1 INTRODUCTION

Vegetables, fruits, flowers, herbs, and other aromatic and medicinal plants are all included in the broad category of horticultural crops. Phytonutrients like phenolic compounds and carotenoids, along with vitamins, minerals, carbs, fats, and amino acids, are essential for human health preservation, and are provided by vegetables and fruits (Willett & Stampfer, 2013; Jacob et al., 2012; Aghdam et al., 2023). Many diverse metabolic pathways in a plant cell communicate with one another, are regulated by one another, and respond to changes in their physiological and environmental conditions via a complex network of signaling molecules. However, poor postharvest management and storage can drastically reduce their quality and shelf life, which in turn can cause infections and consumer rejection, resulting in substantial economic losses (Aghdam et al., 2020, 2023). Fruit ripening is a prolonged system of senescence and is characterized by substantial physiological and biochemical changes in fruit color, surface texture, flavor, and smell (Tayal et al., 2022; Irfan et al., 2014, 2016).

Furthermore, physiological and pathological processes choreograph senescence, which is defined as the abrupt depletion of cellular capabilities and sensitivity by a cell or tissue in an organism (Mishra et al., 2021). Many metabolic and signaling pathways are influenced by endogenous chemicals that govern fruit quality. These include nitric oxide (NO), reactive oxygen species (ROS), elicitors, hydrogen sulfide (H_2S), and ethylene

(Irfan et al., 2014, 2016, 2022), and salicylic acid (SA) (Mukherjee & Bozhkov, 2019). Moreover, simple molecules like H_2S and NO are seen to modulate a significant signaling function in numerous physiological and adverse circumstances via regulation of gene expression, post-translational protein modifications (PTMs), and interaction amongst other plant regulators, but H_2O_2 ought to get its mechanism (Freschi, 2013; Asgher et al., 2016; Prakash et al., 2019; Corpas et al., 2019). It's worth noting that despite their lengthy history of recognition in plant science, NO and H_2S were once thought of as contaminating and dangerous compounds. Animal researchers, however, brought to light the fact that these chemicals are naturally produced in cells and, at healthy amounts, can serve useful purposes like signaling, completely shifting the paradigm. Exogenous NO and H_2S have similar positive effects, such as extending post-harvest fruit shelf life and inducing disease resistance. These results are typically linked to the upregulation of soluble antioxidants like ascorbate and glutathione, together with antioxidant enzymes including catalase (CAT), superoxide dismutase, and peroxidases (PODs) (Corpas et al., 2019). L-arginine-dependent NO synthase-like activity and nitrate reductase (NR) are the two main enzymatic candidate systems in higher plants that produce NO from L-arginine or nitrate, respectively (Corpas & Barroso, 2017). Different parts of plant cells produce H_2S (also called dihydrogen sulfide or sulfane); the cytosol, chloroplasts, and mitochondria are all involved in Cys metabolism (Gotor et al., 2015; Birke et al., 2015). D- and L-cysteine desulfhydrase (LCD), sulfite reductase, cysteine synthase, and cyanoalanine synthase are all enzymes involved in the H_2S metabolism (Hancock & Whiteman, 2016; Calderwood & Kopriva, 2014; Gotor et al., 2015). Moreover, there is a continually expanding market for premium, nutrient-rich meals (Kumar et al., 2016a, 2019). For this reason, it is crucial to both increase yields of high-quality produce and decrease post-harvest losses, especially of perishable fruits and vegetables (Kumari et al., 2022; Ansari et al., 2022). The purpose of this chapter is to give readers a high-level understanding of how H_2S, NO, and H_2O_2 all work together to boost postharvest quality. Here, we focus on how the genetic engineering (GE) of genes involved in the biosynthesis and communication of H_2S, NO, and H_2O_2 can mitigate senescence in fruits and vegetables and cut down on post-harvest waste. We then investigated how these chemicals work together and assessed whether or not they might have a positive impact on post-harvest senescence by priming or limiting damage.

13.2 ENGINEERING OF H_2S BIOSYNTHESIS AND SIGNALING GENES

H_2S is a gaseous molecule that is produced naturally by plants; it has been hypothesized to play a role as a signaling agent in a wide range of cellular functions (Tayal et al., 2022). The main enzyme in the manufacture of H_2S in plants is LCD, which is found in a variety of subcellular locations (Arif et al., 2021). Cyanoalanine synthase, D-cysteine desulfhydrase, LCD, cysteine synthase, and sulfite reductase are key enzymes in maintaining H_2S homeostasis (Li et al., 2013, 2014). Accumulating evidence projects the important role of H_2S in various activities in plants such as postharvest senescence, abiotic stress tolerance and photosynthesis, etc. (Zhang et al., 2011; Chen et al., 2011; Jin et al., 2011). In relation to postharvest senescence, Ni et al. examined the method of H_2S in lessening postharvest aging and deterioration of Kyoho grapes (Ni et al., 2016). Authors elaborated that H_2S controls the redox microenvironment by modulation ROS as they found malondialdehyde (MDA), H_2O_2, and superoxide anion (O_2^-) were all significantly reduced in grape peels and pulp after H_2S treatment, while CAT and ascorbate peroxidase (APX) activity was increased, and lipoxygenase (LOX) activity was decreased (Ni et al., 2016). Hu et al. (2012) found that H_2S increased GR, CAT, APX, and POD activity during strawberry postharvest storage however H_2S inhibits LOX activity, which oxygenates polyunsaturated fatty acids (Hu et al., 2012). Similarly, Hu et al. (2014) revealed the impact of H_2S on the characteristics of mulberry fruit after harvest by fumigating the fruit with a 0.8 mm sodium hydrosulfide (NaHS) solution, an H_2S donor, and found that the endogenous H_2S content was significantly increased due to an uptick in the activities of D- and LCD (Hu et al., 2014). NaHS might thus diminish the anthocyanin level and respiratory activity of mulberry fruit, as well as drastically slow down the rate of ripening (Hu et al., 2014). As a result, H_2S served as a gaseous regulator by enhancing antioxidant enzyme activity and affecting ROS accumulation. Hu et al. (2015) suggested that endogenous H_2S generated by NaHS fumigation could ameliorate the energy status, resulting in increased antioxidant potential and decreased respiration rate, ultimately delaying the senescence of Ipomoea aquatica (Hu et al., 2015). However, fruits' senescence and ripening can be postponed under cold storage during postharvest processing, despite causing chilling stresses due to low non-freezing temperatures. After harvest, fruits have a short shelf life and low quality due to chilling harm. Due to its effects on improving the antioxidant system, H_2S can reduce chilling injury in banana fruit. This is attributed to the increased activity of antioxidative enzymes and P5CS (Δ1-pyrroline-5-carboxylate synthase), both of which elevate the proline content of the fruit (Luo et al., 2015). In addition, H_2S-treated bananas exhibited increased Hue angle and firmness, along with decreased MDA content, electrolyte leakage, and ethylene generation (Li et al., 2016). Moreover, H_2S administration greatly enhanced the activity of enzymes required in energy metabolism, such as cytochrome C oxidase, H(+)-ATPase, succinate dehydrogenase, and Ca(2+)-ATPase. Hence, it can be argued that H_2S can diminish chilling progress in banana fruit by enhancing the energy-metabolizing enzymes' activities to retain energy. In another postharvest study with a 1.5 mM NaHS (H_2S donor), was identified as being helpful for chilling the forbearance of hawthorn fruit by inducing endogenous H_2S augmentation, boosting ROS scavenging enzyme activity, enhancing anthocyanins, flavonoids, and phenols by phenylalanine ammonia-lyase (PAL) enzyme activity, promoting phenols augmentation, which is essential for membrane integrity, and conserving nutritional property (Aghdam et al., 2018). It is crucial to understand the function of H_2S in controlling senescence-related genes in postharvest storage, as senescence is associated with the degradation of chlorophyll and certain proteins. By controlling increased quantities of metabolites such as anthocyanin, ascorbate, and carotenoids, and by reducing MDA buildup, hydrogen peroxide (H_2O_2), and superoxide anion, Li et al. (2014) showed that H_2S, in a dose-dependent manner, delays broccoli's deterioration after harvest (Li et al., 2014). Moreover, postharvest broccoli treatment with H_2S showed downregulated expression of the genes pheophorbide A oxygenase (*BoPaO*), Stay-Green encoding gene (*BoSGR*), red catabolite chlorophyll reductase (*BoRCCR*), chlorophyllase 2 (*BoCLH2*), lipoxygenase 1 gene (*BoLOX1*) and cysteine protease (*BoCP1*). The effects of H_2S on other fruits and vegetables senescence show that this substance has a universal function as a senescence regulator. Inhibition of *BoLSC807* (aspartic protease gene) and *BoCP3* (cysteine protease gene) expression by H_2S was discovered by Li et al. (2015), indicating that H_2S prevented protein degradation throughout the senescence phase in broccoli (Li et al., 2014). By regulating the gene expression engaged in chlorophyll degradation, H_2S prevented the postharvest yellowing of broccoli. During apple storage, H_2S suppressed the expression of *MdPPO* (encoding PPO) and *MdLOX2* (Zheng et al., 2016). APX, CAT, superoxide dismutase, GR, and guaiacol POD activities were found to be significantly higher in H_2S-fumigated apple fruits compared to controls, while lower activities were found in PAL, LOX, protease, and polyphenol oxidase (PPO) (Zheng et al., 2016). In postharvest apple tissue, H_2S increased *MdDHAR* expression and decreased *MdPPO*, *MdLOX2*, *MdACO1*, *MdPG1*, *MdETR1*, and *MdERS1*. Postharvest storage of vegetables and fruits exposed to H_2S reduced the expression of lipid peroxidation genes, thus delaying the onset of senescence (Zheng et al., 2016). Taken together, H_2S modulates the expression of genes involved in senescence and slows the onset of senescence in fruits and vegetables throughout storage. When Aghdam et al. (2018) subjected hawthorn fruit to H_2S donor NaHS, they discovered a striking increase in endogenous H_2S accumulation due to the activation of LCD and d-cysteine desulfhydrase enzymes which, in tandem with increased activity of ROS scavenging enzymes superoxide dismutase, APX and CAT activity to lesser H_2O_2 accumulation (Aghdam et al., 2018). Also, because H_2S stimulates PAL enzyme activity, hawthorn fruit exposed to H_2S accumulates significantly more phenols, flavonoids, and anthocyanins than untreated fruit. H_2S's route of action is unclear, although it appears to include the regulation of energy metabolism (TCA cycle, electron transport chain, glycolysis) to maintain high ATP levels and so postpone fruit senescence (Li et al., 2017). It can either feed electrons into the pathway, resulting in an increase in ATP generation, or it can restrict complex IV, thereby decreasing ATP generation (Hancock & Whiteman, 2016).

Many studies examine the collaborative or antagonistic interaction of H_2S with other compounds, like ethylene and NO, in the postharvest senescence of agricultural crops. The quality of strawberries (as measured by the crust color and firmness) was greatly preserved by a combination of 5 M SNP solution (NO) and 0.8 mM NaHS solution (H_2S), both of which reduced respiration and decay. The co treatment with H_2S and NO maintained the decreased activity of endo—1,4-glucanase (EGase), polygalacturonase (PG), and pectin methyl esterase (PME), all of which are linked to the ripening of fruits (Chang et al., 2014). Ubeed et al. (2017) discovered that the green leafy vegetable pak choy was affected by H_2S and that it reduced ethylene synthesis, chlorophyll loss, respiration, weight loss, and the release of numerous antioxidant components and ions (*Brassicarapa* subsp. *Chinensis*) (Al Ubeed et al., 2017). H_2S and 1-methylcyclopropene (1-MCP) were equally effective in preventing pak choy from losing its green color, respiring, losing ions, and producing ethylene when stored in an ethylene-free atmosphere (Al Ubeed et al., 2018). The research indicated that H_2S's mechanism of action involved inhibiting the effectiveness of ethylene. H_2S may bind to ethylene receptors in agricultural products, but the allosteric action of H_2S and 1-MCP on the protein receptor would be distinct (Pirrung et al., 2008; Lacey & Binder, 2014). Li et al. (2017) demonstrated that H_2S slowed down kiwifruit ripening and senescence when exposed to ethylene (Li et al., 2017). Subsequent studies revealed that H_2S suppressed the expression of the *AdCP1*, cysteine protease genes, and *AdCP3*, as well as the ethylene production genes *AdSAM, AdACO2, AdACO3, AdACS1,* and *AdACS2.* These findings provide further evidence that H_2S delays the ripening and aging of kiwifruit by inhibiting the pathway that generates ethylene. Researchers Ge et al. (2017) found that when ethylene and H_2S were applied together to banana fruit, the expression of genes engaged in ethylene synthesis was suppressed (*MaACS1, MaACS2, MaACO1, and MaPL*), while the expression of genes involved in ethylene receptors (MaETR, MaERS1, and MaERS2) was increased (Ge et al., 2017). Overall, the ethylene signaling pathway can be inhibited by H_2S, which counteracts the action of ethylene and slows down the ripening and senescence of bananas (Ge et al., 2017).

H_2O_2, superoxide anion (O_2^-) generation, and MDA in tomato fruit were all shown to be suppressed by exogenous H_2S in a study by Yao et al. (2018). Further, when compared to their expression following ethylene treatment, the expression of the antioxidant encoding genes *SlCuZnSOD*, *SlAPX2*, *SlPOD12*, and *SlCAT1* was usually enhanced by C_2H_4–H_2S co treatment. The autophagy inhibitor H_2S was discovered by studying the des1 mutant of Arabidopsis, which is defective in the cytosolic synthesis of H_2S from cysteine (Álvarez et al., 2012). Interchangeable post-translational adaptation of enzymes responsible for ubiquitylation or other proteins involved in the beginning and maintenance of the autophagosome may underlie H_2S-mediated signaling in autophagy (Ziogas et al., 2018). Moreover, various molecules in postharvest quality improvement are summarized in Table 13.1.

TABLE 13.1 Various Molecules in Postharvest Quality Improvement

Hydrogen	Reduced Activity	Enhanced Activity	Fruit	References
Sulfide (H2S)	Hydrogen peroxide (H_2O_2) Malondialdehyde (MDA) Superoxide anion (O_2^-) Lipoxygenase	Catalase Ascorbate peroxidase	Grape (*Vitis vinifera*) peels and pulp.	Ni et al. (2016)
	Lipoxygenase	Catalase Glutathione reductase Peroxidase Ascorbate peroxidase	Strawberry (*Fragaria* species)	Hu et al. (2012)
	Anthocyanin level and respiratory activity.	–	Mulberry (*Morus indica*)	Hu et al. (2014)
	Chlorophyll-degrading enzyme and respiration rate.	Antioxidant activity	Spinach (*Ipomoea aquatica*)	Hu et al. (2015)
	Proline dehydrogenase	Δ1-pyrroline-5-carboxylate synthase and antioxidant activity.	Banana(*Musa* spp.)	Luo et al. (2015)
	Senescence-related gene expression.	–	Broccoli (*Brassica oleracea*)	Li et al. (2014)
	Lipid peroxidation-related genes, ethylene biosynthesis, and signaling.	–	Apple (*Malus* × *pumila*)	Zheng et al. (2016)
	AdCP1, cysteine protease genes and *AdCP3*, ethylene production genes *AdSAM*, *AdACO2*, *AdACO3*, *AdACS1*, and *AdACS2*.	–	Kiwifruit (*Actinidia deliciosa*)	Li et al. (2017)

TABLE 13.1 *(Continued)*

Hydrogen	Reduced Activity	Enhanced Activity	Fruit	References
Nitric Oxide (NO)	Ethylene biosynthesis-related genes and signaling associated with *ACO* and *ERS1*.	Ethylene receptors and signaling genes *MiEIN2* and *MiETR1*.	Mango (*Mangifera indica*)	Hong et al. (2014)
	Phenylalanine ammonia-lyase and pectin methyl esterase.	–	Santa Rosa' Japanese plum (*Prunus salicina* Lindell).	Sharma & Sharma (2015)
	Ethylene biosynthesis genes 1-aminocyclopropane-1-carboxylic acid synthase and 1-aminocyclopropane-1-carboxylic acid oxidase.	–	Tomato (*Solanum lycopersicum*)	Gambhir et al. (2023)
	ACC synthase and ACC oxidase.	–	Mango (*Mangifera indica*)	Zaharah & Singh (2011)
	–	Superoxide dismutase, ascorbate peroxidase, glutathione reductase	Zucchini fruit (*Cucurbita pepo*)	Jiménez-Muñoz et al. (2021)
	–	Galactono-1,4-lactone dehydrogenase (GalLDH)	Pepper (*Capsicum annuum* L.)	Rodríguez-Ruiz et al. (2017)
	Ethylene synthesis-related genes.	–	Banana (*Musa* spp.)	Cheng et al. (2009)
Chlorine Dioxide (ClO_2)	Pericarp browning reduces.	–	Longan (*Dimocarpus longan*)	Saengnil et al. (2014)
Ozone (O_3)	–	Improve fruit shelf-life.	Citrus fruits	Karaca (2010)

13.3 ENGINEERING OF NO BIOSYNTHESIS AND SIGNALING GENES

Nitric oxide (NO) has a potential involvement in fruit ripening, maturation, and biotic/abiotic stressors tolerance processes, making it a popular choice for preserving quality and postponing senescence. The genes *NIA1* and *NIA2* code for NR, an enzyme necessary for NO production in higher plants that also uses the amino acid arginine (Mohn et al., 2019; Costa-Broseta et al., 2021) and a NO synthase-like activity dependent on NADPH (Corpas & Barroso, 2017), although it is possible that more enzymatic and non-enzymatic sources exist. It showed that NO has an adverse influence on the fruit's ability to produce ethylene. Some pre-climacteric fruits, such as bananas, mangoes (Hong et al., 2014), and tomato (Steffens et al., 2022) have been observed to have a longer shelf life when exposed to NO after harvest. Sodium-nitroprusside (SNP) and nitrosoglutathione (GSNO) are two types of exogenous NO shown in the study to efficiently postpone fruit ripening and deterioration (Madebo et al., 2022). Moreover, Important cell wall degradation enzymes in many fruits include PG and pectin methyl esterase (PME). It has been reported by Sharma et al. (2014) that treating 'Santa Rosa' plums using sodium nitroprusside (a source of NO) is a simple procedure for applying NO to preserve the quality of the fruit and increase shelf life by preventing fruit softening during refrigeration by restricting pectin methyl esterase (PME) and PAL activities (Sharma & Sharma, 2015). The expression of *ACO* genes including *LeACO1, LeACOH2*, and *LeACO4* is significantly reduced and delayed in NO-treated tomato fruit (Gambhir et al., 2023). The presence of 1-aminocyclopropane carboxylic acid also modifies the NO-ACO binding affinity, leading to a more stable tertiary ACC-ACC oxidase-NO complex (Madebo et al., 2022). Furthermore, NO treatment limits ion leakage and MDA generation by decreasing O_2^- and H_2O_2 accumulation due to increased phenols buildup (Yansheng Wang et al., 2013). Changes in ACC levels and the activity of ACC synthase and ACC oxidase were discovered after applying exogenous NO to unripe *Mangifera indica* L. cv. 'Kensington Pride' fruits (Zaharah & Singh, 2011). Researchers have discovered that NO protects cells from oxidative stress through a coordinated response from the antioxidant system, a reduced concentration of H_2O_2, and the support of elevated levels of phenol and flavonoid metabolites (Jiménez-Muñoz et al., 2021). The outstanding redox-based signaling mechanism that controls a wide range of biological functions is NO and its involvement in altering cysteine residues through protein S-nitrosylation. S-nitrosoglutathione reductase (GSNOR) controls total cellular S-nitrosylated proteins, fine-tunes NO signaling, and regulates NO bioavailability by breaking down S-nitrosoglutathione (GSNO). From *in vitro* culture establishment to plant growth and fruit set, Rasool et al. (2021) demonstrated that SlGSNOR overexpression in tomatoes modifies various developmental programs (Rasool et al., 2021). Hussain et al. (2019) showed that knocking down GSNOR expression in tomatoes (*Solanum lycopersicum* L.) via RNAi altered fruit set, yield, and other aspects of fruit development. Additionally, several other stages of development, such as seed formation, were influenced by the lack of *S. lycopersicum* GSNOR (SlGSNOR). Increased ascorbic acid levels in ripening fruits, such as pepper, are achieved through NO-induced upregulation of galactinol-1,4-lactone dehydrogenase (GalLDH) activity (Rodríguez-Ruiz et al., 2017). NO hyperaccumulation (Short root (shr) mutant) tomatoes have been shown to rely heavily on NO-mediated modulation of phytohormone concentrations and carbon metabolism for the ripening of their fruits (Zhu & Zhou, 2006). Researchers found that applying NO to ripening bananas decreased MA-ACO1 gene expression, which in turn reduced respiration, inhibited hydrolyzing enzymes, and softened the pericarp (Cheng et al., 2009).

13.4 ENGINEERING OF H_2O_2 BIOSYNTHESIS AND SIGNALING GENES

Crop loss primarily occurs in the postharvest process, particularly in developing nations that have little availability of high-quality technologies, affecting not just farmers but also consumers. Many extremely reactive molecules such as H_2O_2, singlet oxygen ($^1O_2$), superoxide anion (O_2•), and the hydroxyl radical (OH) are all examples of ROS (Decros et al., 2019; Apel & Hirt, 2004). Superoxide can either be oxidized to H_2O_2 via chemical reduction or dismutation, or it can directly exercise its signaling role. Key regulators of the plant cell's redox status are superoxide dismutase (SODs; EC 1.15.1.1) that speed up the dismutation of H_2O_2 (Smirnoff & Arnaud, 2019). Critical to the production of H_2O_2 via the activity of peroxisomal glycolate oxidase are the chloroplast, peroxisome, and mitochondria. Specific SODs in the chloroplast quickly convert O_2^- to H_2O_2 (Smirnoff & Arnaud, 2019). Fruits that are damaged mechanically release ROS that weaken the cell walls, leading to spoilage after harvest. Chitosan, a natural and eco-friendly preservative, regulates ROS in both entire fruits and plant cells. Post-harvest chitosan treatment consequences on cell membrane integrity, ROS production, and enzymatic and non-enzymatic antioxidant systems at fruit wounds in apple fruits during storage were investigated by Ackah et al. (2022). Chitosan treatment was found to activate antioxidant mechanisms, which neutralized ROS and protected fruit membranes from damage. Fruit growth can be affected by alterations in oxidative pathways due to the crucial role of ROS signaling in development. For instance, fruits engineered to have higher quantities of antioxidants by boosting chloroplast activities, like tomatoes, stay longer and are firmer (Mehta et al., 2002; Zhang et al., 2013). Disruptions in the redox equilibrium, such as ROS generation, are a primary contributor to postharvest fruit degradation. Since ROS builds up during fruit overripening, enhancing fruit storage conditions is at the forefront of redox signaling's potential (Muñoz & Munné-Bosch, 2018). Ozone treatments in the citrus crop that extend the shelf life of fruit and experimental usage of chlorine dioxide fumigation in longan fruit that slows enzymatic fruit browning (Saengnil et al., 2014) are two examples (Karaca, 2010). Peaches, tomatoes, peppers, and grape berries (Vitis vinifera) all accumulate H_2O_2 pools as the fruit ripens, indicating increased oxidative stress (Decros et al., 2019; Martí et al., 2009; Gambhir et al., 2023). The antioxidative system is crucial to the ripening of tomato fruits, as evidenced by alterations in the activity of CAT, superoxide dismutase, and the enzymes involved in the ascorbate-glutathione cycle (Jimenez et al., 2002). Due to improvements in disease resistance and ROS scavenging capacity, as well as reductions in O_2^- accumulation and membrane lipid peroxidation, Kadozan (a new chitosan formulation) was found in a study by Jiang et al. (2018) to be effective in inhibiting the growth of *P. litchii* in harvested litchis (Jiang et al., 2018). The accumulation (e.g., GABA) or rapid decline (e.g., ASC, GSH) of redox-related metabolites following exposure to low O_2 and/or elevated CO_2 environments is linked to physiological problems in fruits kept under multiple stressful conditions, such as for pome fruit (Lum et al., 2016).

13.5 CROSSTALK OF NO, H_2O_2, AND H_2S PATHWAYS

Further research on the interplay between NO, H_2O_2, and H_2S on the quality and lengthening of storage times in various fruit and vegetable species is warranted. Many studies have demonstrated the importance of signal molecules including ROS, H_2S, and NO in plant resistance to biotic and abiotic stresses. Many endogenous chemicals, including ROS, elicitors, NO, H_2S, and sphingosine (SA), play critical roles in cell death cascade signaling (Figure 13.1).

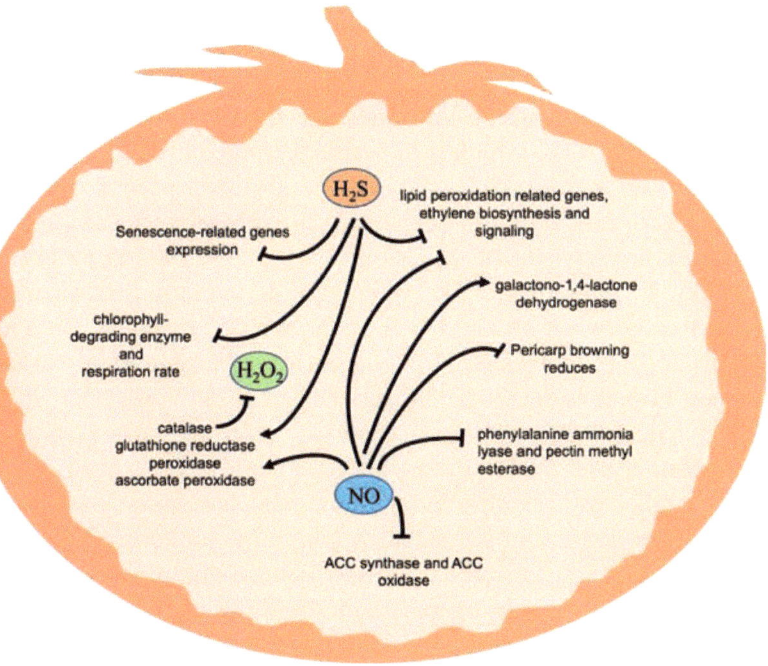

FIGURE 13.1 Crosstalk between H_2S, H_2O_2, and NO.

The improved tolerance of plants to abiotic stress was the result of a crosstalk between NO and H_2S (Wang et al., 2012). Nitrothiol production, a major source of NO reserve in cells, is thought to entail H_2S and NO. Also, the combined effects of NO and H_2S could significantly delay the deterioration of harvested strawberries, extending their shelf life (Chang et al., 2014). The effects of fumigation with NO and H_2S were studied by

Zhu et al. (2019), who reported the presence of collaborative functions in NO and H_2S in blocking ethylene biosynthesis and cell wall metabolism, as well as the softening of postharvest 'Dahong' peach fruit (Zhu et al., 2019). The reaction of the fruits to the ROS burst is the pivotal point of fruit senescence, which is closely related to the regulation of oxidative stress (Ghosh et al., 2013; Kumar et al., 2016b). Selected antioxidant systems, including CAT, SOD, POD, and APX, were effectively enhanced by NO. Increased expression of the SOD, POD, and APX genes (*MaSOD, MaCAT, MaPOD*, and *MaAPX*) is accompanied by an increase in enzyme activity after NO gas fumigation in banana fruit (Wu et al., 2014). Similar improvements in the antioxidant system and increased expression of *VvSOD, VvCAT, VvPOD2*, and *VvGR* were seen in NO-fumigated table grapes (Zhang et al., 2019). Additionally, NO, its donor, includes an improved SOD, and APX improves the tissues' capacity to expel O_2, which produces H_2O_2, which is directly influenced by the ASC-GSH cycle (Jiménez-Muñoz et al., 2021). Postharvest fruit senescence was thus postponed due to their effect in boosting antioxidants and scavenging ROS. The crosstalk between the NO single bond and H_2S has received some major new insights from recent studies of climacteric fruit ripening. The relationship between NO and H_2S under abiotic stress has been reported and H_2S appears to operate both as complementary and opposing to NO signaling (Tayal et al., 2022; Wang et al., 2012). It has been found that H_2S-ethylene interactions are antagonistic in the context of fruit ripening. It is important to note that the nature of the interactions between NO and H_2S crosstalk vary for various physiological processes, despite the fact that they both exhibit positive and negative interactions. Exogenous H_2S and NO application together assisted ripening inhibition by keeping the pericarp stiff and decreasing pectin methyl esterase (PME; EC 3.1.1.11), endo-β-1,4-glucanase (EGase; EC 3.2.1.6), and polygalacturonase (PG; EC 3.2.1.15) activity in strawberry fruits, as described by Chang et al. (2014). Therefore, the post-harvest ripening of fruit is inevitably slowed by the synergistic impacts of H_2S single bond NO treatments' anti-senescence characteristics. The possible participation of H_2S single bond NO crosstalk appears to be useful in post-harvest vegetable and fruit storage. S-nitrosation, persulfidation, tyrosine nitration, and metal nitrosylation are all examples of post-translational modifications (PTMs) that NO and H_2S use to control protein function at the molecular level (Corpas et al., 2020; Corpas & Palma, 2020). Resulting from its interaction with ROS

(H_2O_2) metabolism, NO has been shown to cause cell death in several plant species, but it is not always clear whether H_2O_2 operates downstream or upstream in NO signaling (Mishra et al., 2021). A NO burst occurs in *C. reinhardtii* in response to high light levels via pathways that are independent of NO synthase and NR. Additionally, external exposure to H_2O_2 and NO stimulates higher levels of cell mortality than exposure to NO alone. Using a NO scavenger stops H_2O_2 and NO from interacting, which reduces cell death, demonstrating the essential involvement of NO in these activities and showing that the two molecules work together synergistically to mediate cell death (Kuo et al., 2020). H_2S has been found to only reduce or postpone PCD via ROS-inhibiting pathways through different protein modification processes, especially persulfidation, whereas NO can have a dual effect by functioning as a signaling molecule to either suppress or accelerate PCD (Zhang et al., 2015; Luo et al., 2020; Rather et al., 2020). Exogenous administration of NO and H_2S alleviates Al^{3+}-induced cell death by: (i) enhancing the ability of the antioxidant system to avert oxidative stress; (ii) reducing Al^{3+} buildup in cells; and (iii) controlling the expression of protein transporters, such as Al-activated malate transporter, multidrug, and H^+-ATPase, toxic-extrusion/Al-activated citrate transporter and to promote organic acid secretion and transport (He et al., 2019; Wang et al., 2019). S-nitrosated and persulfidated proteins, as well as antioxidant enzymes such as adenosine polyphosphate synthase and CAT, have been linked to the increase of phosphate membrane (PM) H^+-ATPase-coupled citrate transporter co transport networks (Corpas et al., 2019, 2020). Under nitrate stress, the effects of H_2S and the crosstalk between H_2S, MAPK, and NO on cucumber roots were investigated by Qi et al. (2019), and the authors found that H_2S played a protective function in cucumber seedlings under nitrate stress by modulating the activities of antioxidant enzymes (Qi et al., 2019). However, during senescence and PCD, there is substantial crosstalk between different RNS and ROS. The expression profiles of relevant genes in Litchi (*Litchi chinensis*) plants show that ROS and NO cause senescence in young leaves (Yang et al., 2018). NO-deficient mutant plants (*nos1/noa1*) exhibit early senescence (Niu & Guo, 2012) and are known to scavenge H_2O_2 and relieve ROS toxicity to postpone senescence (Sang et al., 2008). Furthermore, the NO-degrading dioxygenases (NOD) enzyme causes premature senescence in plants when it is expressed ectopically (Mishina et al., 2007). Absence of the cyclic nucleotide-gated ion channel 2 (CNGC2), which is located in the plasma membrane, causes early senescence in plants. For instance, in comparison to their WT counterparts, Arabidopsis thaliana CNGC2 loss of function dnd1 plants show reduced levels of NO, chlorophyll loss, H_2O_2 production, lipid peroxidation, and tissue necrosis (Ma et al., 2010) demonstrating an opposing behavior of ROS and NO during senescence. The antioxidant mechanism, which prevents chlorophyll breakdown and increases thylakoid membrane stability, may play a role in NO's protective properties against ROS-mediated senescence (Liu & Guo, 2013). Catalase enzyme (OsCATC) is encoded by the nitric oxide excess 1 (*NOE1*) gene in rice (Oryza sativa). The rice NO-accumulation mutant noe1 generates more H_2O_2 in the leaves, which stimulates the generation of NO by activating the NR enzyme. Reducing cell death by removing excess NO suggests that NO mediates the effects of H_2O_2 on these cells. Additionally, over-expression of rice GSNOR decreased cell mortality in noe1 rice plants by decreasing cellular SNO levels. Another study found that when NO was applied externally to maize (Zea mays) leaves, it quickly triggered the creation of H_2O_2 and induced cell death (Kong et al., 2013). These findings suggest that NO and NO-mediated protein S-nitrosylation play critical roles in H_2O_2-mediated cell death and senescence (Lin et al., 2012; Yiqin Wang et al., 2013).

13.6 CONCLUSION AND FUTURE PROSPECTS

Protecting fruits and vegetables from H_2S, NO, and H_2O_2, all of which play essential roles in the senescence process, is the focus of this chapter. However, different species of fruits and vegetables require varying quantities of H_2S, NO, and H_2O_2 for postharvest protection. Thus, this chapter illuminated the crucial regulatory role of H_2S, NO, and H_2O_2 in extending the postharvest shelf life of fruits and vegetables. Research on ROS signaling in ripening fruits and vegetables is vast. However, numerous open problems need to be described in order to comprehend the interplay of H_2S, NO, and H_2O_2. More research is needed into the signaling role of H_2S, NO, and H_2O_2 created during post-harvest storage of fruits and vegetables. In this section, it is crucial to investigate whether or how H_2S, NO, and H_2O_2 directly influence post-harvest preservation of fruits and vegetables. The post-harvest effects of H_2S, NO, and H_2O_2 crosstalk remain puzzling, despite the abundance of evidence supporting their signaling role in plants. Therefore, a more comprehensive framework is necessary for future research. Indeed, the interaction of these molecules with other phytohormones and membrane channels is the major function of these molecules in fruit and vegetable development. As a result, biochemical and molecular investigations are needed to decode the complicated mechanism of NO, H_2O_2, and H_2S crosstalk in fruits and vegetables, which is expected to affect plant growth, development, and crop output. Understanding the molecular mechanisms of NO, H_2O_2, and H_2S crosstalk in plants may seem useful in crop management, especially in regard to senescence, and may serve as the foundation for future biotechnology use in agriculture, particularly in light of rising global food shortages.

KEYWORDS

- **bioengineering**
- **biosynthesis**
- **crosstalk**
- **membrane channels**
- **phytohormones**
- **post-harvest**
- **postharvest quality improvement**
- **shelf life**
- **signaling genes**

REFERENCES

Aghdam, M. S., Mahmoudi, R., Razavi, F., Rabiei, V., & Soleimani, A., (2018). Hydrogen sulfide treatment confers chilling tolerance in hawthorn fruit during cold storage by triggering endogenous H_2S accumulation, enhancing antioxidant enzymes activity and promoting phenols accumulation. *Sci. Hortic. (Amsterdam), 238*, 264–271. https://doi.org/10.1016/J.SCIENTA.2018.04.063.

Aghdam, M. S., Mukherjee, S., Flores, F. B., Arnao, M. B., Luo, Z., & Corpas, F. J., (2023). Functions of melatonin during postharvest of horticultural crops. *Plant Cell Physiol., 63*, 1764–1786. https://doi.org/10.1093/PCP/PCAB175.

Aghdam, M. S., Palma, J. M., & Corpas, F. J., (2020). NADPH as a quality footprinting in horticultural crops marketability. *Trends Food Sci. Technol., 103*, 152–161. https://doi.org/10.1016/J.TIFS.2020.07.002.

Al Ubeed, H. M. S., Wills, R. B. H., Bowyer, M. C., & Golding, J. B., (2018). Comparison of hydrogen sulphide with 1-methylcyclopropene (1-MCP) to inhibit senescence of the leafy vegetable, Pak choy. *Postharvest Biol. Technol., 137*, 129–133. https://doi.org/10.1016/J.POSTHARVBIO.2017.11.020.

Al Ubeed, H. M. S., Wills, R. B. H., Bowyer, M. C., Vuong, Q. V., & Golding, J. B., (2017). Interaction of exogenous hydrogen sulphide and ethylene on senescence of green leafy vegetables. *Postharvest Biol. Technol., 133*, 81–87. https://doi.org/10.1016/J.POSTHARVBIO.2017.07.010.

Álvarez, C., García, I., Moreno, I., Pérez-Pérez, M. E., Crespo, J. L., Romero, L. C., & Gotor, C., (2012). Cysteine-generated sulfide in the cytosol negatively regulates autophagy and modulates the transcriptional profile in *Arabidopsis. Plant Cell, 24*, 4621–4634. https://doi.org/10.1105/TPC.112.105403.

Ansari, S., Kumar, V., Bhatt, D. N., Irfan, M., & Datta, A., (2022). N-acetylglucosamine sensing and metabolic engineering for attenuating human and plant pathogens. *Bioeng., 9*, 64. https://doi.org/10.3390/BIOENGINEERING9020064.

Apel, K., & Hirt, H., (2004). *Reactive Oxygen Species: Metabolism, Oxidative Stress, and Signal Transduction, 55,* 373–399. https://doi.org/10.1146/ANNUREV.ARPLANT.55.031903.141701.

Arif, Y., Hayat, S., Yusuf, M., & Bajguz, A., (2021). Hydrogen sulfide: A versatile gaseous molecule in plants. *Plant Physiol. Biochem., 158*, 372–384. https://doi.org/10.1016/J.PLAPHY.2020.11.045.

Asgher, M., Per, T. S., Masood, A., Fatma, M., Freschi, L., Corpas, F. J., & Khan, N. A., (2016). Nitric oxide signaling and its crosstalk with other plant growth regulators in plant responses to abiotic stress. Environ. *Sci. Pollut. Res., 24*, 2273–2285. https://doi.org/10.1007/S11356-016-7947-8.

Birke, H., De Kok, L. J., Wirtz, M., & Hell, R., (2015). The role of compartment-specific cysteine synthesis for sulfur homeostasis during H_2S exposure in *Arabidopsis. Plant Cell Physiol., 56*, 358–367. https://doi.org/10.1093/PCP/PCU166.

Calderwood, A., & Kopriva, S., (2014). Hydrogen sulfide in plants: From dissipation of excess sulfur to signaling molecule. *Nitric Oxide, 41*, 72–78. https://doi.org/10.1016/J.NIOX.2014.02.005.

Chang, Z., Jingying, S., Liqin, Z., Changle, L., & Qingguo, W., (2014). Cooperative effects of hydrogen sulfide and nitric oxide on delaying softening and decay of strawberry. *Int. J. Agric. Biol. Eng., 7*, 114–122. https://doi.org/10.25165/IJABE.V7I6.1582.

Chen, J., Wu, F. H., Wang, W. H., Zheng, C. J., Lin, G. H., Dong, X. J., He, J. X., et al., (2011). Hydrogen sulphide enhances photosynthesis through promoting chloroplast biogenesis, photosynthetic enzyme expression, and thiol redox modification in *Spinacia oleracea* seedlings. *J. Exp. Bot., 62*, 4481–4493. https://doi.org/10.1093/JXB/ERR145.

Cheng, G., Yang, E., Lu, W., Jia, Y., Jiang, Y., & Duan, X., (2009). Effect of nitric oxide on ethylene synthesis and softening of banana fruit slice during ripening. *J. Agric. Food Chem., 57*, 5799–5804. https://doi.org/10.1021/JF901173N.

Corpas, F. J., & Barroso, J. B., (2017). Nitric oxide synthase-like activity in higher plants. *Nitric Oxide, 68*, 5, 6. https://doi.org/10.1016/J.NIOX.2016.10.009.

Corpas, F. J., & Palma, J. M., (2020). H2S signaling in plants and applications in agriculture. *J. Adv. Res., 24*, 131–137. https://doi.org/10.1016/J.JARE.2020.03.011.
Corpas, F. J., González-Gordo, S., & Palma, J. M., (2020). Nitric oxide: A radical molecule with potential biotechnological applications in fruit ripening. *J. Biotechnol., 324*, 211–219. https://doi.org/10.1016/J.JBIOTEC.2020.10.020.
Corpas, F. J., González-Gordo, S., Cañas, A., & Palma, J. M., (2019). Nitric oxide and hydrogen sulfide in plants: Which comes first? *J. Exp. Bot., 70*, 4391–4404. https://doi.org/10.1093/JXB/ERZ031.
Costa-Broseta, Á., Castillo, M., & León, J., (2021). Post-translational modifications of nitrate reductases autoregulates nitric oxide biosynthesis in *Arabidopsis*. *Int. J. Mol. Sci., 22*, 549. https://doi.org/10.3390/IJMS22020549.
Decros, G., Baldet, P., Beauvoit, B., Stevens, R., Flandin, A., Colombié, S., Gibon, Y., & Pétriacq, P., (2019). Get the balance right: ROS homeostasis and redox signaling in fruit. *Front. Plant Sci., 10*, 1091. https://doi.org/10.3389/FPLS.2019.01091/BIBTEX.
Freschi, L., (2013). Nitric oxide and phytohormone interactions: Current status and perspectives. *Front. Plant Sci., 4*, 398. https://doi.org/10.3389/FPLS.2013.00398/BIBTEX.
Gambhir, P., Raghuvanshi, U., Parida, A. P., Kujur, S., Sharma, S., Sopory, S. K., Kumar, R., & Sharma, A. K., (2023). Elevated methylglyoxal levels inhibit tomato fruit ripening by preventing ethylene biosynthesis. *Plant Physiol.* https://doi.org/10.1093/PLPHYS/KIAD142.
Ge, Y., Hu, K. Di., Wang, S. S., Hu, L. Y., Chen, X. Y., Li, Y. H., Yang, Y., et al., (2017). Hydrogen sulfide alleviates postharvest ripening and senescence of banana by antagonizing the effect of ethylene. *PLoS One, 12*, e0180113. https://doi.org/10.1371/JOURNAL.PONE.0180113.
Ghosh, S., Singh, U. K., Meli, V. S., Kumar, V., Kumar, A., Irfan, M., Chakraborty, N., et al., (2013). Induction of senescence and identification of differentially expressed genes in tomato in response to monoterpene. *PLoS One, 8*, e76029. https://doi.org/10.1371/JOURNAL.PONE.0076029.
Gotor, C., Laureano-Marín, A. M., Moreno, I., Aroca, Á., García, I., & Romero, L. C., (2015). Signaling in the plant cytosol: Cysteine or sulfide? *Amino Acids, 47*, 2155–2164. https://doi.org/10.1007/S00726-014-1786-Z/METRICS.
Hancock, J. T., & Whiteman, M., (2016). Hydrogen sulfide signaling: Interactions with nitric oxide and reactive oxygen species. *Ann. N. Y. Acad. Sci., 1365*, 5–14. https://doi.org/10.1111/NYAS.12733.
He, H., Li, Y., & He, L. F., (2019). Role of nitric oxide and hydrogen sulfide in plant aluminum tolerance. *BioMetals, 32*, 1–9. https://doi.org/10.1007/S10534-018-0156-9/METRICS.
Hong, K., Gong, D., Xu, H., Wang, S., Jia, Z., Chen, J., & Zhang, L., (2014). *Effects of Salicylic Acid and Nitric Oxide Pretreatment on the Expression of Genes Involved in the Ethylene Signaling Pathway and the Quality of Postharvest Mango Fruit, 42*, 205–216. http://dx.doi.org/10.1080/01140671.2014.892012; https://doi.org/10.1080/01140671.2014.892012.
Hu, H., Liu, D., Li, P., & Shen, W., (2015). Hydrogen sulfide delays leaf yellowing of stored water spinach (Ipomoea aquatica) during dark-induced senescence by delaying chlorophyll breakdown, maintaining energy status, and increasing

antioxidative capacity. *Postharvest Biol. Technol., 108*, 8–20. https://doi.org/10.1016/J.POSTHARVBIO.2015.05.003.

Hu, H., Shen, W., & Li, P., (2014). Effects of hydrogen sulphide on quality and antioxidant capacity of mulberry fruit. *Int. J. Food Sci. Technol., 49*, 399–409. https://doi.org/10.1111/IJFS.12313.

Hu, L. Y., Hu, S. L., Wu, J., Li, Y. H., Zheng, J. L., Wei, Z. J., Liu, J., et al., (2012). Hydrogen sulfide prolongs postharvest shelf life of strawberry and plays an antioxidative role in fruits. *J. Agric. Food Chem., 60*, 8684–8693. https://doi.org/10.1021/JF300728H/ASSET/IMAGES/MEDIUM/JF-2012-00728H_0006.GIF.

Irfan, M., Ghosh, S., Kumar, V., Chakraborty, N., Chakraborty, S., & Datta, A., (2014). Insights into transcriptional regulation of β-D-N-acetylhexosaminidase, an N-glycan-processing enzyme involved in ripening-associated fruit softening. *J. Exp. Bot., 65*, 5835–5848. https://doi.org/10.1093/JXB/ERU324.

Irfan, M., Ghosh, S., Meli, V. S., Kumar, A., Kumar, V., Chakraborty, N., Chakraborty, S., & Datta, A., (2016). Fruit ripening regulation of α-mannosidase expression by the MADS box transcription factor RIPENING INHIBITOR and ethylene. *Front. Plant Sci., 7*, 10. https://doi.org/10.3389/FPLS.2016.00010/BIBTEX.

Irfan, M., Kumar, P., Kumar, V., & Datta, A., (2022). Fruit ripening specific expression of β-D-N-acetylhexosaminidase (β-Hex) gene in tomato is transcriptionally regulated by ethylene response factor SlERF.E4. *Plant Sci., 323*, 111380. https://doi.org/10.1016/J.PLANTSCI.2022.111380.

Jacob, J. K., Tiwari, K., Correa-Betanzo, J., Misran, A., Chandrasekaran, R., & Paliyath, G., (2012). *Biochemical Basis for Functional Ingredient Design from Fruits, 3*, 79–104. https://doi.org/10.1146/ANNUREV-FOOD-022811-101127.

Jiang, X., Lin, H., Lin, M., Chen, Y., Wang, H., Lin, Y., Shi, J., & Lin, Y., (2018). A novel chitosan formulation treatment induces disease resistance of harvested litchi fruit to *Peronophythora litchii* in association with ROS metabolism. *Food Chem., 266*, 299–308. https://doi.org/10.1016/J.FOODCHEM.2018.06.010.

Jimenez, A., Creissen, G., Kular, B., Firmin, J., Robinson, S., Verhoeyen, M., & Mullineaux, P., (2002). Changes in oxidative processes and components of the antioxidant system during tomato fruit ripening. *Planta, 214*, 751–758. https://doi.org/10.1007/S004250100667/METRICS.

Jiménez-Muñoz, R., Palma, F., Carvajal, F., Castro-Cegrí, A., Pulido, A., Jamilena, M., Romero-Puertas, M. C., & Garrido, D., (2021). Pre-storage nitric oxide treatment enhances chilling tolerance of zucchini fruit (*Cucurbita pepo* L.) by S-nitrosylation of proteins and modulation of the antioxidant response. *Postharvest Biol. Technol., 171*, 111345. https://doi.org/10.1016/J.POSTHARVBIO.2020.111345.

Jin, Z., Shen, J., Qiao, Z., Yang, G., Wang, R., & Pei, Y., (2011). Hydrogen sulfide improves drought resistance in *Arabidopsis thaliana*. *Biochem. Biophys. Res. Commun., 414*, 481–486. https://doi.org/10.1016/J.BBRC.2011.09.090.

Karaca, H., (2010). *Use of Ozone in the Citrus Industry, 32*, 122–129. http://dx.doi.org/10.1080/01919510903520605; https://doi.org/10.1080/01919510903520605.

Kong, X., Zhang, D., Pan, J., Zhou, Y., & Li, D., (2013). Hydrogen peroxide is involved in nitric oxide-induced cell death in maize leaves. *Plant Biol., 15*, 53–59. https://doi.org/10.1111/J.1438-8677.2012.00598.X.

Kumar, V., Chattopadhyay, A., Ghosh, S., Irfan, M., Chakraborty, N., Chakraborty, S., & Datta, A., (2016a). Improving nutritional quality and fungal tolerance in soya bean and grass pea by expressing an oxalate decarboxylase. *Plant Biotechnol. J., 14*, 1394–1405. https://doi.org/10.1111/PBI.12503.

Kumar, V., Irfan, M., & Datta, A., (2019). Manipulation of oxalate metabolism in plants for improving food quality and productivity. *Phytochemistry, 158*, 103–109. https://doi.org/10.1016/J.PHYTOCHEM.2018.10.029.

Kumar, V., Irfan, M., Ghosh, S., Chakraborty, N., Chakraborty, S., & Datta, A., (2016b). Fruit ripening mutants reveal cell metabolism and redox state during ripening. *Protoplasma, 253*, 581–594. https://doi.org/10.1007/S00709-015-0836-Z/METRICS.

Kumari, C., Sharma, M., Kumar, V., Sharma, R., Kumar, V., Sharma, P., Kumar, P., & Irfan, M., (2022). Genome editing technology for genetic amelioration of fruits and vegetables for alleviating post-harvest loss. *Bioeng., 9*, 176. https://doi.org/10.3390/BIOENGINEERING9040176.

Kuo, E. Y. H., Chang, H. L., Lin, S. T., & Lee, T. M., (2020). High light-induced nitric oxide production induces autophagy and cell death in *Chlamydomonas reinhardtii. Front. Plant Sci., 11*, 772. https://doi.org/10.3389/FPLS.2020.00772/BIBTEX.

Lacey, R. F., & Binder, B. M., (2014). How plants sense ethylene gas–the ethylene receptors. *J. Inorg. Biochem., 133*, 58–62. https://doi.org/10.1016/J.JINORGBIO.2014.01.006.

Li, D., Li, L., Ge, Z., Limwachiranon, J., Ban, Z., Yang, D., & Luo, Z., (2017). Effects of hydrogen sulfide on yellowing and energy metabolism in broccoli. *Postharvest Biol. Technol., 129*, 136–142. https://doi.org/10.1016/J.POSTHARVBIO.2017.03.017.

Li, D., Limwachiranon, J., Li, L., Du, R., & Luo, Z., (2016). Involvement of energy metabolism to chilling tolerance induced by hydrogen sulfide in cold-stored banana fruit. *Food Chem., 208*, 272–278. https://doi.org/10.1016/J.FOODCHEM.2016.03.113.

Li, J., Jia, H., Wang, J., Cao, Q., & Wen, Z., (2014). Hydrogen sulfide is involved in maintaining ion homeostasis via regulating plasma membrane Na^+/H^+ antiporter system in the hydrogen peroxide-dependent manner in salt-stress *Arabidopsis thaliana* root. *Protoplasma, 251*, 899–912. https://doi.org/10.1007/S00709-013-0592-X/METRICS.

Li, S. P., Hu, K. Di., Hu, L. Y., Li, Y. H., Jiang, A. M., Xiao, F., Han, Y., et al., (2014). Hydrogen sulfide alleviates postharvest senescence of broccoli by modulating antioxidant defense and senescence-related gene expression. *J. Agric. Food Chem., 62*, 1119–1129. https://doi.org/10.1021/JF4047122/ASSET/IMAGES/MEDIUM/JF-2013-047122_0008.GIF.

Li, T. T., Li, Z. R., Hu, K. Di., Hu, L. Y., Chen, X. Y., Li, Y. H., Yang, Y., et al., (2017). Hydrogen sulfide alleviates kiwifruit ripening and senescence by antagonizing effect of ethylene. *HortScience, 52*, 1556–1562. https://doi.org/10.21273/HORTSCI12261-17.

Li, Z. G., Ding, X. J., & Du, P. F., (2013). Hydrogen sulfide donor sodium hydrosulfide-improved heat tolerance in maize and involvement of proline. *J. Plant Physiol., 170*, 741–747. https://doi.org/10.1016/J.JPLPH.2012.12.018.

Lin, A., Wang, Y., Tang, J., Xue, P., Li, C., Liu, L., Hu, B., et al., (2012). Nitric oxide and protein S-nitrosylation are integral to hydrogen peroxide-induced leaf cell death in rice. *Plant Physiol., 158*, 451–464. https://doi.org/10.1104/PP.111.184531.

Liu, F., & Guo, F. Q., (2013). Nitric oxide deficiency accelerates chlorophyll breakdown and stability loss of thylakoid membranes during dark-induced leaf senescence in *Arabidopsis. PLoS One, 8*, e56345. https://doi.org/10.1371/JOURNAL.PONE.0056345.

Lum, G. B., Shelp, B. J., DeEll, J. R., & Bozzo, G. G., (2016). Oxidative metabolism is associated with physiological disorders in fruits stored under multiple environmental stresses. *Plant Sci., 245*, 143–152. https://doi.org/10.1016/J.PLANTSCI.2016.02.005.

Luo, S., Tang, Z., Yu, J., Liao, W., Xie, J., Lv, J., Feng, Z., & Dawuda, M. M., (2020). Hydrogen sulfide negatively regulates cd-induced cell death in cucumber (*Cucumis sativus* L) root tip cells. *BMC Plant Biol., 20*, 1–13. https://doi.org/10.1186/S12870-020-02687-8/FIGURES/8.

Luo, Z., Li, D., Du, R., & Mou, W., (2015). Hydrogen sulfide alleviates chilling injury of banana fruit by enhanced antioxidant system and proline content. *Sci. Hortic. (Amsterdam), 183*, 144–151. https://doi.org/10.1016/J.SCIENTA.2014.12.021.

Ma, W., Smigel, A., Walker, R. K., Moeder, W., Yoshioka, K., & Berkowitz, G. A., (2010). Leaf senescence signaling: The Ca^{2+}-conducting *Arabidopsis* cyclic nucleotide gated channel2 acts through nitric oxide to repress senescence programming. *Plant Physiol., 154*, 733–743. https://doi.org/10.1104/PP.110.161356.

Madebo, M. P., Ayalew, Y., Zheng, Y., & Jin, P., (2022). *Nitric Oxide and Its Donor Sodium-Nitroprusside Regulation of the Postharvest Quality and Oxidative Stress on Fruits: A Systematic Review and Meta-Analysis*. https://doi.org/10.1080/87559129.2022.2122995. https://doi.org/10.1080/87559129.2022.2122995.

Martí, M. C., Camejo, D., Olmos, E., Sandalio, L. M., Fernández-García, N., Jiménez, A., & Sevilla, F., (2009). Characterization and changes in the antioxidant system of chloroplasts and chromoplasts isolated from green and mature pepper fruits. *Plant Biol., 11*, 613–624. https://doi.org/10.1111/J.1438-8677.2008.00149.X.

Mehta, R. A., Cassol, T., Li, N., Ali, N., Handa, A. K., & Mattoo, A. K., (2002). Engineered polyamine accumulation in tomato enhances phytonutrient content, juice quality, and vine life. *Nat. Biotechnol., 20*, 613–618. https://doi.org/10.1038/nbt0602-613.

Mishina, T. E., Lamb, C., & Zeier, J., (2007). Expression of a nitric oxide degrading enzyme induces a senescence program in *Arabidopsis. Plant. Cell Environ., 30*, 39–52. https://doi.org/10.1111/J.1365-3040.2006.01604.X.

Mishra, V., Singh, P., Tripathi, D. K., Corpas, F. J., & Singh, V. P., (2021). Nitric oxide and hydrogen sulfide: An indispensable combination for plant functioning. *Trends Plant Sci., 26*, 1270–1285. https://doi.org/10.1016/J.TPLANTS.2021.07.016.

Mohn, M. A., Thaqi, B., & Fischer-Schrader, K., (2019). Isoform-specific NO synthesis by *Arabidopsis thaliana* nitrate reductase. *Plants, 8*, 67. https://doi.org/10.3390/PLANTS8030067.

Mukherjee, S., & Bozhkov, P., (2019). Insights into nitric oxide–melatonin crosstalk and N-nitrosomelatonin functioning in plants. *J. Exp. Bot., 70*, 6035–6047. https://doi.org/10.1093/JXB/ERZ375.

Muñoz, P., & Munné-Bosch, S., (2018). Photo-oxidative stress during leaf, flower, and fruit development. *Plant Physiol., 176*, 1004–1014. https://doi.org/10.1104/PP.17.01127.

Ni, Z. J., Hu, K. Di., Song, C. B., Ma, R. H., Li, Z. R., Zheng, J. L., Fu, L. H., et al., (2016). Hydrogen sulfide alleviates postharvest senescence of grape by modulating the antioxidant defenses. *Oxid. Med. Cell. Longev., 2016*. https://doi.org/10.1155/2016/4715651.

Niu, Y. H., & Guo, F. Q., (2012). Nitric oxide regulates dark-induced leaf senescence through EIN2 in *Arabidopsis. Journal of Integrative Plant Biology, 54*(8), 516–525. https://doi.org/10.1111/j.1744-7909.2012.01140.x.

Pirrung, M. C., Bleecker, A. B., Inoue, Y., Rodríguez, F. I., Sugawara, N., Wada, T., Zou, Y., & Binder, B. M., (2008). Ethylene receptor antagonists: Strained alkenes are necessary but not sufficient. *Chem. Biol., 15*, 313–321. https://doi.org/10.1016/J.CHEMBIOL.2008.02.018.

Prakash, V., Singh, V. P., Tripathi, D. K., Sharma, S., & Corpas, F. J., (2019). Crosstalk between nitric oxide (NO) and abscisic acid (ABA) signaling molecules in higher plants. *Environ. Exp. Bot., 161*, 41–49. https://doi.org/10.1016/J.ENVEXPBOT.2018.10.033.

Qi, Q., Guo, Z., Liang, Y., Li, K., & Xu, H., (2019). Hydrogen sulfide alleviates oxidative damage under excess nitrate stress through MAPK/NO signaling in cucumber. *Plant Physiol. Biochem., 135*, 1–8. https://doi.org/10.1016/J.PLAPHY.2018.11.017.

Rasool, G., Buchholz, G., Yasmin, T., Shabbir, G., Abbasi, N. A., & Malik, S. I., (2021). Overexpression of SlGSNOR impairs *in vitro* shoot proliferation and developmental architecture in tomato but confers enhanced disease resistance. *J. Plant Physiol., 261*, 153433. https://doi.org/10.1016/J.JPLPH.2021.153433.

Rather, B. A., Mir, I. R., Sehar, Z., Anjum, N. A., Masood, A., & Khan, N. A., (2020). The outcomes of the functional interplay of nitric oxide and hydrogen sulfide in metal stress tolerance in plants. *Plant Physiol. Biochem., 155*, 523–534. https://doi.org/10.1016/J.PLAPHY.2020.08.005.

Rodríguez-Ruiz, M., Mateos, R. M., Codesido, V., Corpas, F. J., & Palma, J. M., (2017). Characterization of the galactono-1,4-lactone dehydrogenase from pepper fruits and its modulation in the ascorbate biosynthesis. Role of nitric oxide. *Redox Biol., 12*, 171–181. https://doi.org/10.1016/J.REDOX.2017.02.009.

Saengnil, K., Chumyam, A., Faiyue, B., & Uthaibutra, J., (2014). Use of chlorine dioxide fumigation to alleviate enzymatic browning of harvested 'Daw' longan pericarp during storage under ambient conditions. *Postharvest Biol. Technol., 91*, 49–56. https://doi.org/10.1016/J.POSTHARVBIO.2013.12.016.

Sang, J., Jiang, M., Lin, F., Xu, S., Zhang, A., & Tan, M., (2008). Nitric oxide reduces hydrogen peroxide accumulation involved in water stress-induced subcellular anti-oxidant defense in maize plants. *Journal of Integrative Plant Biology, 50*(2), 231–243. https://doi.org/10.1111/j.1744-7909.2007.00594.x.

Sharma, S., & Sharma, R. R., (2015). Nitric oxide inhibits activities of PAL and PME enzymes and reduces chilling injury in 'Santa Rosa' Japanese plum (*Prunus salicina* Lindell). *J. Plant Biochem. Biotechnol., 24*, 292–297. https://doi.org/10.1007/S13562-014-0271-9/METRICS.

Smirnoff, N., & Arnaud, D., (2019). Hydrogen peroxide metabolism and functions in plants. *New Phytol., 221*, 1197–1214. https://doi.org/10.1111/NPH.15488.

Steffens, C. A., Santana, G. R. O., Amarante, C. V. T. Do., Antonovviski, J. L., Miqueloto, T., Anami, J. M., & Fenili, C. L., (2022). Treatment with nitric oxide in controlled atmosphere storage to preserve the quality of 'Laetitia' plums. *LWT, 158*, 113033. https://doi.org/10.1016/J.LWT.2021.113033.

Tayal, R., Kumar, V., & Irfan, M., (2022). Harnessing the power of hydrogen sulphide (H_2S) for improving fruit quality traits. *Plant Biol., 24*, 594–601. https://doi.org/10.1111/PLB.13372.

Wang, H., Ji, F., Zhang, Y., Hou, J., Liu, W., Huang, J., & Liang, W., (2019). Interactions between hydrogen sulphide and nitric oxide regulate two soybean citrate transporters during the alleviation of aluminum toxicity. *Plant. Cell Environ., 42*, 2340–2356. https://doi.org/10.1111/PCE.13555.

Wang, Y., Li, L., Cui, W., Xu, S., Shen, W., & Wang, R., (2012). Hydrogen sulfide enhances alfalfa (*Medicago sativa*) tolerance against salinity during seed germination by nitric oxide pathway. *Plant Soil, 351*, 107–119. https://doi.org/10.1007/S11104-011-0936-2/METRICS.

Wang, Y., Lin, A., Loake, G. J., & Chu, C., (2013). H_2O_2-induced leaf cell death and the crosstalk of reactive nitric/oxygen species. *Journal of Integrative Plant Biology, 55*, 202–208. https://doi.org/10.1111/JIPB.12032.

Wang, Y., Luo, Z., Du, R., Liu, Y., Ying, T., & Mao, L., (2013). Effect of nitric oxide on antioxidative response and proline metabolism in banana during cold storage. *J. Agric. Food Chem., 61*, 8880–8887. https://doi.org/10.1021/JF401447Y/ASSET/IMAGES/MEDIUM/JF-2013-01447Y_0008.GIF.

Willett, W. C., & Stampfer, M. J., (2013). *Current Evidence on Healthy Eating, 34*, 77–95. https://doi.org/10.1146/annurev-publhealth-031811-124646; https://doi.org/10.1146/ANNUREV-PUBLHEALTH-031811-124646.

Wu, B., Guo, Q., Li, Q., Ha, Y., Li, X., & Chen, W., (2014). Impact of postharvest nitric oxide treatment on antioxidant enzymes and related genes in banana fruit in response to chilling tolerance. *Postharvest Biol. Technol., 92*, 157–163. https://doi.org/10.1016/J.POSTHARVBIO.2014.01.017.

Yang, H., Kim, H. J., Chen, H., Lu, Y., Lu, X., Wang, C., & Zhou, B., (2018). Reactive oxygen species and nitric oxide induce senescence of rudimentary leaves and the expression profiles of the related genes in *Litchi chinensis*. *Hortic. Res., 5*, 23. https://doi.org/10.1038/S41438-018-0029-Y/41997884/41438_2018_ARTICLE_29.PDF.

Yao, G. F., Wei, Z. Z., Li, T. T., Tang, J., Huang, Z. Q., Yang, F., Li, Y. H., et al., (2018). Modulation of enhanced antioxidant activity by hydrogen sulfide antagonization of ethylene in tomato fruit ripening. *J. Agric. Food Chem., 66*, 10380–10387. https://doi.org/10.1021/ACS.JAFC.8B03951/ASSET/IMAGES/MEDIUM/JF-2018-03951Z_0007.GIF.

Zaharah, S. S., & Singh, Z., (2011). Mode of action of nitric oxide in inhibiting ethylene biosynthesis and fruit softening during ripening and cool storage of 'Kensington pride' mango. *Postharvest Biol. Technol., 62,* 258–266. https://doi.org/10.1016/J.POSTHARVBIO.2011.06.007.

Zhang, H., Hu, S. L., Zhang, Z. J., Hu, L. Y., Jiang, C. X., Wei, Z. J., Liu, J., et al., (2011). Hydrogen sulfide acts as a regulator of flower senescence in plants. *Postharvest Biol. Technol., 60*, 251–257. https://doi.org/10.1016/J.POSTHARVBIO.2011.01.006.

Zhang, L., Pei, Y., Wang, H., Jin, Z., Liu, Z., Qiao, Z., Fang, H., & Zhang, Y., (2015). Hydrogen sulfide alleviates cadmium-induced cell death through restraining ROS accumulation in roots of *Brassica rapa* L. ssp. pekinensis. *Oxid. Med. Cell. Longev., 2015*. https://doi.org/10.1155/2015/804603.

Zhang, Y., Butelli, E., De Stefano, R., Schoonbeek, H. J., Magusin, A., Pagliarani, C., Wellner, N., et al., (2013). Anthocyanins double the shelf life of tomatoes by delaying overripening and reducing susceptibility to gray mold. *Curr. Biol., 23*, 1094–1100. https://doi.org/10.1016/J.CUB.2013.04.072.

Zhang, Z., Xu, J., Chen, Y., Wei, J., & Wu, B., (2019). Nitric oxide treatment maintains postharvest quality of table grapes by mitigation of oxidative damage. *Postharvest Biol. Technol., 152*, 9–18. https://doi.org/10.1016/J.POSTHARVBIO.2019.01.015.

Zheng, J. L., Hu, L. Y., Hu, K. Di., Wu, J., Yang, F., & Zhang, H., (2016). Hydrogen sulfide alleviates senescence of fresh-cut apple by regulating antioxidant defense system and senescence-related gene expression. *HortScience, 51*, 152–158. https://doi.org/10.21273/HORTSCI.51.2.152.

Zhu, L., Du, H., Wang, W., Zhang, W., Shen, Y., Wan, C., & Chen, J., (2019). Synergistic effect of nitric oxide with hydrogen sulfide on inhibition of ripening and softening of peach fruits during storage. *Sci. Hortic. (Amsterdam), 256*, 108591. https://doi.org/10.1016/J.SCIENTA.2019.108591.

Zhu, S., & Zhou, J., (2006). Effects of nitric oxide on fatty acid composition in peach fruits during storage. *J. Agric. Food Chem., 54*, 9447–9452. https://doi.org/10.1021/JF062451U.

Ziogas, V., Molassiotis, A., Fotopoulos, V., & Tanou, G., (2018). Hydrogen sulfide: A potent tool in postharvest fruit biology and possible mechanism of action. *Front. Plant Sci., 9*, 1375. https://doi.org/10.3389/FPLS.2018.01375/BIBTEX.

CHAPTER 14

METABOLIC ENGINEERING OF PLANT VOLATILES IN FRUITS AND VEGETABLES FOR POSTHARVEST QUALITY ENHANCEMENT

SHIKHA TIWARI[1] and PRANITA BHATELE[2]

[1]*Department of Biochemistry, All India Institute of Medical Sciences, Bhopal, Madhya Pradesh, India*

[2]*Mata Gujri Mahila Mahavidyalaya, Jabalpur, Madhya Pradesh, India*

ABSTRACT

Postharvest losses of fruits and vegetables are a major challenge worldwide, especially in developing countries. Metabolic engineering offers a promising approach to extend shelf life and improve quality by manipulating the production of plant volatiles. This chapter discusses various strategies for metabolic engineering, including overexpression of desired genes, gene silencing, and induced mutations. The chapter also explores the role of different volatiles in biotic and abiotic stress tolerance, delayed ripening, and reduced softening. Examples of successful metabolic engineering of terpenoid, fatty acid-derived, and amino acid-derived volatiles are provided. The chapter concludes by highlighting the limitations of current knowledge and the need for further research to optimize metabolic engineering for postharvest quality enhancement.

Bioengineered Fruit and Vegetables: Trends in Postharvest Quality Improvement.
Mohammad Irfan and Pankaj Kumar (Eds.)

14.1 INTRODUCTION

Post-harvest losses of fruits and vegetables are quite high worldwide, especially in the developing world. Cold chain facilities are an expensive option, and support from companies specializing in cold chain instruments is not quite adequate. In this scenario, metabolic engineering involving recombinant DNA technology to increase the post-harvest shelf life of fruits and vegetables has proved to be an effective approach. Genes related to biochemical and physiological changes during the ripening process have been identified. Regulation of the expression of these genes by modification through engineering could be helpful in preventing post-harvest losses to some extent. The cell wall is composed of cellulose, xyloglucan, and pectin, consisting of polygalacturonides (Fishman et al., 2004). Cell wall metabolism enzymes play an important role in the ripening process. During the ripening process, there occurs modification of the cell wall components. Firm fruits have long pectin chains (>1,000 nm) which are broken down into thinner and shorter chains during the ripening process (Zhang et al., 2008). Fruit softens during ripening due to cell wall degradation and reduction in cell adhesion caused by pectin de-polymerization (Canton et al., 2020). The enzyme PG is involved in pectin de-polymerization and solubilization. Pectin can only be acted upon by PG when it is de-methyl esterified, a reaction catalyzed by pectin methyl esterase (PME) (Brummell & Harpster, 2001). PG and PME have been targeted for extending the shelf life of fruits.

Neutral sugars such as galactose and arabinose are involved in crosslinking pectin. Hydroxylation of these sugars in the cell wall causes the fruits to ripen (Talmadge et al., 1973). Preventing the hydroxylation reaction is beneficial to delay the ripening process. Another target enzyme to delay ripening is pectate lyase (*PL*). It degrades pectin by cleaving linkages between galacturonosyl residues. They also macerate and disrupt the cell wall assembly (Uluisik & Seymour, 2020). In the case of vegetables, approximately 50% of vegetables are lost due to post-harvest damage in developing countries. It has been observed that when there is an increase in the cytokinin level, the endogenous level of ethylene is reduced, which results in a delay in leaf senescence (Van, 1988). A bacterial enzyme IPT catalyzes an important step in cytokinin biosynthesis and can delay leaf senescence. Placing the gene for IPT under the control of a senescence-specific promoter SAG12 leads to the overproduction of this enzyme at the onset of senescence (Gan & Amasino, 1997) (Figure 14.1).

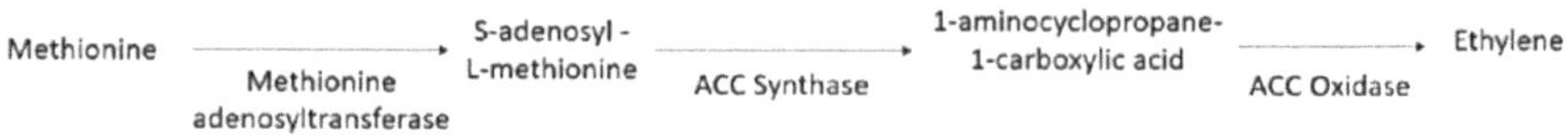

FIGURE 14.1 Ethylene biosynthesis (Imaseki, 1991).

Genes for ACC synthase and ACC oxidase have been identified and cloned. Antisense technology has been used to silence these genes to some extent, thus reducing ethylene production. Enzymes that degrade intermediates of ethylene biosynthesis can also serve as a potential target to prevent ripening. ACC deaminase and SAM hydrolase have been identified in microorganisms for this purpose. Antioxidants such as anthocyanin have also proved to be effective in increasing the shelf-life of fruits and vegetables (Zhang et al., 2013). Cloning antibiotic genes such as Nikkomycin (Kinzel, 1989), Chitinase (Lund et al., 1989), and even peroxidase (POD) (Vanitha & Umesha, 2008) can be an effective approach to protect vegetables and fruits from post-harvest decay by pathogens.

14.2 METABOLIC ENGINEERING APPROACH

Plants can produce various secondary metabolites that are volatile. These volatile compounds have numerous roles in pollination and breeding by attracting insects, dispersal of seeds, improving the strength of the plants, and plant defense. Additionally, maintaining the flavor and aroma of fruits and vegetables (Table 14.1), and control of volatile emissions during postharvest storage can further assist in the reduction of postharvest losses (PHLs). Currently, metabolic engineering is playing a crucial role in modifying the volatiles qualitatively as well as quantitatively during storage. This is done by changing the enzyme activities of certain metabolic pathways involved in secondary metabolism. The basic strategies of metabolic engineering include either gain of function or loss of function as discussed here in detail.

TABLE 14.1 Volatile Organic Compounds Contribution to Aroma of Fruits and Vegetables

Fruits	Volatile Class	Volatile Compounds	References
Citrus fruit, orange, grape, strawberry, tomato, peach, lemon.	Terpenes	R-limonene, valencene, Nerol, nerolidol, S-linalool, S-citronellol, geraniol, geranylacetone, citral, geranic acid, dihydrovomifoliol, α-terpineol, β-ionone, 3-hydroxy-β-ionone, 3-hydroxy-5,6-epoxy-β-ionone, 3-hydroxy-5,6-epoxy-7,8-dihydro-β-ionone nectarine, nectarine, α-sinensal, β-sinensal	Chambers et al. (2012); Zorrilla-Fontanesi et al. (2012)
Passion fruit, strawberry, melon, tomato, peach, grape, cucumber, apple	Fatty acid derived	1-(E, Z)-3,5-Undecatriene, 2-Pentanol, (Z)-3-Hexenol, 3,7-Dimethyl-1,5-octadiene-3,7-diol, (E)-2-hexenal, hexanal, (E)-2-nonenal, (Z)-6-nonenal, (Z)-3-hexenal, (E, Z)-2,6-nonadienal, 2-heptyl butanoate, methyl hexanoate, γ-octalactone, γ-decalactone, (Z, Z)-3,6-nonadienal, δ-decalactone, Ethyl 3-(methylthio)-propanoate, γ-hexalactone, γ-jasmolactone, 2-heptyl hexanoate, 3-(methylthio)-propyl acetate, ethyl (methylthio)-acetate.	Aragüez et al. (2013)
Tomato, melon, strawberry, banana, apple, pineapple pear, grape, bell pepper	Amino acid-derived	Isoamyl alcohol, 2-methyl butanol, 3-hydroxy-2-butanone, isoamyl acetate, 2-methyl butanoic, 2-methylpropyl acetate, 2-methyl butyl acetate, butyl acetate, hexyl acetate, ethyl-2-methyl propanoate, methyl-2-methyl butanoate, ethyl-2-methyl butanoate, 3-methyl butyl butanoate, 2-methoxy-3-isobutylpyrazine, 1-nitro-2-phenylethane, benzyl alcohol, isoeugenol, 2-phenyl ethanol, 3-phenyl-1-propanol, 2-phenylacetaldehyde.	Aubert et al. (2005)
Strawberry, pineapple, peach	Carbohydrate derived	2,5-dimethyl-4-hydroxy-3 (2H)-furanone, 6-pentyl α-pyrone.	Luan et al. (2005)

14.2.1 OVERPRODUCTION OF DESIRED GENE PRODUCTS BY TRANSGENESIS

The insertion of new genes into living plants for over-expression or additional expression is the process of transgenesis. The inserted gene may provide new features to the plant (additional expression) or improve the habitual quality of the plant (over-expression) (Holme et al., 2013). Cisgenia and intragenia are two main components of transgenesis. Cisgenia is the insertion of exogenous genes from the same plant, and intragenia is the insertion of exogenous genes from different but related plants (Holme et al., 2013). As the genetic code in all eukaryotic living organisms is the same, this maintains the functionality of exogenous genes in multiple plants at the species or genus level. A simple example is *Agrobacterium tumefaciens* (*A. tumifaciens*), a bacterium that develops "crown gall" disease in more than 150 species of plants by inserting a gene into the plant genome. This quality makes it a vector to carry exogenous genes inside the plant genome. The main mechanism is the replacement of the tumor-causing gene (T-DNA) by our gene of interest and infecting the plant cells with transformed (having our gene of interest) *A. tumefaciens*. This allows the natural insertion of desired genes and expression under the plant's regular life cycle (Gelvin et al., 2003). Other approaches include gene gun/biolistic gun, electroporation, and the use of vectors other than *A. tumefaciens*. Several commercial fruits and vegetables are available that have been modified by this technique, for example, soybean, canola, cotton, maize, tomato, banana, and citrus fruits, etc. (Quiro-Iturra et al., 2017).

14.2.2 GENE SILENCING (DOWN-REGULATION OR NO EXPRESSION)

The natural defense mechanism and proliferation of transposons in plants involve the mechanism of RNA interference (RNAi). RNAi can be defined as post-transcriptional gene silencing (PTGS) or translational inhibition by small non-coding RNAs (ncRNAs). These ncRNAs can be miRNA, siRNA, shRNA, and lncRNA. All non-coding RNAs are small, 20–25 nt in length, except lncRNA, which is 1,000–10,000 nt long (Ali et al., 2008). These RNAs can also be synthesized synthetically and act as anti-sense RNA. The hybridization of anti-sense non-coding RNA with sense mRNA strand activates the host plant RNA-induced silencing complex (RISC) and leads to degradation or inhibition of mRNA expression (Mocellin et al., 2004). According to the central dogma, each gene is expressed finally in the form of proteins. This mechanism of inhibiting mRNA expression ultimately results in gene silencing, leading to no or reduced expression of silenced genes. Many genes responsible for the ripening and discoloration of fruits and vegetables can be silenced to increase postharvest storage output.

14.2.3 INDUCED MUTATION IN PLANTS

It has been observed that sometimes mutation (change in the sequence of the gene) is beneficial, and mutated plants are evolutionarily adapted better than wild plants. This generates the hypothesis of induced mutation to improve the quality and shelf life of fruits and vegetables. Traditionally, there are two methods: Physical and Chemical mutagenesis. Physical mutagenesis includes the use of physical agents like ultraviolet rays, heat shock, electromagnetic radiation, and gamma rays. However, these methods are not preferred and cause damage to the host plant. Chemical methods are simpler and more effective ways of induced mutation, and ethyl-methane sulfonate (EMS) is the most widely used chemical mutagen (Munoz et al., 2004). EMS is one of the most reliable mutagens and has been used for the last 60 years in crop improvement programs. EMS either generates single nucleotide polymorphisms (SNPs; changes in a single nucleotide) or deletions (deletion of single or multiple nucleotides from the gene sequence), causing a change in expression at the protein level. This is visualized phenotypically and makes induced mutation commercially more feasible than transgenesis (Shu et al., 2012). Around 3,000 plant species have been modified with EMS-based methods such as wheat, rice, carrot, brinjal, tomato, banana, etc. (Quiroz-Iturra et al., 2017).

The current approach of induced mutation is clustered regularly interspaced short palindromic repeats-CRISPR associated proteins 9 (CRISPR/Cas9) based gene editing system. Physical and chemical methods mutate the gene non-specifically at many sites; this drawback can be overcome by CRISPR-Cas9 technology. CRISPR-Cas9 is the natural immune mechanism of many archaea and bacteria against phages and viruses. The main requirement of this editing technique is the presence of a protospacer adjacent motif (PAM) in the upstream region of the target gene sequence. PAM acts as a recognition and cutting site for Cas9 and inserts about 1–10 nucleotide long deletion. CRISPR/Cas9 system has been effectively used for many plants like citrus, wheat, tomato, etc., and considered as the most versatile and sensitive method for induced mutation (Gao et al., 2015).

14.3 POSTHARVEST QUALITY IMPROVEMENT BY METABOLIC ENGINEERING OF VOLATILES IN FRUITS AND VEGETABLES

Decay of fruits and vegetables during postharvest storage is a major challenge worldwide and is well documented with the degree of PHLs. In developing countries, more than 20–35% of harvested vegetables and fruits are spoiled by biotic as well as abiotic stress during postharvest handling and storage (Sharma et al., 2009; Ghaouth et al., 2004). The lack of facilities for proper transport, storage, processing, and the use of synthetic microbicides and chemical fertilizers are the real causes of postharvest damage to fruits and vegetables (Zhang et al., 2007). There are various health concerns about using these microbicides and objections by the scientific community. Therefore, metabolic engineering plays a crucial role in improving fruits and vegetables postharvest storage. Advanced biotechnology technologies can be implemented to develop fruits and vegetables with prolonged shelf life and enhanced quality, with engineering secondary metabolism being one way to make them more resistant to stress-induced spoilage. However, advances in biotechnology can be employed to develop fruits and vegetables with improved quality and shelf life. Principally, the biochemical, physiological, and molecular behavior of stored crops contribute to maintaining quality during storage (Lers, 2012). These phenotypic traits can be manipulated by genetic changes that may improve postharvest storage with an improvement in food supply to humans. There are various types of volatiles based on their chemical structure and biosynthetic pathways. Terpenes are the largest class, followed by aromatics, nitrogen-containing compounds. The terpenes are further classified into monoterpenes (biosynthetic pathway is methylerythritol phosphate pathway), and sesquiterpenes (synthesized by mevalonate pathway) (Phillips, Leon, Boronat, & Rodriguez-Concepcion, 2008). Isoprenoids, dimethylallyl diphosphate (DMADP), and isopentenyl diphosphate (IDP) are the end products of terpene biosynthesis. Aromatic volatiles are emitted during plant developmental processes, such as young leaves and flowering (Dudareva & Pichersky, 2006). Nitrogen-containing volatiles are synthesized from numerous plant primary metabolites with diverse origins. The role of volatile metabolic engineering in plants is discussed in subsections.

14.3.1 METABOLIC ENGINEERING FOR BIOTIC STRESS TOLERANCE

During postharvest storage, biotic stress is the result of the growth of microbial contamination. However, the source of microbes mainly belongs to soil but postharvest handling and storage conditions can also be sources of contamination. Bacteria, fungi, and viruses are the main biotic stress contributors, and these viral contaminations focus more on investigation. Viral infection is completely internal, in which the viral genome is inserted in the host genome, and the virus utilizes host machinery for replication and development. Some bacteria like *Erwinia carotovora* are also present in the internal sites of fruits and vegetables, enabling them to escape washing and surface sterilization procedures before postharvest storage.

Volatile organic compounds (VOCs) are metabolites of secondary metabolism and are affected by the presence of microbial contamination. There is a crosstalk between microbes and horticultural crops, and manipulation in the internal metabolism of plants also affects microbial growth. Some VOCs also fight against microbial attack, and up-regulation of their expression may reduce the biotic stress faced during storage. In the ancient history of humans, the antimicrobial nature of plants remains an integral part of pharmacology. For example, essential oils have been used in Indian Ayurvedic medicine to overcome stress, heal infections, especially against drug-resistant microbial infections (Dima & Dima, 2015). These volatiles act by changing the permeability of the microbial cell membrane and breaking microbial cell integrity. Terpenes can break the cell membranes of microbes by affecting the phospholipid bilayer. Terpenes integrate between the phospholipid acyl chains, which increases cell permeability and causes leakage of internal contents (Lambert et al., 2001). The aromatic volatiles can bind with proteins of the cell membrane and change their conformation (Bennis et al., 2004). Some Green Leaf Volatiles, like (E)-2-hexenal, also change the conformation of microbial extracellular proteins (Myung et al., 2007). Other volatiles, such as indolic volatiles, interrupt cytoskeleton structure (Mei et al., 2019). Some volatiles induce apoptosis (Chen et al., 2014), respiratory distress (Fry & Munch, 1975), change enzyme function (Wendakoon & Sakaguchi, 1995), and break quorum sensing to avoid microbial communication (Joshi et al., 2016) and microbial growth. Therefore, an increase in concentration of this plant defense volatile may lead to the inhibition of microbial growth during storage. Metabolic engineering of genes from the biosynthetic pathways of these volatiles by any of the above-mentioned approaches can prove beneficial to combat biotic stress during postharvest storage. Table 14.2 summarizes some important volatiles having a role in boosting fruits and vegetables' immunity against microbial infection. Transgenic tomatoes that over-express GLVs have been made and found to be resistant to sp. *lycopersici* (Xin et al., 2014) compared to previous wild-type tomatoes.

TABLE 14.2 Volatile Organic Compounds Against Microbial Infection

Volatile Organic Compound	Pathogen/Disease	References
Sesquiterpene, (E)-nerolidol, monoterpene,(–)-limonene	*Plasmopara viticola*	Algarra et al. (2015)
C6 aldehydes	Huanglongbing disease	Hijaz et al. (2016)
(–)-limonene	*Mangnaporthe oryzae*	Chen et al. (2018)
Green-leaf volatiles; C6 aldehydes and alcohols.	*P. syringae*, *Colletotrichum* sp., *Botrytis cinerea*	Mastui et al. (2006)
(Z)-hexenal and (Z)-decenal	*Colletotrichum lindemuthianum*	Quintana-Rodriguez et al. (2015)
(E)-β-caryophyllene	*P. syringae*	Huang et al. (2012)
(Z)-3-hexenyl acetate	*Fusarium graminearum*	Melotto et al. (2017)

There are some volatiles released during viral infection that attract various insects, resulting in postharvest losses. For instance, a cucumber infected with *cucumber mosaic virus* (CMV) secretes volatiles that attract insects. Silencing the genes responsible for these volatile compounds can help prevent insects from causing cucumber loss during storage. Further research and understanding of volatile crop defenses against microbial infections can lead to sustainable postharvest decay management.

14.3.2 METABOLIC ENGINEERING FOR ABIOTIC STRESS TOLERANCE

Fruits and vegetables are generally stored in cold storage at temperature ranges from 20°C to 15°C with varying humidity levels. Chilling injury, humidity pressure, and mechanical injury are the main sources of spoilage during storage. These injuries lead to wounds, making fruits and vegetables more susceptible to microbial infections and discoloration. Abiotic stresses trigger signaling pathways such as ethylene production, resulting in premature ripening. Efforts have been made to develop abiotic stress-resistant fruits and vegetables by manipulating genes involved in metabolic, regulatory, and signaling pathways, as well as enzymes within these pathways. Isoprene, a volatile compound, aids in temperature stress tolerance, but its concentration decreases with rising CO_2 levels. Up-regulating isoprene expression, even in the presence of CO_2, can reduce the risk of chilling injury and extend the shelf life of stored produce. Studies have shown that inhibitors of isoprene biosynthesis lead to increased reactive oxygen species (ROS) and lipid peroxidation in stored fruits, while transgenic plants with enhanced isoprene production exhibit reduced ROS and lipid peroxidation levels. These findings highlight the role of volatile isoprenoids in mitigating oxidative damage under abiotic stress conditions. Apart from isoprene, nitric oxide (NO) also serves as an oxidative stress scavenger during cold storage. Wang et al. (2015) observed that chilling injury during cold storage leads to a decrease in certain key volatiles such as aldehydes, ketones, alcohols, esters, terpenes, and acids, while increasing others: hexanal, 6-methyl-5-hepten-2-one, β-ionone, trans-2-hexenal, guaiacol, 2-phenyl ethanol, 2-isobutylthiazole, and 2-methyl butanal, thereby diminishing the flavor of tomatoes. Preceding cold storage with heat treatment was shown to preserve the concentration of flavor volatiles, thereby maintaining the tomato flavor (Wang et al., 2015). However, heat treatment may reduce shelf life and could potentially be substituted by genetic modification of enzymes involved in the biosynthesis of key volatiles. Similarly, red mango fruits (exposed directly to sunlight and located on the outer part of the tree) were found to be more cold-tolerant than green ones (positioned on the inner part of the mango tree) (Sudheeran et al., 2018). The red color of mangoes is attributed to their anthocyanin and flavonoid content. Engineering mango fruits to enhance anthocyanin and flavonoid expression can delay premature ripening. Cold injury in plums (*Prunus domestica L.*) can be mitigated by phenylalanine treatment prior to storage (Sogvar et al., 2020), while methyl jasmonate (JA) treatment in eggplants (*Solanum melongena L.*) (Shi et al., 2019) can confer resistance to chilling injury during cold storage. JAs, derived from C-18 fatty acid derivatives, can be upregulated in fruits and vegetables to combat both biotic and abiotic stress.

14.3.3 DELAYED RIPENING

The ripening of fruits and vegetables under storage is generally promoted by volatile compounds, especially ethylene. Ethylene production can be reduced by metabolic engineering successfully by down-regulating genes that encode enzymes of the biosynthetic pathway for ethylene or its precursor synthesis. This can be achieved by shifting metabolic flux from ethylene biosynthesis by over-expression of an enzyme that degrades its precursors. The 1-aminocyclopropane-1-carboxylic acid (ACC) is the most favorable precursor for the above-mentioned approach. RNAi technique has been implemented to silence the gene coded for 1-amino-cyclopropane-1-carboxylate (ACC) synthase. Three homologs were used to silence the ACC synthase gene that delays the tomato ripening up to 45 days. Engineered fruits and vegetables showed prolonged life because of low levels of ethylene but compromised flavor quality (Gupta et al., 2013). Likewise, in papaya, watermelon, and muskmelon fruit ripening inhibition and extension of shelf life can be attained by either stopping ethylene production by direct inhibition of enzymes ethylene synthesis or its precursors (Nuñez-Palenius et al., 2006).

Mutation has also been successfully implemented to reduce ethylene synthesis during storage. The target-induced local lesions in genomes approach has been used to introduce missense mutations at 194 positions of the ACC oxidase 1 enzyme in ethylene biosynthesis.

14.3.4 REDUCE RIPENED FRUITS AND VEGETABLES SOFTENING

Postharvest handling, shipping, and processing cause damage to fruits and vegetables, leading to extra softening. This excessive reduction in firmness results in around 35–40% loss of fruits and vegetables produced in India. India is the second largest producer of horticulture crops, and this type of loss results in significant commercial failure. Ethanol, (E)-β-caryophyllene, α-pinene, (Z)-3-hexanol, and hexadecanoic acid are the main volatile compounds associated with softening of fruits and vegetables (Lourdes Cardeal et al., 2005). The genes encoding enzymes of their biosynthetic pathway can be silenced or mutated to inhibit the emission of these volatiles. This can increase the shelf life and also reduce the chances of softening during storage.

14.4 METABOLIC ENGINEERING OF VOLATILE COMPOUNDS OF FRUITS AND VEGETABLES

Now that we understand the role of volatiles in the nutrient availability, ripening, softening, stress tolerance, and shelf life of fruits and vegetables, we will discuss various metabolic engineering evidence for the upregulation/downregulation of volatiles (Table 14.3). However, the biochemistry of horticultural crops under storage is still unexplored. In the past few years, several enzymes have been investigated, especially terpenoids, furanones, and phenylpropanoids synthesis pathways. For example, sesquiterpene synthase produces monoterpene S-linalool and sesquiterpene nerolidol, methyl ketone synthase, phenylacetaldehyde reductases catalyze the synthesis of 2-phenylethanol, and alcohol acyltransferases, etc. (Aharoni et al., 2004; Günther et al., 2011; Cumplido-Laso et al., 2012; Mageroy et al., 2012). Based on previous research and genome sequencing information, several volatiles have been modified to date.

14.4.1 METABOLIC ENGINEERING OF TERPENOID CONTENT OF FRUITS AND VEGETABLES

Tomato is the most challenging fruit under storage to combat, therefore numerous experiments have been performed on the improvement program for enhancement of flavor and aroma in tomatoes (*Solanum lycopersicum*) using a metabolic engineering approach. The most important volatile is monoterpene S-linalool which determines the aroma of tomato. Lewinsohn et al. (2001) successfully increased the concentration of S-linalool by metabolic engineering. S-linalool synthase gene isolated from *C. breweri* (CbLIS) under the promoter specific for late ripening causes an increase in the level of S-linalool and 8-hydroxylinalool in ripened tomato fruit (Lewinsohn et al., 2001). Similarly, the gene encoding for S-linalool or (3S)-E-nerolidol synthase was over-expressed in potato (*Solanum tuberosum*) constitutively. But these modifications lead to the formation of undesired non-volatile compounds. This problem was overcome by silencing SlCCD1B and CCDs SlCCD1A which decreased synthesis of undesired non-volatiles significantly.

In another experiment, Davidovich-Rikanati et al. (2007) inserted a gene encoding for geraniol synthase in tomatoes under the control of a promoter specific for ripening. This led to the biosynthesis of monoterpenes: nerol, geraniol, geranic, citronellol, neric acid, and citronellic acid, which were absent in control tomato fruit (Davidovich-Rikanati et al., 2007). This engineered tomato was very rich in flavor; however, the tomato did not develop a natural red color. This was due to lower levels of phytoene and

TABLE 14.3 Metabolic Engineering of Some Plants and its Effects

Fruit/ Vegetable	Genes	Expression of Gene	Effects	References
Tomato	*CbLIS, FaNES1, ObGES, ObZIS, AmGDPS-SSU, ShzFDPS, SlADH2, SlAADC1A, SlSAMT1*	Overexpression	Overproduction of S-linalool, 8-hydroxy linalool, geraniol, nerol, citronellol, citronellic acid, geranic acid, α-zingiberene, resistance to biotic stress, modify short-chain aldehyde and alcohols.	Prestage et al. (1999); Moummou et al. (2012)
Tomato	*SlCCD1A, SlCCD1B, SlscADH1, TomloxC, SlAADC1B, SlCXE1-5*	Lower expression/ no expression	Reduction of β-ionone,2-phenylacetaldehyde, 2-phenylethanol, geranyl acetone, and 1-nitro-2-phenylethane, increase C5 and C6 aldehydes.	Chen et al. (2004); Zawirska-Wojtasiak (2009)
Potato	*FaNES1*	Overexpression	Increase production of S-linalool, aphids, and resistance.	León et al. (2002); Torregrosa et al. (2008)
Potato	*StLOX H1*	Lower expression/ no expression.	Decrease C6 aldehydes	Cheng et al. (2007)
Grapevines	*VvADH2*	Overexpression	Lower level of benzyl alcohol and 2-phenylethanol in grapefruit.	Rodríguez et al. (2011)
Grapevines	*VvADH2*	Lower expression/ no expression	No modification at volatile level.	Torregrosa et al. (2008)
Strawberry	*FaCHS, ObEGS1, PhIGS1, FaOMT*	Lower expression/ no expression	Increase the level of eugenol, chavicol, isoeugenol, and methoxyeugenol.	Lunkenbein et al. (2006)
Apple	*MdACS, MdACO, MdACO1*	Lower expression/ no expression	Decrease ethylene production and prolong shelf life.	Dandekar et al. (2004)
Kiwi fruit	*AcACO1*	Lower expression/ no expression	Decrease ethylene production and prolong shelf life.	Günther et al. (2011)

lycopene in the tomato fruit. Sequentially, α-zingiberene synthase of *O. basilicum* and the small subunit (SSU) of GDP synthase of *Antirrhinum majus* that catalyzes the synthesis of sesquiterpenes, overexpression leads to loss of color in tomato fruit under storage (Davidovich-Rikanati et al., 2008). GDP synthase of *Antirrhinum majus* and α-zingiberene synthase coexpression increase monoterpenes expression threefold, while coexpression of GDP synthase-SSU of *Antirrhinum majus* and α-zingiberene synthase increases both sesquiterpenes and monoterpenes level (Gutensohn et al., 2013).

Regarding the development of resistance towards biotic stress, tomato fruits were engineered with 7-epizingiberene synthase and Z-Z farnesyl diphosphate synthase that coded for the enzyme of biosynthesis of 7-epizingiberene. These genes were under the control of trichome-specific promoters. This transgenic tomato with engineered genes showed accumulation of 7-epizingiberene, and the tomato fruits developed resistance towards several pests (Bleeker et al., 2012). The storage of citrus fruit faces the problem of fungal and bacterial contamination, which can be resolved with metabolic engineering. RNAi-based down-regulation of the limonene synthase gene reduced the accumulation of limonene (aroma compound in the peel of citrus fruits) (Rodríguez et al., 2011). This makes citrus fruits resistant to bacterial and fungal infection remarkably. Other crops like oregano (*Origanum vulgare*) with up-regulation of (E)-β-caryophyllene synthases increase sesquiterpenes and improve their defense mechanism (Cheng et al., 2007).

14.4.2 METABOLIC ENGINEERING OF FATTY ACID DERIVED VOLATILES

In fruits and vegetables, alcohol and aldehydes play important roles in ripening, aroma development, and softening. Key enzymes involved in alcohol and aldehyde biosynthetic pathways include alcohol dehydrogenase (ADH2). Overexpression of ADH2 has been shown to impact alcohol and aldehyde levels (Prestage et al., 1999). Conversely, silencing the short-chain ADH1 in tomatoes with RNAi did not result in significant changes in fruit, except for the accumulation of LOX pathway C5 and C6 volatiles (Moummou et al., 2012). In grapevines (*Vitis vinifera*), ADH2 overexpression and ADH1 downregulation in grape berries did not lead to morphological changes, but increased levels of 2-phenyl ethanol and benzyl alcohol were observed (Torregrosa et al., 2008). (Z)-3-hexenal is considered a freshness indicator in various fruits, vegetables, and grains. Another approach involves controlling the biosynthesis of short-chain aldehydes through the LOX pathway. Up-regulation of tomato LOX C has been shown to reduce the expression of (Z)-3-hexenal, hexanal, and (Z)-3-hexenol (Chen et al., 2004). Similarly, in potatoes, a reduction in linolenic acid, a precursor of hexanal, was observed (Leóneta et al., 2002). Up-regulation of Δ9-desaturase, a yeast enzyme, in tomatoes was found to influence the emission of short-chain aldehydes and alcohols, enhancing flavor and aroma (Domínguez et al., 2010). Overexpression of certain ω-3 fatty acid desaturases from potatoes increases the (Z)-3-hexenal/hexanal ratio and linolenic/linoleic acid ratio, aiding in chilling injury tolerance (Domínguez et al., 2010). In cucumbers, (E,Z)-2,6 nonadienal contributes to aroma and appears to be up-regulated by overexpression of preprothaumatin II (Zawirska-Wojtasiak et al., 2009).

14.4.3 METABOLIC ENGINEERING OF VOLATILES ORIGINATES FROM AMINO ACID

Some amino acid decarboxylases such as 1A and 1B constitutively express higher levels causing the decarboxylation of amino acids and elevating the levels of 2-phenylethanol, 2-phenylacetaldehyde, and 1-nitro2-phenylethane (Tieman et al., 2006). These processes mainly involve the decarboxylation of phenylalanine, which is not always responsible for amino acid-derived volatiles. The PhODORANT1 MYB gene from petunia introduced into tomato was discovered to enhance the production of phenylpropanoid volatiles other than those derived from phenylalanine (Dal Cin et al., 2011). Another enzyme, Adenosyl L-methionine: salicylic acid (SA) carboxy methyltransferase, transfers a methyl group from S-adenosyl-L-methionine to SA, forming methyl salicylate (Tieman et al., 2006). Methyl salicylate contributes to the tomato's flavor, and its increased expression through metabolic engineering enhances the flavor. In certain transgenic plants, methyl salicylate has been linked to delayed pathogenic infections. Interestingly, studies have revealed that the gene silencing of carboxylesterase (CXE1) in tomatoes reduces the expression of another related gene (CHE2-CXE5) and boosts the release of acetate esters (Goulet et al., 2012). Typically, acetate esters are present in green tomatoes, while in red tomatoes, retrotransposons elevate esterases that break down acetate esters.

In strawberries, wild types are rich in some Phenylpropenes that were not found in cultivated varieties. Phenylpropenes are aroma and flavor enhancing volatiles; therefore, many metabolic engineering attempts have been made to restore wild characteristics. Chalcone synthase (FaCHS) was down-regulated, ultimately increasing the precursor for the phenylpropanoids pathway (Lunkenbein et al., 2006). In some studies, O-methyltransferase overexpression or down-regulation was found to affect the aroma of strawberries by mediating the biosynthesis of mesifurane (Lunkenbein et al., 2006). ACO (ACC oxidase) and ACS (ACC synthase; ACC: 1-aminocyclopropane-1-carboxylic acid) gene silencing in apples reduces ethylene production, and simultaneously esters are also reduced. This indirectly increases the shelf life of apples under storage (Atkinson et al., 2011). A similar observation was found in kiwi fruit. Metabolic engineering of sulfur-containing amino acids is also a remarkable step towards postharvest storage improvement of fruits and vegetables.

14.5 CONCLUSION

Metabolic engineering of fruits and vegetables has been an inseparable part of the prolonged storage of perishable crops. As secondary metabolites, i.e., volatiles, play an important role in shelf life, aroma, and flavor. These can be a good target for manipulation to improve storage. Overexpression or down-regulation of these metabolites ensures variation in certain biosynthetic pathways, leading to morphological and physiological changes in crops. However, engineering a single gene alone cannot result in overall advancement in postharvest storage. The engineered metabolites discussed in this chapter are from various research experiments, but limited knowledge of volatile function may hinder crop improvement. This limitation may result in aroma or flavor variation, and sometimes color changes as well. Further research on key volatiles and their roles may enhance the success rate of metabolic engineering.

KEYWORDS

- **amino acid**
- **carboxylesterase**
- **key volatiles**
- **limonenesynthasegene**
- **mesifurane**
- **metabolic engineering**
- **pathogenic infection**
- **Phenylpropenes**
- **shelf life**

REFERENCES

Aharoni, A., Giri, A. P., Verstappen, F. W., Bertea, C. M., Sevenier, R., Sun, Z., & Bouwmeester, H. J., (2004). Gain and loss of fruit flavor compounds produced by wild and cultivated strawberry species. *The Plant Cell, 16*(11), 3110–3131.

Algarra, A. A., Lazazzara, V., Cappellin, L., Bianchedi, P. L., Schuhmacher, R., Wohlfahrt, G., & Perazzolli, M., (2015). Emission of volatile sesquiterpenes and monoterpenes in grapevine genotypes following *Plasmopara viticola* inoculation *in vitro*. *Journal of Mass Spectrometry, 50*(8), 1013–1022.

Ali, I., Husnain, T., & Riazuddin, S., (2008). RNA interference: The story of gene silencing in plants and humans. *Biotechnology Advances, 26*(3), 202–209.

Aragüez, I., & Valpuesta, F. V., (2013). Metabolic engineering of aroma components in fruits. *Biotechnology Journal, 8*(10), 1144–1158.

Atkinson, R. G., Gunaseelan, K., Wang, M. Y., Luo, L., Wang, T., Norling, C. L., & Schaffer, R. J., (2011). Dissecting the role of climacteric ethylene in kiwifruit (*Actinidia chinensis*) ripening using a 1-aminocyclopropane-1-carboxylic acid oxidase knockdown line. *Journal of Experimental Botany, 62*(11), 3821–3835.

Aubert, C., Baumann, S., & Arguel, H., (2005). Optimization of the analysis of flavor volatile compounds by liquid– liquid microextraction (LLME). Application to the aroma analysis of melons, peaches, grapes, strawberries, and tomatoes. *Journal of Agricultural and Food Chemistry, 53*(23), 8881–8895.

Bennis, S., Chami, F., Chami, N., Bouchikhi, T., & Remmal, A., (2004). Surface alteration of *Saccharomyces cerevisiae* induced by thymol and eugenol. *Letters in Applied Microbiology, 38*(6), 454–458.

Bleeker, P. M., Mirabella, R., Diergaarde, P. J., VanDoorn, A., Tissier, A., Kant, M. R., & Schuurink, R. C., (2012). Improved herbivore resistance in cultivated tomato with the sesquiterpene biosynthetic pathway from a wild relative. *Proceedings of the National Academy of Sciences, 109*(49), 20124–20129.

Brummell, D. A., & Harpster, M. H., (2001). Cell wall metabolism in fruit softening and quality and its manipulation in transgenic plants. *Plant Molecular Biology, 47*(1, 2), 311–340.

Calfapietra, C., Pallozzi, E., Lusini, I., & Velikova, V., (2013). Modification of BVOC emissions by changes in atmospheric [CO_2] and air pollution. In: *Biology, Controls, and Models of Tree Volatile Organic Compound Emission*s (pp. 253–284). Springer, Dordrecht.

Canton, M., Drincovich, M. F., Lara, M. V., Vizzotto, G., Walker, R. P., Famiani, F., & Bonghi, C., (2020). Metabolism of stone fruits: Reciprocal contribution between primary metabolism and cell wall. *Frontiers in Plant Science*, 11.

Chambers, A., Whitaker, V. M., Gibbs, B., Plotto, A., & Folta, K. M., (2012). Detection of the linalool-producing NES1 variant across diverse strawberry (*Fragaria* spp.) accessions. *Plant Breeding, 131*(3), 437–443.

Chen, G., Hackett, R., Walker, D., Taylor, A., Lin, Z., & Grierson, D., (2004). Identification of a specific isoform of tomato lipoxygenase (TomloxC) involved in the generation of fatty acid-derived flavor compounds. *Plant Physiology, 136*(1), 2641–2651.

Chen, X., Chen, H., Yuan, J. S., Köllner, T. G., Chen, Y., Guo, Y., & Chen, F., (2018). The rice terpene synthase gene OsTPS 19 functions as an (S)-limonene synthase in planta, and its overexpression leads to enhanced resistance to the blast fungus *Magnaporthe oryzae*. *Plant Biotechnology Journal, 16*(10), 1778–1787.

Chen, Y., Zeng, H., Tian, J., Ban, X., Ma, B., & Wang, Y., (2014). Dill (*Anethum graveolens* L.) seed essential oil induces *Candida albicans* apoptosis in a metacaspase-dependent manner. *Fungal Biology, 118*(4), 394–401.

Cheng, A. X., Xiang, C. Y., Li, J. X., Yang, C. Q., Hu, W. L., Wang, L. J., & Chen, X. Y., (2007). The rice (E)-β-caryophyllene synthase (OsTPS3) accounts for the major inducible volatile sesquiterpenes. *Phytochemistry, 68*(12), 1632–1641.

Cumplido-Laso, G., Medina-Puche, L., Moyano, E., Hoffmann, T., Sinz, Q., Ring, L., & Blanco-Portales, R., (2012). The fruit ripening-related gene FaAAT2 encodes an acyl transferase involved in strawberry aroma biogenesis. *Journal of Experimental Botany, 63*(11), 4275–4290.

Dal Cin, V., Tieman, D. M., Tohge, T., McQuinn, R., De Vos, R. C., Osorio, S., & Klee, H. J., (2011). Identification of genes in the phenylalanine metabolic pathway by ectopic expression of a MYB transcription factor in tomato fruit. *The Plant Cell, 23*(7), 2738–2753.

Dandekar, A. M., Teo, G., Defilippi, B. G., Uratsu, S. L., Passey, A. J., Kader, A. A., & James, D. J., (2004). Effect of down-regulation of ethylene biosynthesis on fruit flavor complex in apple fruit. *Transgenic Research, 13*(4), 373–384.

Davidovich-Rikanati, R., Lewinsohn, E., Bar, E., Iijima, Y., Pichersky, E., & Sitrit, Y., (2008). Overexpression of the lemon basil α-zingiberene synthase gene increases both mono-and sesquiterpene contents in tomato fruit. *The Plant Journal, 56*(2), 228–238.

Davidovich-Rikanati, R., Sitrit, Y., Tadmor, Y., Iijima, Y., Bilenko, N., Bar, E., & Lewinsohn, E., (2007). Enrichment of tomato flavor by diversion of the early plastidial terpenoid pathway. *Nature Biotechnology, 25*(8), 899–901.

Dima, C., & Dima, S., (2015). Essential oils in foods: Extraction, stabilization, and toxicity. *Current Opinion in Food Science, 5*, 29–35.

Domínguez, T., Hernández, M. L., Pennycooke, J. C., Jiménez, P., Martínez-Rivas, J. M., Sanz, C., & Sanmartín, M., (2010). Increasing ω-3 desaturase expression in tomato results in altered aroma profile and enhanced resistance to cold stress. *Plant Physiology, 153*(2), 655–665.

Dudareva, N., & Pichersky, E., (2006). *Biology of Floral Scent*. CRC press.

Fishman, M. L., Cooke, P. H., & Coffin, D. R., (2004). Nanostructure of native pectin sugar acid gels visualized by atomic force microscopy. *Biomacromolecules, 5*, 334–341.

Gan, S., & Amasino, R. M., (1997). Making sense of senescence (molecular genetic regulation and manipulation of leaf senescence). *Plant. Physiol., 113*(2), 313.

Gao, J., Wang, G., Ma, S., Xie, X., Wu, X., Zhang, X., & Xia, Q., (2015). CRISPR/Cas9-mediated targeted mutagenesis in *Nicotiana tabacum. Plant Molecular Biology, 87*(1), 99–110.

Gelvin, S. B., (2003). *Agrobacterium*-mediated plant transformation: The biology behind the "gene-jockeying" tool. *Microbiology and Molecular Biology Reviews, 67*(1), 16–37.

Ghaouth, A. E., Wilson, C., & Wisniewski, M., (2004). Biologically-based alternatives to synthetic fungicides for the control of postharvest diseases of fruit and vegetables. In: *Diseases of Fruits and Vegetables* (Vol. II, pp. 511–535). Springer, Dordrecht.

Goulet, C., Mageroy, M. H., Lam, N. B., Floystad, A., Tieman, D. M., & Klee, H. J., (2012). Role of an esterase in flavor volatile variation within the tomato clade. *Proceedings of the National Academy of Sciences, 109*(46), 19009–19014.

Günther, C. S., Chervin, C., Marsh, K. B., Newcomb, R. D., & Souleyre, E. J., (2011). Characterization of two alcohol acyltransferases from kiwifruit (*Actinidia* spp.) reveals distinct substrate preferences. *Phytochemistry, 72*(8), 700–710.

Gupta, A., Pal, R. K., & Rajam, M. V., (2013). Delayed ripening and improved fruit processing quality in tomato by RNAi-mediated silencing of three homologs of 1-aminopropane-1-carboxylate synthase gene. *Journal of Plant Physiology, 170*(11), 987–995.

Gutensohn, M., Orlova, I., Nguyen, T. T., Davidovich-Rikanati, R., Ferruzzi, M. G., Sitrit, Y., & Dudareva, N., (2013). Cytosolic monoterpene biosynthesis is supported by plastid-generated geranyl diphosphate substrate in transgenic tomato fruits. *The Plant Journal, 75*(3), 351–363.

Hijaz, F., Nehela, Y., & Killiny, N., (2016). Possible role of plant volatiles in tolerance against huanglongbing in *Citrus. Plant Signaling & Behavior, 11*(3), e1138193.

Holme, I. B., Wendt, T., & Holm, P. B., (2013). Intragenesis and cisgenesis as alternatives to transgenic crop development. *Plant Biotechnology Journal, 11*(4), 395–407.

Huang, M., Sanchez-Moreiras, A. M., Abel, C., Sohrabi, R., Lee, S., Gershenzon, J., & Tholl, D., (2012). The major volatile organic compound emitted from *Arabidopsis thaliana* flowers, the sesquiterpene (E-β-caryophyllene, is a defense against a bacterial pathogen. *New Phytologist, 193*(4), 997–1008.

Imaseki, H., (1991). The biochemistry of ethylene biosynthesis. *Plant Hormone Ethylene*, 1–20.

Joshi, J. R., Khazanov, N., Senderowitz, H., Burdman, S., Lipsky, A., & Yedidia, I., (2016). Plant phenolic volatiles inhibit quorum sensing in pectobacteria and reduce their virulence by potential binding to ExpI and ExpR proteins. *Scientific Reports, 6*(1), 1–15.

Kinzel, B., (1989). *Nikkomycin: Antibiotic for Plants*. Agricultural Research-US Department of Agriculture, Agricultural Research Service (USA).

Lambert, R. J. W., Skandamis, P. N., Coote, P. J., & Nychas, G. J., (2001). A study of the minimum inhibitory concentration and mode of action of oregano essential oil, thymol and carvacrol. *Journal of Applied Microbiology, 91*(3), 453–462.

León, J., Royo, J., Vancanneyt, G., Sanz, C., Silkowski, H., Griffiths, G., & Sánchez-Serrano, J. J., (2002). Lipoxygenase H1 gene silencing reveals a specific role in supplying fatty acid hydroperoxides for aliphatic aldehyde production. *Journal of Biological Chemistry, 277*(1), 416–423.

Lers, A., (2012). Potential application of biotechnology to maintain fresh produce postharvest quality and reduce losses during storage. In: *Plant Biotechnology and Agriculture* (pp. 425–441). Academic Press.

Lewinsohn, E., Schalechet, F., Wilkinson, J., Matsui, K., Tadmor, Y., Nam, K. H., & Pichersky, E., (2001). Enhanced levels of the aroma and flavor compound S-linalool by metabolic engineering of the terpenoid pathway in tomato fruits. *Plant Physiology, 127*(3), 1256–1265.

Lourdes, C. Z. D., Guimarães, E. M., & Vilela, P. F., (2005). Analysis of volatile compounds in some typical Brazilian fruits and juices by SPME-GC method. *Food Additives and Contaminants, 22*(6), 508–513.

Luan, F., Mosandl, A., Münch, A., & Wüst, M., (2005). Metabolism of geraniol in grape berry mesocarp of *Vitis vinifera* L. cv. Scheurebe: Demonstration of stereoselective reduction, E/Z-isomerization, oxidation, and glycosylation. *Phytochemistry, 66*(3), 295–303.

Lund, P., Lee, R. Y., & Dunsmuir, P., (1989). Bacterial chitinase is modified and secreted in transgenic tobacco. *Plant Physiol., 91*(1), 130–135.

Lunkenbein, S., Coiner, H., De Vos, C. R., Schaart, J. G., Boone, M. J., Krens, F. A., & Salentijn, E. M., (2006). Molecular characterization of a stable antisense chalcone synthase phenotype in strawberry (*Fragaria× ananassa*). *Journal of Agricultural and Food Chemistry, 54*(6), 2145–2153.

Mageroy, M. H., Tieman, D. M., Floystad, A., Taylor, M. G., & Klee, H. J., (2012). A *Solanum lycopersicum* catechol-O-methyltransferase involved in synthesis of the flavor molecule guaiacol. *The Plant Journal, 69*(6), 1043–1051.

Matsui, K., Minami, A., Hornung, E., Shibata, H., Kishimoto, K., Ahnert, V., & Feussner, I., (2006). Biosynthesis of fatty acid derived aldehydes is induced upon mechanical wounding and its products show fungicidal activities in cucumber. *Phytochemistry, 67*(7), 649–657.

Mei, X., Liu, Y., Huang, H., Du, F., Huang, L., Wu, J., & Yang, M., (2019). Benzothiazole inhibits the growth of *Phytophthora capsici* through inducing apoptosis and suppressing stress responses and metabolic detoxification. *Pesticide Biochemistry and Physiology, 154*, 7–16.

Melotto, M., Zhang, L., Oblessuc, P. R., & He, S. Y., (2017). Stomatal defense a decade later. *Plant Physiology, 174*(2), 561–571.

Mocellin, S., & Provenzano, M., (2004). RNA interference: Learning gene knock-down from cell physiology. *Journal of Translational Medicine, 2*(1), 1–6.

Moummou, H., Tonfack, L. B., Chervin, C., Benichou, M., Youmbi, E., Ginies, C., & Van, D. R. B., (2012). Functional characterization of SlscADH1, a fruit-ripening-associated short-chain alcohol dehydrogenase of tomato. *Journal of Plant Physiology, 169*(15), 1435–1444.

Munoz, M., Cistue, L., Castillo, A. M., Romagosa, I., & Valles, M. P., (2004). A retrotransposon sequence is related to DNA instability in barley culture. In: *17th Eucarpia General Congress: Genetic Variation for Plant Breeding* (pp. 8–11) Tullin, Austria.

Myung, K., Hamilton-Kemp, T. R., & Archbold, D. D., (2007). Interaction with and effects on the profile of proteins of *Botrytis cinerea* by C6 aldehydes. *Journal of Agricultural and Food Chemistry, 55*(6), 2182–2188.

Nuñez-Palenius, H. G., Cantliffe, D. J., Huber, D. J., Ciardi, J., & Klee, H. J., (2006). Transformation of a muskmelon 'Galia' hybrid parental line (*Cucumis melo* L. var. reticulatus Ser.) with an antisense ACC oxidase gene. *Plant Cell Reports, 25*(3), 198–205.

Phillips, M. A., León, P., Boronat, A., & Rodríguez-Concepción, M., (2008). The plastidial MEP pathway: Unified nomenclature and resources. *Trends in Plant Science, 13*(12), 619–623.

Prestage, S., Linforth, R. S., Taylor, A. J., Lee, E., Speirs, J., & Schuch, W., (1999). Volatile production in tomato fruit with modified alcohol dehydrogenase activity. *Journal of the Science of Food and Agriculture, 79*(1), 131–136.

Quintana-Rodriguez, E., Morales-Vargas, A. T., Molina-Torres, J., Ádame-Alvarez, R. M., Acosta-Gallegos, J. A., & Heil, M., (2015). Plant volatiles cause direct, induced, and associational resistance in common bean to the fungal pathogen *Colletotrichum lindemuthianum*. *Journal of Ecology, 103*(1), 250–260.

Quiroz-Iturra, L., Rosas-Saavedra, C., & Klein, C. S., (2017). Functional fruits through metabolic engineering. In: *Superfood and Functional Food-the Development of Superfoods and Their Roles as Medicine*. IntechOpen.

Rodríguez, A., San, A. V., Cervera, M., Redondo, A., Alquézar, B., Shimada, T., & Peña, L., (2011). Terpene down-regulation in orange reveals the role of fruit aromas in mediating interactions with insect herbivores and pathogens. *Plant Physiology, 156*(2), 793–802.

Sharma, R. R., Singh, D., & Singh, R., (2009). Biological control of postharvest diseases of fruits and vegetables by microbial antagonists: A review. *Biological Control, 50*(3), 205–221.

Shi, J., Zuo, J., Xu, D., Gao, L., & Wang, Q., (2019). Effect of low-temperature conditioning combined with methyl jasmonate treatment on the chilling resistance of eggplant (*Solanum melongena* L.) fruit. *Journal of Food Science and Technology, 56*(10), 4658–4666.

Shu, Q. Y., Forster, B. P., Nakagawa, H., & Nakagawa, H., (2012). *Plant Mutation Breeding and Biotechnology.* CABI.

Sogvar, O. B., Rabiei, V., Razavi, F., & Gohari, G., (2020). Phenylalanine alleviates postharvest chilling injury of plum fruit by modulating antioxidant system and enhancing the accumulation of phenolic compounds. *Food Technology and Biotechnology, 58*(4), 433–444.

Sudheeran, P. K., Feygenberg, O., Maurer, D., & Alkan, N., (2018). Improved cold tolerance of mango fruit with enhanced anthocyanin and flavonoid contents. *Molecules, 23*(7), 1832.

Talmadge, K. W., Keegstra, K., Bauer, W. D., & Albersheim, P., (1973). The structure of plant cell walls: I. The macromolecular components of the walls of suspension-cultured sycamore cells with a detailed analysis of the pectic polysaccharides. *Plant Physiol., 51*(1), 158–173.

Tieman, D., Taylor, M., Schauer, N., Fernie, A. R., Hanson, A. D., & Klee, H. J., (2006). Tomato aromatic amino acid decarboxylases participate in synthesis of the flavor volatiles 2-phenylethanol and 2-phenylacetaldehyde. *Proceedings of the National Academy of Sciences, 103*(21), 8287–8292.

Tietel, Z., Porat, R., Weiss, K., & Ulrich, D., (2011). Identification of aroma-active compounds in fresh and stored 'Mor' mandarins. *International Journal of Food Science & Technology, 46*(11), 2225–2231.

Torregrosa, L., Pradal, M., Souquet, J. M., Rambert, M., Gunata, Z., & Tesniere, C., (2008). Manipulation of VvAdh to investigate its function in grape berry development. *Plant Science, 174*(2), 149–155.

Uluisik, S., & Seymour, G. B., (2020). Pectate lyases: Their role in plants and importance in fruit ripening. *Food Chemistry*, 309.

Van, S. J., (1988). Cytokinins and auxins in carnation senescence as related to chemical treatments. In: *IV International Symposium on Postharvest Physiology of Ornamental Plants* (Vol. 261, pp. 69–80.

Vanitha, S. C., & Umesha, S., (2008). Variations in defense related enzyme activities in tomato during the infection with bacterial wilt pathogen. *J. Plant Interact., 3*(4), 245–253.

Velikova, V., Sharkey, T. D., & Loreto, F., (2012). Stabilization of thylakoid membranes in isoprene-emitting plants reduces formation of reactive oxygen species. *Plant Signaling & Behavior, 7*(1), 139–141.

Velikova, V., Várkonyi, Z., Szabó, M., Maslenkova, L., Nogues, I., Kovács, L., & Loreto, F., (2011). Increased thermostability of thylakoid membranes in isoprene-emitting leaves probed with three biophysical techniques. *Plant Physiology, 157*(2), 905–916.

Wang, L., Baldwin, E. A., Zhao, W., Plotto, A., Sun, X., Wang, Z., & Yu, Z., (2015). Suppression of volatile production in tomato fruit exposed to chilling temperature and alleviation of chilling injury by a pre-chilling heat treatment. *LWT-Food Science and Technology, 62*(1), 115–121.

Wendakoon, C. N., & Sakaguchi, M., (1995). Inhibition of amino acid decarboxylase activity of Enterobacter aerogenes by active components in spices. *Journal of Food Protection, 58*(3), 280–283.

Zawirska-Wojtasiak, R., Gośliński, M., Szwacka, M., Gajc-Wolska, J., & Mildner-Szkudlarz, S., (2009). Aroma evaluation of transgenic, thaumatin II-producing cucumber fruits. *Journal of Food Science, 74*(3), C204–C210.

Zhang, H., Zheng, X., & Yu, T., (2007). Biological control of postharvest diseases of peach with *Cryptococcus laurentii*. *Food Control, 18*(4), 287–291.

Zhang, L., Chen, F., An, H., Yang, H., Sun, X., Guo, X., et al., (2008). Physicochemical properties, firmness, and nanostructures of sodium carbonate-soluble pectin of 2 Chinese cherry cultivars at 2 ripening stages. *J. Food Sci., 73*, N17–N22.

Zhang, Y., Butelli, E., Stefano, R. D., Schoonbeek, H., Magusin, A., Pagliarani, C., & Wellner, N., (2013). Anthocyanins double the shelf life of tomatoes by delaying overripening and reducing susceptibility to gray mold. *Curr. Biol., 23*(12), 1094–1100.

Zorrilla-Fontanesi, Y., Rambla, J. L., Cabeza, A., Medina, J. J., Sánchez-Sevilla, J. F., Valpuesta, V., & Amaya, I., (2012). Genetic analysis of strawberry fruit aroma and identification of O-methyltransferase FaOMT as the locus controlling natural variation in mesifurane content. *Plant Physiology, 159*(2), 851–870.

INDEX

D

E

F

H

M

N

Z